MW00617902

Band Theory of S

An Introduction from the Point of View of Symmetry

QC
176.8
.E4
A45
1994

Band Theory of Solids
An Introduction from the Point of View of Symmetry

Simon L. Altmann
Brasenose College
Oxford

CLARENDON PRESS · OXFORD

GOSHEN COLLEGE LIBRARY
GOSHEN, INDIANA

Oxford University Press, Walton Street, Oxford OX2 6DP
Oxford New York
Athens Auckland Bangkok Bombay
Calcutta Cape Town Dar es Salaam Delhi
Florence Hong Kong Istanbul Karachi
Kuala Lumpur Madras Madrid Melbourne
Mexico City Nairobi Paris Singapore
Taipei Tokyo Toronto
and associated companies in
Berlin Ibadan

Oxford is a trade mark of Oxford University Press

Published in the United States by
Oxford University Press Inc., New York

© *Simon L. Altmann 1991*

First published 1991
First published as paperback 1994 (with corrections)
Reprinted 1995

All rights reserved. No part of this publication may be reproduced, stored in a retrieval system, or transmitted, in any form or by any means, without the prior permission in writing of Oxford University Press. Within the UK, exceptions are allowed in respect of any fair dealing for the purpose of research or private study, or criticism or review, as permitted under the Copyright, Designs and Patents Act, 1988, or in the case of reprographic reproduction in accordance with the terms of licences issued by the Copyright Licensing Agency. Enquiries concerning reproduction outside those terms and in other countries should be sent to the Rights Department, Oxford University Press, at the address above.

This book is sold subject to the condition that it shall not, by way of trade or otherwise, be lent, re-sold, hired out, or otherwise circulated without the publisher's prior consent in any form of binding or cover other than that in which it is published and without a similar condition including this condition being imposed on the subsequent purchaser.

A catalogue record for this book is available from the British Library

Library of Congress Cataloging in Publication Data
Altmann, Simon L., 1924–
Band theory of solids: an introduction from the point of view of symmetry/Simon L. Altmann.
p. cm.
1. Energy-band theory of solids. 2. Group theory. 3. Symmetry (Physics). I. Title.
QC176.8.E4A45 1991 91–7360
530.4'1—dc20 CIP
ISBN 0 19 855866 X

Printed and bound by Biddles Ltd
Guildford and King's Lynn

GOSHEN COLLEGE LIBRARY
GOSHEN, INDIANA

Preface

Whereas no chemist would these days treat, say, molecular vibrations without using the vocabulary of group theory, there is, unfortunately, no elementary introduction to the band theory of solids which adopts this approach. There are, of course, excellent books on the group theory of the solid state, but they are all at a rather rigorous level and are not suitable as an introduction to band theory. Almost twenty years ago I wrote such an introduction in which I made much use of the concept of symmetry, but I could not employ the vocabulary of group theory since undergraduates in the physical sciences were not exposed to this subject in those days. Finite group theory is now the *lingua franca* of chemistry and many undergraduate courses in physics include some group theory. I felt, therefore, that the time had now come to have a book which would bridge the gap between the current elementary texts on band theory and the more advanced treatments of the group theory of solid state.

I am sure I do not have to justify the use of group theory in the band theory of solids. There is hardly any subject in physics in which so much of the structure comes straight from group theory and a lot of important ideas become obscure if this language cannot be used: to have to understand band theory without realizing that the Brillouin zone is a graphical depiction of the irreducible representations of the space group of a crystal, and of the structure of their bases, is an almost intolerable limitation by present standards.

I hope that this work will not be pigeon-holed with texts on the group theory of solid state. In the present book, as a difference, only the simplest and most basic of group-theoretical ideas are used and no attempt is made at the great problems of the irreducibility and completeness of the space group representations which take so much room in the group-theoretical treatments. I have made no attempt at absolute rigour but, rather, I made a serious effort to convey a feeling for the way in which symmetry determines many of the essential ideas in solid-state theory, thus, I hope, illuminating some of the major concepts in the band theory of solids.

In writing this book I was very much aware of the present very strong interest in solid-state chemistry and it was clear to me that there was no quicker entry into the subject for chemists, with their training in the use of group theory, than by means of the approach used here. The choice of many topics was also made taking the needs of chemists into account and many of the examples used have a chemical flavour. I hope, however, that the general approach of this book will appeal to all physical scientists who would like to go a little beyond the elementary treatments so far available.

PHY/CH 9.22.00

The book started as a set of lectures which I gave in 1982 at the Department of Chemistry of the University of Perugia at the invitation of Professor Antonio Sgamellotti and I am very grateful to him for having given me this opportunity. When I was a visiting professor at the Department of Chemistry in Rome (University La Sapienza) in 1985, I was presented with the possibility of greatly improving the lecture notes which I had provided at Perugia. Indeed, the typescript that was circulated at Rome contained, albeit in rudimentary form, most of the material in the present volume. I am most grateful to Professor Piero Porta who not only invited me to give those lectures but also provided numerous useful suggestions. It was my experience during these lectures that the material of this book, suitably simplified where necessary, can be covered in some twenty hours and that audiences including undergraduate students in both physics and chemistry were perfectly able to follow the material.

There is, of course, some overlap between many topics treated in this book and those in my former *Band theory of metals*. Naturally, Bloch functions and Brillouin zones have to re-appear, although the form of presentation is generally different. On the other hand, many new subjects are included. Space groups and their symmetry operations are given in some detail. There is a chapter on phonons in which the phonon spectrum of silicon is discussed as an example. The larger part of a chapter is devoted to Peierls instabilities and their relation to the Jahn–Teller effect, including a detailed treatment of quasi one-dimensional chains. A uniform treatment of Löwdin and Wannier functions is provided through a discussion of equivalent functions. A chapter on surface and impurity states contains a detailed treatment of the Koster and Slater method, as an introduction to the use of Green functions, for which a full but simple example is provided. Numerous problems throughout the book, for which complete solutions are given in its last chapter, provide an opportunity not only for applying the major ideas discussed but also for completing some parts of the treatment.

I am acutely aware, of course, not only that there are many subjects missing in this book but also that for much of the modern work in solid-state theory a different approach to the one used here is desirable. It goes without saying, for instance, that if one wants to move towards the study of bands in non-crystalline solids, the geometrical and crystallographic approach of this book is not the most useful. Likewise, in the study of many structural properties, a much more detailed study of the potential field in the solid than the one provided here is necessary. Many properties of some solids are best understood from a far more chemical point of view, and so on. I took the view, however, that what is done extremely well in other books there was no point for me to emulate. I did try instead to take a very central and significant part of the subject and then to treat it in a way that would not be easy to find in the literature. I hope that with the firm basis provided by this book readers will find it much easier to move forward in the subject.

Many kind friends have helped me to improve this book and I am most grateful to them. Through the kind intervention of Dr Graham Richards, Mr Jonathan Essex checked my approximate calculations for eqn (**14**-2.12). I am particularly indebted to Professor Roald Hoffmann who not only provided me with a critical reading of the manuscript but also arranged for two of his collaborators at Cornell, Dr Christian Kollmar and Mr Yat-Ting, to do the same. Their critical comments were enormously useful in improving the text. Drs Harry Rosenberg and Terry Willis read selected chapters and I am sure that it is thanks to them that, amongst other things, my crystallographic terminology was greatly improved. Dr Peter Herzig, of the University of Vienna, read the whole text and his critical eye spotted a number of mistakes that would have certainly made the book more difficult to read. Finally, I can hardly thank sufficiently Dr Tony Cox who read the manuscript not just once but twice and who never failed to find weak arguments in need of improvement.

In order to save on the cost of the book I have drawn all the illustrations myself on a Macintosh computer and I am very grateful to my sons, Drs Daniel and Paul Altmann for their help and advice with the necessary software.

An author is indeed fortunate who has such good friends to help him improve his work. I only hope that my readers will also approach this book in the spirit in which I wrote it. It is harder, in my experience, to provide persuasive rather than complete and rigorous argument and the reader would probably do better in trying to grasp the gist of the discussions presented here rather than attempting to analyse them to exhaustion. Within these limits, nevertheless, I have tried to make my arguments as easy to follow as possible, partly by providing a very comprehensive cross-referencing system. I have also attempted to foresee possible misreadings and I have given here and there warnings to that effect. Where I may have failed, however, it is not for want of trying!

Oxford S.L.A.
1990

Contents

0 Notation 1

1 References 1
General cross-references 1
Cross-references on left margins of displayed formulae 1
Literature references 1
2 Symbols used 2

1 The free-electron picture 7

1 Free-electron model 7
2 Born–von Karman boundary conditions 8
3 Quantum-mechanical treatment 9
Boundary conditions 11
The **k** space 12
4 Physical consequences: Fermi energy and Fermi surface 13
5 The density of states 14
6 The Fermi energy and the total energy 16
7 Bands 17
8 Electron waves 18
9 Problems 21

2 Symmetry and group theory: a revision 23

1 Group definitions 23
Conjugates and classes 24
Subgroups 25
Direct product 25
2 Symmetry operators 26
Transformation of functions 28
Transformation of operators 30
3 The Schrödinger group 30
Degeneracy 31
The conjugator 32
4 Representations 34
Direct sum. Reduction of the representations 35
Similarity. Irreducibility 36
Characters 37
5 Irreducible representations 37
Representations of the direct product 38

6 An example: Benzene 39
7 Problems 42

3 Space groups 46

1 Translations and unit cells 47
2 Centred primitive cells 50
3 The Bravais lattice 51
4 Space groups and their symmetry operations 53
5 The point group and the Bravais lattice 56
6 The Seitz operators 59
7 The space group and its point group 60
8 The translation subgroup 60
Commutation of translations 61
Direct product form of T 61
Invariance 61
9 Classification of space groups 62
10 Problems 63

4 The reciprocal lattice and the Fourier series 64

1 Periodic functions 65
2 Reciprocal vectors and general periodic functions 67
3 Construction of the reciprocal vectors 69
4 Reciprocal vectors and crystal planes 69
5 Reciprocal lattice and reciprocal space. An example 73
6 Summary of vectors 75
7 A three dimensional example of a reciprocal lattice 76
8 Plane waves in crystal structures 77
9 The Fourier series 78
Notation and definitions 79
Orthonormality 80
Expansion of a periodic function 81
Fourier series for a lattice with basis 82
An example of the structure factor: hexagonal close-packed structure 83
10 Plane-wave notation and properties 84
11 Problems 84

5 Bloch functions and Brillouin zones 86

1 The translation group T 86
2 The representations of T 89
3 Equivalent $\mathbf{k}$ vectors and the Brillouin zone 91
4 Quantization of $\mathbf{k}$ 93
5 The number of states in the Brillouin zone 94

6 The bases for the representations of T: the Bloch functions 97
Physical considerations. Bands 98
7 Problems 100

6 Space group representations 102

1 Effect of the space group operators on the Bloch functions 102
2 The bases of the space group 103
3 The star, the group of the **k** vector, and the small representation 105
The group of the **k** vector 107
The small representation 108
4 The irreducible representations of the space group and the energy as a function of **k** 111
Band trends: guesswork 114
Properties of the $E(\mathbf{k})$ surfaces over the Brillouin zone 115
The symmetry of the Brillouin zone and of the reciprocal lattice: a warning 116
5 Symmetry of $E(\mathbf{k})$ and $E(-\mathbf{k})$: The conjugator operator 118
6 Problems 119

7 The representation of space groups: an example 120
1 The space group model 120
2 The irreducible representations of T 123
3 The irreducible representations of G 124
The bands 127
4 Problems 128

8 Brillouin zones and energy bands 130

1 The Brillouin zone for the face-centred cubic lattice 130
2 Properties of the Brillouin zone faces 131
Conditions on the **k** vectors 131
Condition for $\partial E/\partial \mathbf{k}_n$ 132
3 Properties of the $E(\mathbf{k})$ surfaces and curves. The gradient 134
4 Energy bands: a one-dimensional example 137
Filling the bands. Conductors, semiconductors, and insulators 140
5 Reduced and extended zone schemes: Brillouin zones of higher order 141
6 Construction of the Brillouin zones of higher order 143
7 The energy level surfaces and the Fermi surface 144
Holes, semiconductors, and semimetals 148
Effective mass and holes 149
8 Problems 151

9 Bands and Fermi surfaces in metals and semiconductors 153

1 Silicon 153
The silicon structure (diamond crystal pattern) 153
The orbital basis 155
The Brillouin zone and **k**-vector symmetries 156
The band structure of silicon 157
Guessing band trends 160
2 Bands and Fermi surfaces in copper 161
3 Noble and transition metals 164
4 Problems 165

10 Methods of calculation of band structures 166

1 The nearly free-electron method 167
The band gap 169
Band gap classification 172
2 The tight-binding method 173
An example: the linear chain 175
3 The orthogonalized plane waves method (OPW) 178
Orthogonalization to the core 178
The crystal eigenfunctions 179
4 Pseudopotential method 180
The Austin–Heine–Sham pseudopotential 181
5 Problems 183

11 Phonons and conductivity 184

1 Conductivity and the Fermi surface 184
2 Crystal normal modes 186
Kinetic and potential energies in a crystal 187
Translations and their effect on $\boldsymbol{T}$ and V 189
The eigenvectors of the translations. Normal coordinates 190
3 Electron–phonon scattering 192
Scattering by a rigid crystal: Bragg scattering 193
The phonon drag 194
4 Lattice with basis. The silicon phonon spectrum 195
5 Problems 197

12 Phase stability: Brillouin zone effects 198

1 Effect of the Brillouin zone faces on the Fermi energy of alloys 198
2 Phase change in a linear chain: The Peierls instability 200
The electronic energy stabilization 202
Electron–phonon interactions 203

3 The Peierls instability for a quasi one-dimensional chain 205
Sticking together of the bands at the Brillouin zone edge 206
The double Brillouin zone 208
Reconstruction 209
4 Problems 210

13 Wannier functions and Löwdin orbitals 213

1 Equivalent functions 213
Definition of equivalent functions 214
Properties of cyclic groups and of their direct products 215
Construction of the equivalent functions 217
Orthogonality of the equivalent functions 218
2 The Wannier functions 219
The Wannier functions for free electrons 220
3 Löwdin orbitals 222
Appendix. The matrix $\boldsymbol{N}$ 224
4 Problems 225

14 Surface and impurity states 227

1 States in the energy gap. Tamm states 227
2 Shockley states 229
The bulk states 230
Breaking a strong bond 232
Breaking a weak bond 234
Comments. Shockley and Tamm states 234
3 The Koster and Slater method for impurities and surface states 235
Definitions and basic formulae 235
The Green function 237
Explicit form of the Green function 238
Localized perturbation 239
4 Surface states in a linear chain 240
The surface state 243
5 Problems 244

15 Solutions to the problems 248

1 Problems **1**-9 248
2 Problems **2**-7 250
3 Problems **3**-10 252
4 Problems **4**-11 253
5 Problems **5**-7 254
6 Problems **6**-6 256
7 Problems **7**-4 257

8 Problems **8**-8 258
9 Problems **9**-4 260
10 Problems **10**-5 262
11 Problems **11**-4 264
12 Problems **12**-4 265
13 Problems **13**-4 270
14 Problems **14**-5 270

References 274

Index 279

0

Notation

1. References

General cross-references

§ **3**-5	Section 5 of Chapter **3**. Chapters are numbered with bold-face numerals which are dropped in all cross-references, formulae, etc. corresponding to the current chapter.
§ 5	Section 5 of the current chapter.
(**6**-3.5)	equation (5) of § **6**-3.
(3.5)	equation 5 of § 3 of the current chapter.
(5)	equation 5 of the current section.
(L3), (R3)	left-hand side and right-hand side, respectively, of eqn (3).
2.1	after Problem, Fig., or Table, refers to the corresponding item serially numbered in Section 2. The section number is dropped within the current section.

Cross-references on left margins of displayed formulae

3	equation (3) is used to derive the equation on the right.
3′	equation (3), in a changed notation, is used to derive the equation on the right.
,3	equation (3) is used, but not immediately, to derive the equation on the right.
2,3	equations (2) and (3) are applied in that order to obtain the equation on the right.
2|3	equation (2) applied on equation (3) gives the equation on the right.
F, T, P	on any of the above, indicate a Figure, Table, or Problem, respectively.

Literature references

Ono (1945) identifies a paper or book under that name in the alphabetic list of references at the end of the book.

2. Symbols used

A', (A'')	one-dimensional irreducible representations of a point group symmetrical (antisymmetrical) with respect to a symmetry plane, §**6**-4.
$\mathbb{A}$	some arbitrary operator.
$\forall$	for all.
$\mathbf{a}, \mathbf{b}, \mathbf{c}$	unit vectors, mostly chosen to be primitive vectors.
$\mathbf{a}^{\#}, \mathbf{b}^{\#}, \mathbf{c}^{\#}$	reciprocal vectors, (**4**-2.6), (**4**-2.7); (**4**-3.3).
a^*, $\mathbf{a}^*$	complex conjugates. *Notice that in order to avoid confusion, the asterisk is never used in this book to denote reciprocal vectors.*
α	Coulomb integral, (**14**-2.1).
$\mathscr{B}$	Bravais lattice, §**3**-5.
β	resonance integral, (**10**-2.15), (**14**-2.2).
$C(g_i)$	class of the element g_i, §**2**-1.
$\lvert C(G) \rvert$	number of classes in G, §**15**-2, Problem 2.
$\mathscr{C}$	electron concentration per atom, §**12**-1.
$\chi(g \mid u)$	character of operation g in the basis $\boldsymbol{\langle u \rvert}$, (**12**-4.1).
$\chi(\mathring{g} \mid \hat{G})$	character of the operation $\mathring{g}$ (written as g when unambiguous) in the representation $\hat{G}$ of the group G, (**2**-4.17).
D	interplanar distance between planes of a stack, §**4**-4.
det A	determinant of the matrix A.
$\Delta\alpha$	perturbation potential at a lattice site, §**14**-3.
$\Delta\mathscr{E}$	band gap, §**10**-1.
δ_{ij}, $\delta(\mathbf{g}, \mathbf{g}')$	Kronecker's delta, (**2**-5.2), (**4**-9.10).
δ	distortion parameter for a linear lattice, §**12**-2.
$\boldsymbol{\delta}$	distortion vector for a linear lattice, §**12**-2.
$\partial E/\partial \mathbf{k}_{\mathbf{n}}$	derivative of the energy with respect to **k** for **k** varying along **n** normal to a Brillouin zone face, §**8**-2.
E	energy.
e, E	identity element of a group, (**2**-1.3)
$\mathscr{E}$	electric field.
E_0	eigenvalue of an atomic orbital at a given site, (**10**-2.17).
E_{F}	Fermi energy, §**1**-4.
E_T	total electronic energy, (**1**-6.7).
$E^j(\mathbf{k})$	energy as a function of **k** for the j-th band, §**6**-4.
e	electron charge.
$\varepsilon_{\mathbf{q}}$	translation eigenvalues labelled by the N discrete propagation vectors **q** in the Brillouin zone, (**11**-2.16).
$\in$	belongs to.
$\phi_{\mathbf{t}}(\mathbf{r})$	atomic orbitals, §**13**-3.

φ, $\varphi(\mathbf{r})$	momentum eigenfunction, (**1**-3.3), (**1**-3.13).
$\varphi_{\mathbf{k}}(\mathbf{r})$	momentum eigenfunction in a more complete notation, (**1**-8.1); free-electron waves.
φ_i	equivalent functions, (**13**-1.8).
$\varphi_{\mathbf{t}}(\mathbf{r})$	Wannier functions, (**13**-2.1); Löwdin orbitals, § **13**-3.
G	group of operations g; a space group.
$\lvert G \rvert$	order (number of elements) of G, (**15**-2.4).
$G(E)_{pm}$	Green function, (**14**-3.17).
$\hat{G}(\mathring{g})$, $\hat{G}(g)$	matrix representative of the operator $\mathring{g}$ (written as g when unambiguous), (**2**-4.5).
${}^{i}\hat{G}$	i-th irreducible representation of G, § **15**-2, Problem 2.
$\lvert {}^{i}\hat{G} \rvert$	dimension of the above, § **15**-2, Problem 2.
g	configuration space operator, § **2**-2.
$\mathring{g}$	function space operator, written as g when unambiguous, (**2**-2.7).
g_i^{-1}	inverse of element g_i, (**2**-1.4).
$\mathbf{g}$	vector *of* the reciprocal lattice, (**4**-2.10), (**4**-6.4).
$\mathrm{grad}_{\mathbf{k}}$	gradient in **k** space, (**8**-3.1).
Γ	centre of the Brillouin zone.
γ	resonance integral, (**14**-2.3).
γ	glide plane, (**12**-3.1).
$H \subset G$	H is a subgroup of G, (**2**-1.5).
$H \lhd G$	H is an invariant subgroup of G, (**2**-1.6).
$\mathbb{H}$	quantum mechanical Hamiltonian.
$\mathbb{H}$ (italic)	pseudo Hamiltonian, (**10**-4.6).
$\mathbb{H}$ (outline italic)	Austin–Heine–Sham pseudo Hamiltonian, (**10**-4.15).
$\mathbb{H}$ (outline)	perturbed Hamiltonian, (**14**-3.2).
$H_{\mathbf{kk}'}$	Hamiltonian matrix element, (**10**-1.8).
$\hbar$	Planck's constant h divided by 2π.
I	transition probability integral, (**12**-2.9).
$\lvert I \rvert$	number of irreducible representations in G, § **15**-2, Problem 2.
i	imaginary unit.
$\mathring{\mathbb{i}}$	inversion at the origin of coordinates.
$\mathbf{J}$	current density, (**11**-1.2).
$\mathring{\mathbb{j}}$	conjugator operator, (**2**-3.11).
$\mathbf{k}$	vector of orthogonal components $[k_x, k_y, k_z]$, (**1**-3.15), or more usually a vector $[k_x, k_y, k_z]^{\#}$ *in* the reciprocal lattice, (**4**-6.3).
$\lvert \mathbf{k} \rvert$	circular wave number, (**1**-8.6).
k_{B}	Boltzmann's constant.
κ	oscillator force constant, § **11**-2.
$\boldsymbol{\kappa}$	small vector normal to a Brillouin zone face, (**14**-1.9).
κ_x	integer; likewise κ_y, κ_z, (**1**-3.19).

L	length of one-dimensional crystal.
$L \otimes M$	direct product, (**2**-1.8).
λ	wave length.
m	electron mass.
μ	mass of an oscillator particle, § **11**-2.
$\boldsymbol{\mu}$	real factor of the imaginary vector $\boldsymbol{\kappa}$, § **14**-1.
N	total number of primitive cells in a crystal, (**5**-1.3); number of atoms in a piece of metal.
N_x	number of primitive cells of a crystal in the x direction, § **5**-1.
$\mathcal{N}$	electron concentration or density, § **1**-5, § **12**-1.
$\mathbf{N}$	number of electrons, § **1**-4.
$\boldsymbol{N}$	normalization constant, (**15**-1-6).
$\boldsymbol{N}_{\mathbf{k}}$	normalization constant, (**13**-3.10).
n	order of the set $\{g_i\}$.
$n(E)$	density of states, (**1**-5.1).
Ω	volume of the unit (primitive) cell of the crystal, (**4**-9.1).
ω	circular frequency (**1**-8.7); oscillator frequency, § **11**-2.
$\omega_{\mathbf{q}}$	normal frequency, § **11**-2.
P	point group of a space group G, § **3**-5.
$P_{\mathscr{B}}$	point group of the Bravais lattice $\mathscr{B}$, § **3**-5.
$\mathbb{P}$	projection operator, (**10**-3.10).
$\mathbf{p}$	momentum (linear).
$\mathbb{p}$	momentum operator.
$\mathbf{q}$	propagation vector with discrete values (N in number) which label vibrational states, (**11**-2.19).
$\mathbf{r}$	position vector of components $[xyz]$, (**4**-4.3); position vector *in* the lattice (**4**-1.12), (**4**-6.1).
$\boldsymbol{\rho}$	displacement vector of an atom from its equilibrium position, (**11**-2.6).
$\boldsymbol{\rho}_n$	position vector of the n-th atom of the basis, § **4**-9.
$\boldsymbol{\rho}_{\mathbf{t}}$	displacement vector corresponding to the site $\mathbf{t}$, § **11**-2.
$S_{\mathbf{kk}'}$	overlap integral, (**10**-1.9).
S	Schrödinger group, § **2**-3.
$\mathbf{s}_{\mathbf{t}}$	reduced coordinates, (**11**-2.13).
$\boldsymbol{\sigma}$	electrical conductivity, conductivity tensor, (**11**-1.2).
σ	reflection plane.
σ_v, σ_d	reflection planes (vertical, that is parallel to a principal rotation axis), the σ_d planes bisecting the angle between two binary axes normal to the principal rotation axis.
$\mathbb{T}$	kinetic energy operator.
T	translation subgroup, § **3**-8.
$\|T\|$	order of the translation group T.
$\boldsymbol{T}$	kinetic energy, (**11**-2.10).

T	period (as inverse of frequency).
$_{\mathbf{k}}\hat{T}$	**k**-th irreducible representation of the translation group T, § **5**.2.
$_{\mathbf{k}}\hat{T}\{E\|\mathbf{t}\}$	the **k**-th irreducible representation of $\{E\|\mathbf{t}\}$, (**5**-2.9).
$_{k_x}\hat{T}\{E\|m\mathbf{a}\}$	the k_x irreducible representation of the translation $\{E\|m\mathbf{a}\}$, § **5**.2.
t	translation in configuration space by the translation vector **t**, (**2**-2.3); translation $\{E\|\mathbf{t}\}$, (**11**-2.16).
$\mathbf{t}$	a translation vector, in particular a translation vector of the Bravais lattice $\mathscr{B}$ or vector *of* the lattice, (**3**-8.1), (**4**-6.2).
τ	relaxation time, (**11**-1.3).
$u^j_{\mathbf{k}}(\mathbf{r})$	cell function in the j-th band, (**5**-6.11).
$u_{\mathbf{k}}(\mathbf{r})$	cell functions, (**5**-6.4).
$\mathbf{u}, \mathbf{v}, \mathbf{w}$	primitive vectors.
$\mathbb{V}$	potential energy operator, (**10**-1.1).
V	crystal volume, volume of a lump of a solid, (**1**-3.24).
V	potential energy, (**11**-2.11).
$V(\mathbf{r})$	crystal potential field.
$V_{\mathbf{g}}$	Fourier coefficient, (**4**-9.19).
$\mathscr{V}_{\mathbf{g}}$	structure factor, (**4**-9.26).
v	volume in **k** space occupied by one electron state, spin included, (**1**-3.25).
v'	volume in **k** space occupied by one electron state, spin not included, (**1**-3.24).
v_p	phase velocity, (**1**-8.8).
$\mathbf{v}$	fractional translation vector, (**3**-4.3).
$\mathbf{v}$	velocity, § **11**-1 only.
$\mathbf{v}_g$	group velocity, (**1**-8.14).
$\mathbb{W}$	pseudopotential, (**10**-4.8).
$\mathbf{w}$	either a translation vector **t** or a fractional translation vector, § **3**-6.
X	bisector point of one of the unit vectors of the reciprocal lattice; belongs to a Brillouin zone edge, (**6**-4.2).
ξ_i	normal coordinates, (**11**-2.4).
$\Theta(x)$	step function, (**14**-4.10).
Ψ	perturbed wave function, (**14**-3.12).
ψ	wave function.
ψ^i	unperturbed wave function, (**14**-3.3).
$\psi^j_k(\mathbf{r})$	Bloch function of the j-th band, (**5**-6.11).
ψ^k	molecular orbital (linear combination of atomic orbitals), (**14**-4.1).
$\psi_{\mathbf{k}}(\mathbf{r})$	translation eigenfunctions, (**5**-5.1); Bloch functions, (**5**-6.3), (**5**-6.9).

$\mathbf{z_q}$	eigenvectors of the translations $\{E\,\vert\,\mathbf{t}\}$, (**11**-2.16); normal coordinates, (**11**-2.18).
(hkl)	Miller indices of a plane of a stack, § **4**-4.
$[a\ b\ c]$	a vector in direct components.
$[a\ b\ c]^{\#}$	a vector in reciprocal components.
$\{g_i\}$	set of all elements g_i, $i = 1, 2, \ldots, n$.
$\{E\,\vert\,\mathbf{t}\}$	translation group operation, § **3**-8.
$\{\mathbb{i}\,\vert\,\mathbf{0}\}$	inversion operator, § **6**-1.
$\{p\,\vert\,\mathbf{0}\}$	point group operation, (**3**-7.2).
$\{p\,\vert\,\mathbf{w}\}$	Seitz space group operator, (**3**-6.2).
$\vert\mathbf{g}\rangle$	plane wave *of* the lattice, (**4**-10.1), (**4**-9.6).
$\vert\mathbf{k}\rangle$	free-electron plane wave *in* the lattice, (**4**-10.2).
$\vert\varphi(\mathbf{r})\rangle$	ket, § **4**-9.
$\vert\varphi_{\mathbf{k}}\rangle$	pseudo wave function, (**10**-4.2).
$\vert m\rangle$	orbital $\phi(m)$ at lattice site m, § **10**-2.
$\langle\varphi(\mathbf{r})\vert$	bra, (**4**-9.2).
$\langle\varphi(\mathbf{r})\vert\psi(\mathbf{r})\rangle$	bra-ket, (**4**-9.3).
$\langle\varphi_1, \varphi_2, \ldots, \varphi_n\vert$	row vector of components $\varphi_1, \varphi_2, \ldots, \varphi_n$; basis of a representation, § **2**-4.
$\langle\varphi\vert$	abbreviated form of the above symbol. A basis of the space group G, (**6**-2.1).
$\bar{\otimes}$	symmetrized direct product, (**12**-4.4).
$\rightarrow$	mapping: the set on the left of this symbol maps into the set on the right.
$\mapsto$	mapping: the element on the left of this symbol maps to the element on the right.
$\Rightarrow$	if then: the statement on the left of this symbol implies the statement on the right.
$=_{\text{def}}$	the corresponding equality entails a definition.

1

The free-electron picture

We shall be concerned in this chapter with a simple model of solid materials which is particularly adapted for the description of metallic crystals. The discussion of this picture, however, will serve as an introduction to some of the ideas of the band formalism which will be developed in later chapters and which will also allow us to describe covalent crystals, such as silicon.

1 Free-electron model

Consider as an example sodium metal: the sodium atoms in their ground states have the configuration $1s^2$, $2s^2$, $2p^6$, $3s$, in which the first ten electrons (*core electrons*) are tightly bound, whereas the $3s$ electron (*valence electron*) is easily ionizable, this property being a characteristic of metal atoms. We shall construct a model of a metal based on two assumptions. First, we assume that the valence electrons are ionized, leaving behind positive ion cores at the atomic sites. The valence electrons can thus move over the crystal and they are often called the *metallic electrons* because it is they that give the crystal its characteristic metallic properties. The second assumption is not so easily acceptable, although as we shall see the results of the theory provide an eventual justification. We shall imagine that each metallic electron moves quite freely, independently of all the other electrons and of the positive ion cores, very much like each particle of an ideal gas moves, except of course that it will follow a different energy-distribution function. The picture which supports this approximation is the following one. First, we take the repulsive interaction between any two electrons to be sufficiently weak with respect to all other interactions so that it can be neglected. Secondly, if we consider for the sake of an example a piece of sodium metal containing N atoms, each electron will move in the field of N singly-charged positive ions plus $N - 1$ metallic electrons. If we assume, as it is compatible with the neglect of electron–electron interactions, that the $N - 1$ metallic electrons are uniformly distributed over the crystal, they will therefore cancel out (or *screen*, as it is said), fairly precisely (for N large) the field of the N positive ions, so that any one electron will move in an approximately vanishing field. In order to solve quantitatively for the energy eigenvalues in the free-electron picture which we are constructing it is necessary to discuss the boundary conditions to be used.

2 Born–von Karman boundary conditions

Since we want to describe the electrons as being able to travel freely through the finite piece of crystal to be studied, it is sensible to consider the latter to be part of an infinite crystal. This is acceptable since, if our crystal piece is not too small, its properties should not depend on its size and we can assume just as well that it is infinitely large. Also, for a reasonably large crystal, its surface is negligible in extent in relation to its bulk, so that the effect of the surface on the bulk states, in which we are interested, can be neglected. The way in which the proposed extension of the crystal up to infinity is to be carried out can thus be chosen with complete freedom as to the conditions attached to the surface of the given piece, since these conditions cannot affect the bulk states. This allows us to make an assumption as regards the manner of this extension which was first proposed by Born and von Karman and which considerably simplifies the work, since it avoids the difficulties of dealing with an infinite crystal.

For simplicity, we shall consider a one-dimensional crystal of length L lying along the x axis between the origin O and the point of abscissa L, as shown in Fig. 1*a*. We extend this piece of metal periodically to $L, 2L, 3L, \ldots, -L, -2L, -3L, \ldots$, meaning by this statement that when the original piece of metal is translated so that O goes to L and L to $2L$, say, then absolutely nothing can change. We have thus embedded our crystal in an infinite one, but, as we have seen, this embedding is done periodically. In this way, rather than being presented with awkward boundary conditions at plus and minus infinity, all that we require is that any physically significant function in the crystal must have the same value at O as at L, that is, must be *periodic* in the period from O to L.

Another way in which this periodic condition can be brought about is by means of a different but equivalent picture. Imagine that the original piece of metal is bent to form a circle as in Fig. 1*b*. (If the metal piece is long enough, the curvature introduced on bending is negligible and cannot affect physical properties.) The periodic translation by $L, 2L, 3L, \ldots$, in Fig. 1*a* corresponds to rotation by $2\pi, 4\pi, 6\pi, \ldots$, in Fig. 1*b*, which, obviously, cannot affect physical properties.

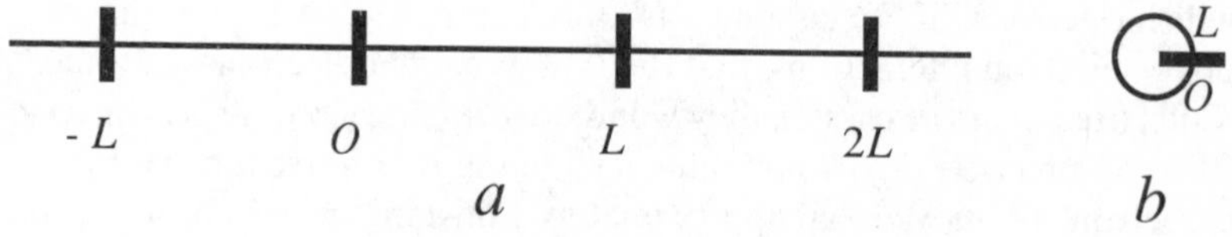

Fig. 1-2.1. Periodic boundary conditions. The circle in *b* has precisely the same length as the segment in *a* from 0 to *L*.

In three dimensions, the crystal must be extended periodically along three directions which we take for convenience to be orthogonal: we are ignoring the crystal structure for the time being and we can therefore assume for simplicity that our finite lump of the solid is a cube of side L. Although this conceptual operation is now much more difficult (if not impossible) to visualize, all that it means is that we require all our physical functions in the crystal to be invariant under translation along x or y or z by L.

3 Quantum-mechanical treatment

Remember that in our free-electron model we deal with one electron at a time moving in a potential field which vanishes throughout the crystal. We must therefore solve the Schrödinger equation for a single electron

$$\mathbb{H}\psi = E\psi, \tag{1}$$

where $\mathbb{H}$ is the quantum-mechanical Hamiltonian operator, ψ the wave function, and E the one-electron energy. The Hamiltonian $\mathbb{H}$ is the operator that corresponds to the energy and the quantum mechanical prescription required in order to write it is as follows. We must first write the classical energy expression in terms of the momentum and position of the particle. This expression is called the *classical Hamiltonian H*:

$$H = \frac{1}{2m}\mathbf{p}^2 + V(\mathbf{r}). \tag{2}$$

Here $\mathbf{p}$ is the momentum of the electron, that is the vector $m\mathbf{v}$, where m and $\mathbf{v}$ are the mass and velocity, respectively, of the electron. The first term in (R2) is the *kinetic energy* of the electron, whereas the second term $V(\mathbf{r})$ is the *potential energy* at the point $\mathbf{r}$ of the crystal, $\mathbf{r}$ denoting the position vector from the origin of coordinates to the point in question. For free electrons, $V(\mathbf{r})$ vanishes and can of course be ignored. The quantum-mechanical Hamiltonian $\mathbb{H}$ is obtained by replacing $\mathbf{p}$ by its corresponding momentum operator $\mathbb{p}$, for which the prescription is well known (see below). Before we do this, it will be useful for later purposes to study the eigenfunctions of $\mathbb{p}$ on their own. These must satisfy the eigenvalue equation

$$\mathbb{p}\varphi = \mathbf{p}\varphi, \tag{3}$$

in which the eigenvalue is the momentum vector $\mathbf{p}$. The momentum eigenfunctions φ are functions of position $\varphi(\mathbf{r})$, and they are most important, since we shall now prove that, in the free-electron case, they are also eigenfunctions of the Hamiltonian, which is, from (2),

$$\mathbb{H} = \frac{1}{2m}\mathbb{p}^2. \tag{4}$$

It follows in fact, by first acting on (3) with $\mathbb{p}$ and then using eqn (3) once more, that

$$\mathbb{p}\mathbb{p}\varphi = \mathbf{p}\mathbb{p}\varphi = \mathbf{p}^2\varphi, \tag{5}$$

whence, from (4),

$$\mathbb{H}\varphi = \frac{1}{2m}\mathbf{p}^2\varphi, \tag{6}$$

which proves, on comparison with (1), that the momentum eigenfunctions φ are the eigenfunctions ψ of the Hamiltonian, corresponding to the energy

$$E = \frac{1}{2m}\mathbf{p}^2 = \frac{1}{2m}|\mathbf{p}|^2, \tag{7}$$

as one should expect, since this is the kinetic energy of the electron.

In order to solve (3), and thus the Schrödinger equation (6), we consider first the component equation of (3) along the x axis:

$$\mathbb{p}_x\varphi(x) = p_x\varphi(x). \tag{8}$$

It is well known from quantum mechanics that the momentum operator $\mathbb{p}_x$ is $-\hbar\mathrm{i}\partial/\partial x$, where $\hbar$ is Planck's constant h divided by 2π:

$$-\hbar\mathrm{i}\partial\varphi(x)/\partial x = p_x\varphi(x). \tag{9}$$

This equation indicates that $\varphi(x)$ must be an exponential, $\exp(\alpha x)$, say. For real α, however, this goes to infinity for infinite x, which will not do (otherwise, the electron would not be confined to the region in x, from O to L, where the metal crystal lies). We thus take α imaginary, $\mathrm{i}k_x$, say. On taking $\varphi(x)$ as $\exp(\mathrm{i}k_x x)$ in (9), we have

$$-\hbar\mathrm{i}\mathrm{i}k_x\exp(\mathrm{i}k_x x) = \hbar k_x\exp(\mathrm{i}k_x x) = p_x\exp(\mathrm{i}k_x x), \tag{10}$$

so that

$$p_x = \hbar k_x, \qquad E = \frac{1}{2m}p_x^2 = \frac{1}{2m}\hbar^2 k_x^2. \tag{11}$$

If we now consider the three orthogonal directions x, y, z, along the edges of the piece of metal under consideration, which can always be taken to be a cube, as discussed in § 2, it easily follows (see, for example, Problem 9.1) that the eigenvalue in (3) is the sum of the eigenvalues (in the vectorial sense) and that the eigenfunction is the product of the eigenfunctions:

$$\mathbf{p} = \hbar[k_x, k_y, k_z] =_{\mathrm{def}} \hbar k, \tag{12}$$

(here, as always in this book, the square bracket denotes a vector, in this case on the basis of the three unit vectors along the orthogonal directions x, y, z);

$$\varphi(\mathbf{r}) = \exp(\mathrm{i}k_x x)\exp(\mathrm{i}k_y y)\exp(\mathrm{i}k_z z) \tag{13}$$

$$= \exp\{\mathrm{i}(k_x x + k_y y + k_z z)\} = \exp(\mathrm{i}\mathbf{k}\cdot\mathbf{r}), \tag{14}$$

where, as in (12),

$$\mathbf{k} = [k_x, k_y, k_z], \qquad \mathbf{r} = [x\ y\ z]. \tag{15}$$

The functions (14) are eigenfunctions of the energy with eigenvalue E equal to $\hbar\mathbf{k}^2/2m$. It should be noticed that they depend on the $\mathbf{k}$ vector which, except for the constant $\hbar$, coincides with the linear momentum of the free electron. For this reason (see also § 8) the $\mathbf{k}$ vector is also called the *propagation vector*. From the point of view of the energy eigenfunctions $\varphi(\mathbf{r})$, this $\mathbf{k}$ vector acts as a label for the distinct eigenstates of the energy, which should thus more properly be rewritten as $\varphi_{\mathbf{k}}(\mathbf{r})$. It is also important to realize that the energy eigenvalues just quoted are *doubly degenerate*. This is so because it is not only the eigenfunction $\exp(\mathrm{i}\mathbf{k}\cdot\mathbf{r})$ of (14), corresponding to the momentum $+\hbar\mathbf{k}$, which belongs to this energy eigenvalue, but also, as it can be seen from (9) and (10), the eigenfunction $\exp(-\mathrm{i}\mathbf{k}\cdot\mathbf{r})$, corresponding to the momentum $-\hbar\mathbf{k}$. It is clear that this must be so since the two functions mentioned differ only in the sign of $\mathbf{k}$, whereas the energy eigenvalue depends only on $\mathbf{k}^2$. It can be seen in a different way that, although these functions correspond to equal but opposite momenta, they must belong to the same energy: the energy is a *scalar* quantity which, by symmetry of free space, cannot depend on the direction of propagation of a free electron but only on the modulus of its momentum.

Boundary conditions

We must satisfy periodic boundary conditions as in § 2, but now with respect to three periods along orthogonal directions, L_x, L_y, L_z, say, all equal to L. (Remember that the lump of metal studied is always assumed to be a cube, since its shape cannot affect the properties of the free electrons inside it.) Thus:

$$\varphi(\mathbf{0}) = \varphi([L\,0\,0]) = \varphi([0\,L\,0] = \varphi([0\,0\,L]). \tag{16}$$

14|16
$$\exp(0) = \exp(\mathrm{i}k_x L) = \exp(\mathrm{i}k_y L) = \exp(\mathrm{i}k_z L) = 1. \tag{17}$$

17
$$k_x L = 2\pi\kappa_x, \qquad k_y L = 2\pi\kappa_y, \qquad k_z L = 2\pi\kappa_z, \tag{18}$$

$$\kappa_x, \kappa_y, \kappa_z = 0, \pm 1, \pm 2, \ldots . \tag{19}$$

Equations (18) and (19) mean that the $\mathbf{k}$ vector (and therefore the momentum) is *quantized*,

18|15
$$\mathbf{k} = \frac{2\pi}{L}[\kappa_x\,\kappa_y\,\kappa_z], \tag{20}$$

and so also is the energy:

12|7, 20
$$E = \frac{1}{2m}\hbar^2|\mathbf{k}|^2 = \frac{1}{m}\hbar^2\frac{2\pi^2}{L^2}(\kappa_x^2 + \kappa_y^2 + \kappa_z^2). \tag{21}$$

Notice that the energy eigenstates can be very highly degenerate. This degeneracy arises through the fact that each **k** corresponds to one eigenfunction $\exp(i\mathbf{k}\cdot\mathbf{r})$ whereas the energy depends only on $\mathbf{k}^2$. We have already seen, for example, that **k** and $-\mathbf{k}$ lead to the same energy but we shall now consider a more general example. It follows at once from (21) that all values of **k** of the form displayed below lead to the energy eigenvalue shown:

19|20, 21
$$\mathbf{k} = \frac{2\pi}{L}[\pm 1, \pm 1, \pm 1] \quad \Rightarrow \quad E = \frac{1}{m}\hbar^2 \frac{2\pi^2}{L^2} 3. \tag{22}$$

Since there are eight values of **k** of the form given in (22), all corresponding to the same energy, the *orbital degeneracy* is eight, and the *total degeneracy*, including spin, is sixteen.

The **k** space

Notice that the propagation vector **k** (which, from eqn 12, is effectively the momentum vector, except for a factor of $1/\hbar$) determines fully the energy, as shown in (21). Thus, the energy can best be plotted as a function of the **k** vector in a *space* with k_x, k_y, k_z along orthogonal axes. This space is called the **k** *space.* All that we are saying here is that if we take k_x, k_y, k_z along orthogonal axes then, for each point of the space so defined, a value of the energy can in principle be given. The fact that such a function is not easy to visualise should not worry us in interpreting the meaning of this space: the temperature as a function of position in ordinary space is just as difficult to represent.

It must be remembered that, from (20), this space is quantized, which means that only discrete points in this space correspond to physical states. (The **k** vector, in fact, is a sort of three-dimensional quantum number which labels the momentum and energy eigenstates.) It is easy to see that these acceptable values of **k** form a simple cubic lattice, as we shall now verify. The only values of **k** permitted along the x direction are, from (19) and (20),

$$\ldots, \frac{-2\pi}{L}3, \frac{-2\pi}{L}2, \frac{-2\pi}{L}, 0, \frac{2\pi}{L}, \frac{2\pi}{L}2, \frac{2\pi}{L}3, \ldots \tag{23}$$

and exactly the same sequence follows along the y and z directions. We thus have a simple cubic lattice of lattice constant equal to $2\pi/L$, the points of which are the permitted values of **k**.

Notice a very useful feature of **k** space in this quantized form: each point of it corresponds to one permitted energy state of one electron in the system. In practice, however, we are allowed by the Pauli principle to place two electrons with opposite spins in such a state, so that, when counting states, we must carefully consider whether spin is included or not.

It will be useful for later purposes to find the volume v' occupied by one state (no spin included) in **k** space. Each elementary cube in this space has

8 permitted states at its vertices, but each of these vertices is shared by 8 such cubes of the simple cubic lattice. It follows therefore that each elementary cube corresponds to one state (no spin included) in **k** space and thus that v' is the volume of this cube, the side of which, from (23), is $2\pi/L$. Therefore,

$$v' = 8\pi^3/L^3 = 8\pi^3/V, \tag{24}$$

where V is the crystal volume. A more important quantity than v' is v, the volume occupied in **k** space by one electron state, spin included. Clearly, it is a half of v':

24 $$v = 4\pi^3/V. \tag{25}$$

Notice that, because V is large (of macroscopic dimension), v is very small. That is, although the **k** space is discrete, i.e. 'grained', this graining is so dense that for most purposes the **k** space can be regarded as quasi-continuous.

4 Physical consequences: Fermi energy and Fermi surface

It is clear from (3.21) that, in one dimension, the energy E is a parabola as a function of **k**, as shown in Fig. 1*a*, where we also take account of the quantization of E required by (3.19). Notice that the two distinct states for **k** and $-$**k** are degenerate and also that the electrons are *fermions*, that is particles for which the Pauli principle requires that no more than two electrons, of opposite spins, may occupy the same *orbital state* which is a state entirely defined by the space coordinates alone. If we have **N** electrons in the crystal we must start filling states from the lowest energy level upwards, until all **N** electrons have been accommodated. The energy just reached, that is the energy of the highest occupied state, is called the *Fermi energy*, E_F. The effect

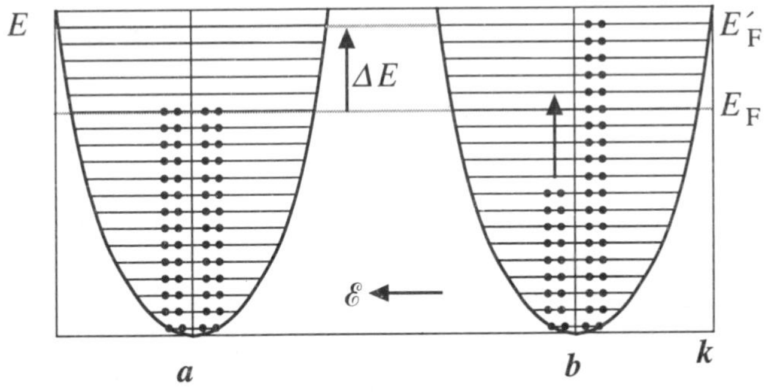

Fig. 1-4.1. The energy E as a function of **k** in the free-electron case, (*a*), and in the presence of an electric field $\mathscr{E}$, (*b*). Notice that the Fermi energy in *b* has been raised by ΔE to the new value E'_F. In the absence of the field each level is occupied by two pairs of electrons of opposite momenta, the electrons in each pair being of opposite spin. The energy levels shown symbolize a quasi-continuous distribution.

of an electric field $\mathscr{E}$ on this electron distribution is illustrated in Fig. 1*b*. Naturally, we no longer have free (and therefore isotropic) space, so that the **k**, −**k** degeneracy is destroyed. In the example shown, the field will raise the energy of each state with momentum $\mathbf{p} = -\hbar\mathbf{k}$ by a positive value ΔE, that is, it will make more favourable the states with positive **k**. (Remember that the electron charge is negative!) As a result, all the states with negative **k** for which $E + \Delta E$ is larger than the Fermi energy E_F will become empty, since states with positive **k** exist which now have lower energy. The result is that there are now more electrons with positive than with negative **k**, and there is thus a flow of electrons to the right. We have in this way a simple explanation of the conductive properties of metals.

The generalization of these ideas to three dimensions is easy. If we represent the energy as a function of **k** in **k** space with k_x, k_y, k_z along orthogonal axes, then it follows from (3.21) that the surfaces of constant energy are spheres (for which $|\mathbf{k}|^2$ is constant), with the energy increasing outwards from the origin. Thus the states near the origin must fill first and when all the **N** electrons have been accommodated then a sphere of energy E_F is reached: this sphere is called the *Fermi sphere*. More generally, the isoenergetic surfaces are not spheres, as will be the case when the electrons are not entirely free, and then the surface in **k** space outside which all the states are unoccupied is called the *Fermi surface*.

5 The density of states

Each quantized value of **k** determines an (orbital) electron state, the degeneracy of which is doubled by spin. In order to avoid confusion we shall agree that *electron states will always be assumed to include the spin degeneracy*. Likewise, when actual electrons are counted it will be assumed that *electrons of either spin are always included*. In order to avoid our quantities depending on the size of the metal sample under study, we must first of all define the *electron density* $\mathscr{N}$, which is the total number of electrons *per unit volume*. We known that at 0 K all electrons in the crystal are at an energy equal to or lower than the Fermi energy. (At higher temperatures, of course, some electrons will be thermally excited above the Fermi energy.) That is, at 0 K, all the states up to the Fermi level are occupied, but the number of electrons in each state varies a great deal because, as shown in § 3, these states are highly degenerate and the degeneracy varies steeply with the energy. Given the degeneracy (always including spin) of each state, we know the number of electrons which have the corresponding energy E and thus the electron density (number of electrons per unit volume) of the electrons with any given energy E. This quantity is nevertheless of litttle significance, as we shall now discuss. First, we notice that although E is quantized, the distribution of its values is practically continuous, since E is incremented by incrementing in unit steps the integers κ in (3.21), which leads to small energy increments of the order of $\hbar^2$. Secondly, we must remember that given a continuous distribution in E the probability of determining a single precise value of E is nil, so that it does not make much sense to ask for the number of electrons that occupy such a state. We define for this reason a new and very important quantity called the *density of states* $n(E)$. This is the density of electrons with energy in the range $\mathrm{d}E$ around the stated energy E, per unit range of energy. This means that the stated density of electrons is divided by the interval $\mathrm{d}E$ in order that the defined quantity be independent of $\mathrm{d}E$. From this definition, $n(E)\,\mathrm{d}E$ is the density of electrons in the range $\mathrm{d}E$, so that we can write

$$n(E)\,\mathrm{d}E = \text{density of electrons in range } \mathrm{d}E \text{ around } E. \tag{1}$$

It is not difficult to calculate (R1). If we consider $E(\mathbf{k})$, the energy as a function in $\mathbf{k}$ space, in order to keep to a constant E as in (R1), we must keep to the surface of a sphere of the radius $|\mathbf{k}|$ given by (3.21) for the chosen value of E. The electrons in the range $\mathrm{d}E$ given in (R1) will thus form a spherical shell in $\mathbf{k}$ space of thickness $\mathrm{d}|\mathbf{k}|$, which we write as $\mathrm{d}k$. Their number will be given by

$$\text{volume of shell/volume occupied by each state} =$$

3.25

$$= (\text{area of sphere radius } |\mathbf{k}|)\,\mathrm{d}k/(4\pi^3/V) \tag{2}$$

$$= V4\pi|\mathbf{k}|^2\,\mathrm{d}k/4\pi^3. \tag{3}$$

The *density* of the electrons in question required in (R1) is clearly obtained on dividing (3) by V. Thus, in (1):

$$n(E)\,\mathrm{d}E = |\mathbf{k}|^2\,\mathrm{d}k/\pi^2, \tag{4}$$

where, from (3.21),

$$|\mathbf{k}| = (2mE)^{1/2}/\hbar. \tag{5}$$

Equation (5) is now used twice, first to replace $|\mathbf{k}|$ in (4) and then to obtain $\mathrm{d}k$ in terms of $\mathrm{d}E$:

$$n(E)\,\mathrm{d}E = \{(2m)^{3/2}\,E^{1/2}/(2\pi^2\hbar^3)\}\,\mathrm{d}E. \tag{6}$$

The density of states follows at once from (6):

$$n(E) = (2m)^{3/2}\,E^{1/2}/(2\pi^2\hbar^3). \tag{7}$$

This is a parabola as a function of E, shown in Fig. 1. Since at $\mathrm{T} = 0\,\mathrm{K}$ there are no occupied levels for $E > E_\mathrm{F}$, the density of states has the form shown by the thick curve in the figure. Thermal excitation of the free-electron gas, which is $k_\mathrm{B}\mathrm{T}$, or about 0.03 eV at normal temperatures, will blur somewhat the Fermi edge, as shown by the grey line in the figure.

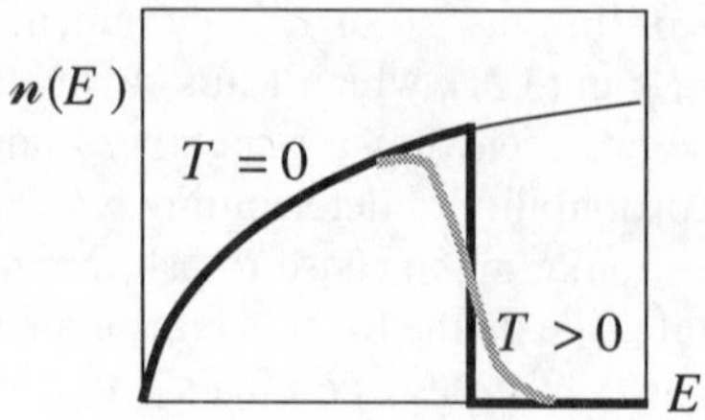

Fig. 1-5.1. Density of states for a free-electron gas.

6 The Fermi energy and the total energy

It is clear from (5.1) that, if we integrate that equation in the energy range from 0 to E_F, the right-hand side must become the total electron density $\mathcal{N}$, since no electrons occupy states above the Fermi energy:

$$\int_0^{E_F} n(E)\,\mathrm{d}E = \mathcal{N}. \tag{1}$$

This relation will allow us to obtain E_F.

5.7|1
$$\mathcal{N} = (2m)^{3/2}(2\pi^2\hbar^3)^{-1}\int_0^{E_F} E^{1/2}\,\mathrm{d}E. \tag{2}$$

$$= (2m)^{3/2}(2\pi^2\hbar^3)^{-1}\frac{2}{3}E_F{}^{3/2}. \tag{3}$$

3
$$E_F = \frac{1}{2m}\hbar^2(3\pi^2\mathcal{N})^{2/3}. \tag{4}$$

As an example of the use of this equation, we know in sodium that there is only one electron per atom, which allows us to compute $\mathcal{N}$ quite easily. The free-electron Fermi energy then turns out to be about 3.2 eV. (See Problem 9.5.)

The Fermi energy is particularly significant for the free-electron system since, as we shall now see, the *total energy* E_T of all electrons occupying states up to E_F is proportional to E_F:

,5.6
$$E_T = \int_0^{E_F} E n(E)\,\mathrm{d}E = (2m)^{3/2}(2\pi^2\hbar^3)^{-1}\int_0^{E_F} E^{3/2}\,\mathrm{d}E \tag{5}$$

$$= (2m)^{3/2}(2\pi^2\hbar^3)^{-1}\frac{2}{5}(E_F)^{5/2}. \tag{6}$$

On introducing here the value of $(2m)^{3/2}(2\pi^2\hbar^3)^{-1}$ from (3), we get

$$E_T = \frac{3}{5}\mathcal{N}E_F. \tag{7}$$

7 Bands

The description of sodium metal provided by the free-electron model is depicted in Fig. 1. We show on the left here the energy levels of an isolated sodium atom. We assume in the metal that the core levels 1*s*, 2*s*, 2*p* are unaltered, but that the 3*s* level spreads out into the free-electron levels which form a *band* of energy levels occupied up to E_F. Because, as we have seen, E_F is 3.2 eV, the *band width* of this band is also 3.2 eV. This band is often called the 3*s band* in order to indicate the provenance of its energy levels. When an electron from the 3*s* band decays to a core level it will emit X-rays. If, as shown in the figure, the core level is high in energy, the X-rays emitted must be of low energy and such X-rays are called *soft X-rays.* Measurement of their intensity as a function of the energy (that is as a function of the band level from which the electrons decay) gives some indication of the band width and of the form of the density of states curve. The intensity of the line shown in the figure, for example, must depend on the occupancy of the emitting level, that is on the value of $n(E)$ shown in grey. Therefore, the intensity of the soft X-ray spectrum should in principle reproduce the shape of the $n(E)$ curve. In practice, however, this is not precisely the case, because the intensity of the emitted radiation depends on factors other than the occupancy of the levels, and these factors are not by any means constant along a band. The shape of

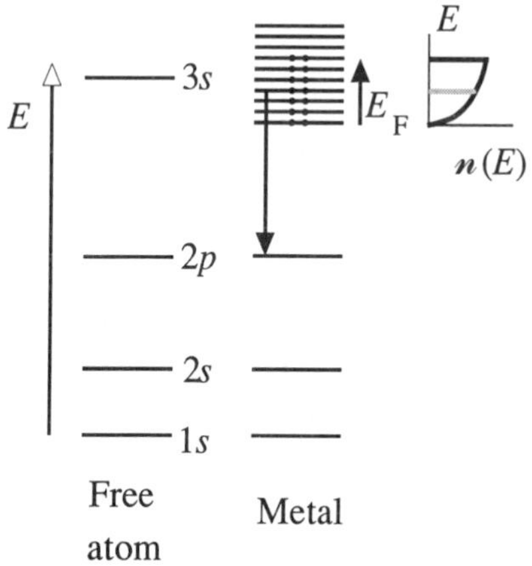

Fig. 1-7.1. The free-electron picture of sodium metal.

the soft X-ray spectrum, however, agrees semi quantitatively with the free-electron density of states curves for some simple metals like sodium and aluminium, and provides strong experimental support for the validity of the free-electron model at least as a first approximation.

Before we go further in preparing our background for the study of more precise approximations in band theory, it will be useful to revise briefly the properties of the electron waves which are described by the free-electron eigenfunctions, which we shall do in the next section.

8 Electron waves

We wrote the free-electron eigenfunctions in (3.14) and their corresponding energies are given underneath (3.15):

$$\varphi_{\mathbf{k}}(\mathbf{r}) = \exp(\mathrm{i}\mathbf{k}\cdot\mathbf{r}), \qquad E = \hbar^2\mathbf{k}^2/2m. \tag{1}$$

These functions turn out to be *plane waves*, so that it is worthwhile reviewing the definition and properties of such waves, for which the **k** vector has a very clear physical meaning, as we shall see, which will explain why it is called the propagation vector. In the most general notation, a plane wave $\varphi(\mathbf{r}t)$ is written as:

$$\varphi_{\mathbf{k}}(\mathbf{r}t) = \exp\{\mathrm{i}(\mathbf{k}\cdot\mathbf{r} - \omega t)\}. \tag{2}$$

This is a *wave* because it is doubly periodic in **r** and t and it is a *plane wave* because it has a constant value over all planes normal to **k**, as we shall now verify. Consider in Fig. 1 the two vectors $\mathbf{r}_1$ and $\mathbf{r}_2$ which denote arbitrary points of the plane normal to **k**: it is clear that $\mathbf{k}\cdot\mathbf{r}_1$ and $\mathbf{k}\cdot\mathbf{r}_2$ are equal, thus verifying our assertion. As regards periodicity of the space part of $\varphi(\mathbf{r}t)$,

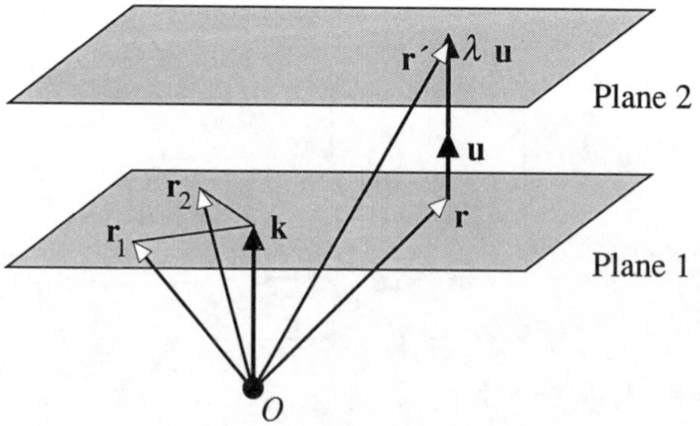

Fig. 1-8.1. Plane waves. The vectors **k** and **u** (a unit vector) are normal to the two planes shown.

consider the value of this function at an arbitrary point $\mathbf{r}'$ of a plane parallel to the first plane and at a distance λ from it. We have:

$$\mathbf{r}' = \mathbf{r} + \lambda\mathbf{u} = \mathbf{r} + \lambda\mathbf{k}/|\mathbf{k}|, \tag{3}$$

3

$$\mathbf{k}\cdot\mathbf{r}' = \mathbf{k}\cdot\mathbf{r} + \lambda|\mathbf{k}|^2/|\mathbf{k}| = \mathbf{k}\cdot\mathbf{r} + \lambda|\mathbf{k}|. \tag{4}$$

If follows from (2) and (4) that if

$$\lambda = 2\pi/|\mathbf{k}|, \tag{5}$$

then $\mathbf{k}\cdot\mathbf{r}'$ differs from $\mathbf{k}\cdot\mathbf{r}$ by 2π. This means, from (2), that $\varphi_{\mathbf{k}}(\mathbf{r}t)$ has the same value over plane 2 of the figure as over plane 1, and that these two are the nearest planes for which this is the case. Thus λ is the *wave length*. The vector $\mathbf{k}$ in (2) thus carries two pieces of information: its *direction* is normal to the planes over which the wave function is constant, and its *modulus* is 2π times the inverse of the wave length. It is for this reason that the $\mathbf{k}$ vector is called the *propagation vector* of the plane wave. Its modulus, which is given by

$$|\mathbf{k}| = 2\pi/\lambda, \tag{6}$$

is called the *circular wave number* ($1/\lambda$ being the *wave number*).

It can be seen from (2), in the same manner, that

$$\omega = 2\pi/T, \tag{7}$$

where T is the *period* of the wave and ω its *circular frequency*. The reader should notice that it is a traditional convention in solid-state theory to write plane waves in the form (2) rather than with a 2π in the exponent, which means that ordinary wave numbers and frequencies are not used and are replaced by their circular counterparts.

The *wave velocity*, also called the *phase velocity* v_p, is given by the well-known expression

,5,6

$$v_p = \lambda/T = \omega/|\mathbf{k}|. \tag{8}$$

From comparison of (8) with (2), it is useful to remember the phase velocity as *the coefficient of t in the wave function divided by the coefficient of* $\mathbf{r}$.

It should be noticed that the free-electron eigenfunctions $\exp(i\mathbf{k}\cdot\mathbf{r})$ of (3.14) agree with the plane waves (2): the missing factor $\exp(-i\omega t)$ gives no trouble because the wave function in quantum mechanics can always be multiplied by a constant, and this factor is just such a constant in eqns (3.1) or (3.3), since we are dealing there with time-independent operators.

We have thus established that the free-electron eigenfunctions, that is the eigenfunctions of the momentum operator, are plane waves. Because these waves must describe electrons in the solid as endowed with some particle-like behaviour, it is well known that these waves must appear in *wave packets*, that is as sums of plane waves with slightly different $\mathbf{k}$'s and ω's.

We shall therefore consider the superposition of waves with **k** vectors and frequencies defined as follows.

$$\mathbf{k}_0 + \delta\mathbf{k}, \qquad \omega = \omega_0 + \delta\omega, \tag{9}$$

with

$$-\Delta\mathbf{k} \le \delta\mathbf{k} \le \Delta\mathbf{k}, \qquad -\Delta\omega \le \delta\omega \le \Delta\omega. \tag{10}$$

(Notice that, whereas the δ are infinitesimal, the Δ are finite, defining finite ranges.) The resultant wave is then written as

$$\psi(\mathbf{r}t) = \sum_{\delta\mathbf{k},\delta\omega} \exp i\{(\mathbf{k}_0 + \delta\mathbf{k})\cdot\mathbf{r} - (\omega_0 + \delta\omega)t\}, \tag{11}$$

where the summation is over all values in the ranges given in (10). Clearly, from (11),

$$\psi(\mathbf{r}t) = \{\exp i(\mathbf{k}_0\cdot\mathbf{r} - \omega_0 t)\} \sum_{\delta\mathbf{k},\delta\omega} \exp i(\delta\mathbf{k}\cdot\mathbf{r} - \delta\omega t). \tag{12}$$

The first factor in (R12) is the plane wave with the wave vector $\mathbf{k}_0$ and the frequency ω_0 at the centre of the range (see eqns 9 and 10) and, by definition, the second factor is then the *amplitude* of this wave. The important result, of course, is that this amplitude is not constant but *modulated*. (See Fig. 2). The velocity of the electron must be identified not with the wave velocity but rather with the velocity of the modulated amplitude, since this is the velocity with which the *localized* perturbation propagates. This velocity is called the *group velocity* $\mathbf{v}_g$. On applying to the modulated amplitude in (12) the definition of the velocity given in italics below eqn (8), we have:

$$\mathbf{v}_g = \frac{\delta\omega}{\delta\mathbf{k}} \qquad \Rightarrow \qquad \mathbf{v}_g = \frac{\partial\omega}{\partial\mathbf{k}}, \tag{13}$$

where, on the right, we have taken limits. Since $E = \hbar\omega$, eqn (13) gives

$$\mathbf{v}_g = \hbar^{-1}\frac{\partial E}{\partial\mathbf{k}}. \tag{14}$$

(See § **8**-3 for the meaning of the derivative here.)

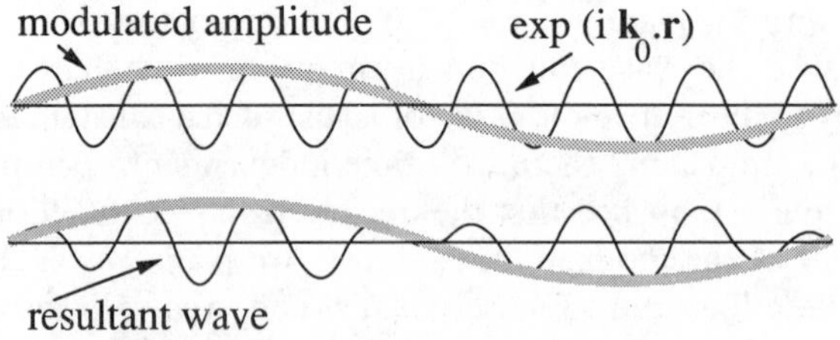

Fig. 1-8.2. Superposition of waves.

It is instructive to check eqn (14) with the free-electron expression for the energy:

$$E = \hbar^2 \mathbf{k}^2/2m \qquad \Rightarrow \qquad \mathbf{v}_g = \hbar\mathbf{k}/m = \mathbf{p}/m. \tag{15}$$

This is as we should expect, the momentum **p** being by definition the mass of the electron times its velocity $\mathbf{v}_g$. It must be clearly understood, however, that this relation is valid *only in the free-electron picture* and that grievous mistakes can be made by ignoring this constraint. For bound electrons the momentum **p** and the velocity $\mathbf{v}_g$ are not related by the classical condition which appears on (R15). It cannot even be assumed, in fact, that the momentum and the velocity are parallel! (See § **8**-3.)

9 Problems

1. An operator $\mathbb{A}$ is of the form $\mathbb{A}_1 + \mathbb{A}_2$. (The suffices denote different independent variables.) The eigenvalue equations of $\mathbb{A}_1$ and $\mathbb{A}_2$ are

$$\mathbb{A}_1\psi_1 = a_1\psi_1, \qquad \mathbb{A}_2\psi_2 = a_2\psi_2. \tag{1}$$

   Prove that

$$\mathbb{A}\psi = a\psi \qquad \Rightarrow \qquad a = a_1 + a_2, \qquad \psi = \psi_1\psi_2. \tag{2}$$

   Extend this result for an operator $\mathbb{A}$ of the form $\sum \mathbb{A}_i$.
2. Prove that, if V is the volume of the lump of metal under study, the normalized free-electron eigenfunctions (3.14) are

$$\varphi(\mathbf{r}) = V^{-1/2} \exp(\mathrm{i}\mathbf{k}\cdot\mathbf{r}). \tag{3}$$

   (Normalization is assumed over the volume V.)
3. Prove that the degeneracy (including spin) of the first four eigenstates (3.21) is 2, 12, 24, and 16, respectively.
4. Use eqn (3.21) to prove that, in the free-electron approximation,

$$E = 0.038\,|\mathbf{k}|^2\,(\mathrm{nm})^2\,\mathrm{eV}, \tag{4}$$

   where $|\mathbf{k}|$ must be measured in nm^{-1}.
5. Sodium and copper are body-centred and face-centred cubic respectively and their lattice constants a are 0.428 and 0.361 nm respectively. Assume in both cases that each atom contributes one electron to the free-electron gas and show that their Fermi energies are 3.2 eV for sodium and 7.1 eV for copper. (Remember that the number of atoms per unit cell is 2 and 4, respectively, for the body-centred and face-centred cubic structures.)
6. Consider a one-dimensional metal of length L with *box boundary conditions* (in which the wave function $\varphi(x)$ vanishes when x equals 0 or L). Show that the wave functions normalized over L are

$$\varphi(x) = (2/L)^{1/2} \sin(\kappa\pi x/L), \quad \kappa = 1, 2, 3, \ldots. \tag{5}$$

7. Show that, under the conditions of Problem 6, the wave functions for a cube of length L and volume V, normalized over V, are:

$$\varphi(\mathbf{r}) = (8/V)^{1/2} \sin(k_x x) \sin(k_y y) \sin(k_z z), \tag{6}$$

$$k_x = \pi\kappa_x/L, \quad k_y = \pi\kappa_y/L, \quad k_z = \pi\kappa_z/L, \tag{7}$$

$$\kappa_x, \kappa_y, \kappa_z = 1, 2, 3, \ldots . \tag{8}$$

   Notice: (i) That under box boundary conditions only the first octant of **k** space is significant. (Compare eqns 8 and 3.19). (ii) That eqns (7) and (3.18) differ by a factor of 2, whence the graining of the new **k** space is denser.
8. Calculate the volume v occupied by one state (spin included) in the **k** space with box boundary conditions and show that it is $\pi^3/2\,V$, where V is the crystal volume. Notice that this is 1/8 of the value in (3.25) and explain the meaning of this result.
9. Determine the density of states in the **k** space with box boundary conditions, in the same manner as in § 5, and show that it has the same value as for travelling waves.

Further reading

A very clear and simple introduction to the properties of the free-electron gas is given in Ziman (1963) and in Cottrell (1988). The book by Rosenberg (1988) contains an admirably concise introduction to this subject. This book is very much to be recommended as an introduction to many of the topics treated in further chapters here, as well as, at a higher level (of length as well as difficulty), the excellent work of Ashcroft and Mermin (1976). The textbook by Blakemore (1985) is clear and useful and Kittel (1986) is probably the most popular solid-state book in print. Chemists will find the books by Hoffmann (1988) and Cox (1987) invaluable in providing a chemical insight into many of the topics discussed in the present book.

2

Symmetry and group theory: a revision

A lump of metal was considered in Chapter **1** as if it had no internal structure at all, the surface of the metal acting as the walls of a container inside which a free-electron gas is to be found. Even in this case we did find a degeneracy which arises from the isotropy of the space in which the electrons move: the pair of free-electron states corresponding to the values **k** and $-$ **k** of the wave vector must be degenerate. (See § **1**-3.) We shall want in the rest of this book, of course, to introduce the effect of the crystal structure on the electron states and we must expect many more, and more complicated, symmetries which, as we shall see, will lead also to more complicated degeneracies. The study of these degeneracies is essential: if we did not know, for example, that the level $2p$ of the free sodium atom is six-fold degenerate, we would get the electronic structure of this atom hopelessly wrong. As we shall remind ourselves in this chapter, the study of degeneracy is very much simplified by using some simple ideas of group theory. We shall expect the reader to have some familiarity with these ideas, as treated for instance in the well-known textbook of physical chemistry by Atkins (1990, Chapter 15).

1 Group definitions

This chapter starts with a number of rather abstract definitions and I hope that the reader will not be unnecessarily put off by them. Examples are provided in § 7 at the end of this chapter and the reader can then satisfy himself or herself that the abstract definitions given here describe in fact fairly straightforward properties of the symmetry operations with which this book will be concerned. The reader must also appreciate that these definitions are mainly given here in order to establish the notation that we shall need. Otherwise, they are not going to be much used in our work, since these concepts are mostly required in the proofs of theorems, which we shall not provide, since we shall be much more interested in understanding and exploiting the major ideas of the subject.

We shall use the symbol $\{g_i\}$ to denote the *set* of all elements g_i with i ranging over the values 1, 2, . . . , n. (In practice, for us, such elements will be the symmetry operations of a molecule or a crystal.) The number n of elements in this set is called the *order* of the set. We assume that by some means (of which more later), the *product* of any two elements of the set, $g_i g_j$, is defined, although the result may not always be a member of the set $\{g_i\}$. Also, this product may not be commutative, that is, $g_i g_j$ may not be the same as $g_j g_i$.

A set $\{g_i\}$ is a *group* G with respect to a given product if the following four properties are satisfied:

(1) The result of the product $g_i g_j$ is always an element of the set. We write this as follows:

$$g_i g_j \in G, \qquad \forall g_i \in G, \qquad \forall g_j \in G, \tag{1}$$

where $\in$ means '*belongs to*' and $\forall$ means '*for all*'. If (1) is satisfied we say that the set $\{g_i\}$ *closes*, whence this condition is called the *closure* condition.

(2) The product must be *associative*, which means that when triple products arise they can be effected as pairs of products at will:

$$(g_i g_j) g_k = g_i (g_j g_k). \tag{2}$$

(3) An element of the set exists, to be called the *identity element* e (sometimes also written as E in which case it must not be confused with the energy) such that its product with any element of the set gives always this same element:

$$g_i e = e g_i = g_i. \tag{3}$$

(4) An *inverse element*, written as g_i^{-1}, exists for every element g_i of $\{g_i\}$, such that its product with g_i gives always the identity:

$$g_i g_i^{-1} = g_i^{-1} g_i = e. \tag{4}$$

Conjugates and classes

Given an operation $g_i \in G$, the product $g g_i g^{-1}$ can always be formed for any arbitrary operation g of the group and it is called the *conjugate element* of g_i by g. Again, what is important in this definition arises from a result which is valid when the elements of G are symmetry operations: $g g_i g^{-1}$ must always be an operation of the same type as g_i, for any operation g. Thus, if we take g_i to be one of the eight ternary rotations of a cube (about one of its diagonal axes), then its conjugate, under any other operation of the group of the cube, must always be one of the eight ternary rotations.

Given any operation g_i, its *class* $C(g_i)$ is defined as the set of all its conjugates under all the operations g of the group. (In the cubic group, for example, we have a class consisting of the eight ternary rotations.) A most important property of classes is that no two classes can have any element in common. As a result, every group G can always be partitioned as a sum of classes. This partitioning will give us, as we shall see, important information.

Subgroups

Symmetry groups of crystals are very large and considerable simplification can be obtained by recognizing that they contain smaller sets of elements which also form groups on their own. These sets are called *subgroups*. In general, if $H = \{h_i\}$ is a group such that all its elements belong to another group G, we say that H is a *subgroup* of G,

$$h_i \in G, \qquad \forall\, h_i \in H \qquad \Rightarrow \qquad H \subset G, \tag{5}$$

a relation which is denoted as shown on the right of the above line.

If H is a subgroup of G such that the conjugate of any element of H under any element of G is always an element of H, then H is said to be an *invariant subgroup* of G:

$$g h_i g^{-1} \in H, \quad \forall\, h_i \in H, \quad \forall\, g_i \in G, \qquad \Rightarrow \qquad H \lhd G. \tag{6}$$

The symbol for this relation is given on the right of (6). Invariant subgroups play a major part in solid-state theory, as we shall see. Notice that, if H is invariant, the conjugate of h_i under any g must belong to H, that is, H must contain the whole of the class of G to which h_i belongs: an invariant subgroup of G must therefore be a sum of classes of G.

Direct product

We have already said that the symmetry group of a crystal will contain lots of subgroups. Because operations of different subgroups are nevertheless operations of the same total group, we know how to multiply such operations. (Remember that products are defined, in general, only within a group.) If we have a subgroup L of three elements, say, and another M of four elements, we can form a set of twelve elements by multiplying the operations of one subgroup with all the operations of the other. The question we have in mind is this: will this new set be a group? If the answer were yes, we would then have a very good way of organizing the subgroup structure of the total group, since, starting from a pair of subgroups we can immediately recognize the existence of another subgroup that comprises that pair. This is done by means of the direct product.

Consider two groups L and M with no common element except the identity, and such that the product of any two operations of L and M is defined and always commutes:

$$l_i m_j = m_j l_i, \qquad \forall\, l_i \in L, \quad \forall\, m_j \in M. \tag{7}$$

Then the set $\{l_i m_j\}$, which contains all the possible products of the form $l_i m_j$, is a group H, say, which is called the *direct product* of L and M and which is denoted with the following symbol:

$$H = L \otimes M. \tag{8}$$

We shall verify only that the closure property is satisfied. Consider the product of two elements of $\{l_i m_j\}$:

7 $$l_i m_i \, l_r m_s = l_i l_r \, m_j m_s = l_p m_q \in \{l_i m_j\}. \tag{9}$$

We use here the fact that the product of any pair of elements of L must be another element of L, because of the closure property of L, and similarly for M.

If the operations of a third group N commute with all the operations of H, it is clear that we can form a new group G as the direct product $H \otimes N$, whence

$$G = L \otimes M \otimes N. \tag{10}$$

The reader who finds all this rather abstract can return to this subject after the example treated in §**3**-8.

2 Symmetry operators

The definition of a symmetry operation is delicate and we shall discuss it by means of an example illustrated in Fig. 1. Our *configuration space* (the space in which the physical objects under study are defined) contains an *infinite square* grid of entirely equivalent points, which *underlies* Fig. 1*a*. (We shall soon explain why we use the italicized words shown.) The whole of this space is referred to axes **x**, **y** which are *fixed in space*. With respect to these axes twelve grid points have been labelled with black numerals from 1 to 12. In order to perform a symmetry operation, the first thing we need is a *copy of the configuration space*. In the two-dimensional case of Fig. 1, this is easy to obtain: all we need is a transparent sheet on which a tracing of the system is made. This sheet is assumed in Fig. 1 to be superimposed on the original system, which thus *underlies* the figure. Although the copy must be absolutely identical to the system we differentiate it a little, purely so as to distinguish it. First, we number the sites with open numerals. (The numbering of the sites, of course, is not in any case part of either the system or its copy!) Secondly, we have tinted a square portion of the copy.

We are now ready to carry out a symmetry operation. In order to perform a *translation* by the vector **t** shown in Fig. 1*a*, the *whole* of the copy is translated parallel to **t** by a length equal to $|\mathbf{t}|$ so that it lands on top of the original system in the position shown in Fig. 1*b*. Notice that the axes **x**, **y** have not moved: as we have said they are fixed in space. Thus the system is related to these axes precisely as before, since it has not moved either: the only thing we have done to the system in Fig. 1*b* is to show a somewhat larger part of its infinite extent.

There are several things which must now be noted. First, the translation effected is a *covering operation*, an expression by which we mean that every

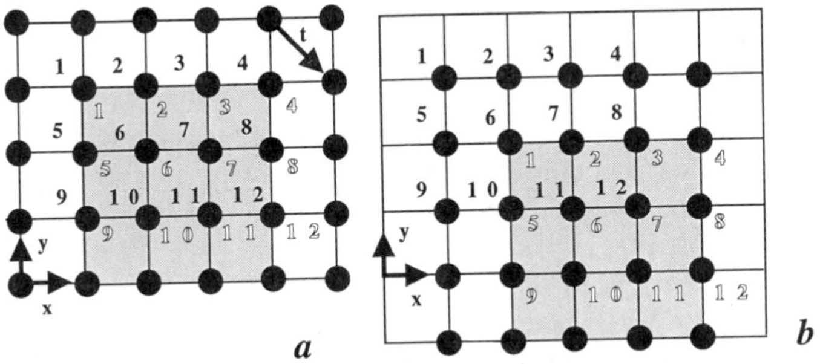

Fig. 2-2.1. Definition of a symmetry operation: a translation by the vector **t**.

grid point in the (infinite) copy overlaps an entirely equivalent grid point of the system. (This would not be the case, of course, if the copy and system were finite.) Because of this any physical scalar quantities computed for the copy must have identical values to those computed for the system: this is the essence of a symmetry operation. Secondly, we must reiterate the very important fact that the coordinate axes are not changed at all by the transformation.

If we effect a covering operation g_2 after a covering operation g_1 the result must clearly be a covering operation, which shall be denoted with the symbol $g_2 g_1$ and which is called the *product* of the two operations. Notice that this product is always read *from right to left*, that is that the operation on the right of it is always the first one to be performed.

We must now understand what happens to the coordinates of points (or, what is the same, to position vectors) during the transformation. The position vector $\mathbf{r}_1$ of the grid point 1, which is [1,3,0] in units of the grid spacing, (see Fig. 1*a*) is transformed after the transformation into the grid point 𝟙, with position vector [2,2,0] in Fig. 1*b*, which is identical to $\mathbf{r}_6$. If we call $\mathbf{r}'$ the transformed vector (always referred to the original axes because no others are used), what we have is this:

$$\mathbf{r}'_1 = \mathbf{r}_6 = \mathbf{r}_1 + \mathbf{t}, \qquad \mathbf{t} = [1,\bar{1},0], \tag{1}$$

as can readily be verified from the figure and the numerical values quoted above. It is clear that the transformation of $\mathbf{r}$ into $\mathbf{r}'$ induced by the translation by $\mathbf{t}$ is given by

$$\mathbf{r}' = \mathbf{r} + \mathbf{t}, \tag{2}$$

for all $\mathbf{r}$.

We can now summarize our definition of a symmetry operation as a transformation of the configuration space in which the coordinate axes are kept fixed and the configuration space is transformed by a covering operation which changes the coordinates of points of the configuration space in such a way that the coordinates of particles are always transformed into the coordinates of identical particles (as, in eqn 1, the coordinates of particle 1 are transformed into the coordinates of particle 6). This picture of a symmetry operation in which the *whole* of the configuration space is rotated, reflected or translated, as the case might be, while the coordinate axes are kept fixed in space, is called the *active picture* and it is now standard in solid-state work. An alternative picture, in which the configuration space is never changed or displaced in any way but, instead, the coordinate axes are transformed, is called the *passive picture*, but it will never be used in this book.

We must now return to eqn (2). The notation $\mathbf{r}'$ which we used in it is not convenient, because it does not make it explicit which is the operation responsible for the change of $\mathbf{r}$ into $\mathbf{r}'$. For this reason, we shall replace the symbol $\mathbf{r}'$ by a new symbol $t\mathbf{r}$. Just in the same manner as we read $\mathbf{r}'$ as a single symbol, $t\mathbf{r}$ must also be read as a *single* symbol, in which t (as the dash before) is an *operator* (or symbol modifier), which in our case is called the *symmetry operator* or *configuration space operator*, its particular form being always chosen so as to distinguish the symmetry operation in question. We thus rewrite (2) as follows:

$$t\mathbf{r} = \mathbf{r} + \mathbf{t}. \tag{3}$$

In general, the set of all symmetry operations g which leave a system invariant (that is, that cover it) form a group G. (It is clear, for instance, that the product of two symmetry operations is a symmetry operation, and the other group properties are easily established.) In this case, for each $g \in G$, the transform of $\mathbf{r}$ by the operation g will be written as $g\,\mathbf{r}$.

Transformation of functions

We get now nearer to the heart of things. However much we might divert ourselves transforming away configuration spaces we must ask ourselves: why do we do it? For our real interest is in energy eigenvalues, degeneracies, and energy eigenfunctions. So, in practice, we are not so much interested in configuration space as in the functions defined over it. We can see quite easily, however, that transformations of the configuration space effect transformations of the functions defined over it. We illustrate this in Fig. 2, where the configuration space is the x axis, over which the function

$$y = f(x) = x^2, \tag{4}$$

is defined. The transformation of the configuration space which we consider is its translation by a, which changes each point x into the point x', equal to $x + a$. In our active picture the parabola is also translated with the whole of this space in such a way that the value y of the function at the point x changes into a value y' at the point x'. The first thing we must notice is that, of course, this value y' is numerically the same as y. On the other hand, before the transformation we knew the relation between y and x: it was $y = x^2$. What we

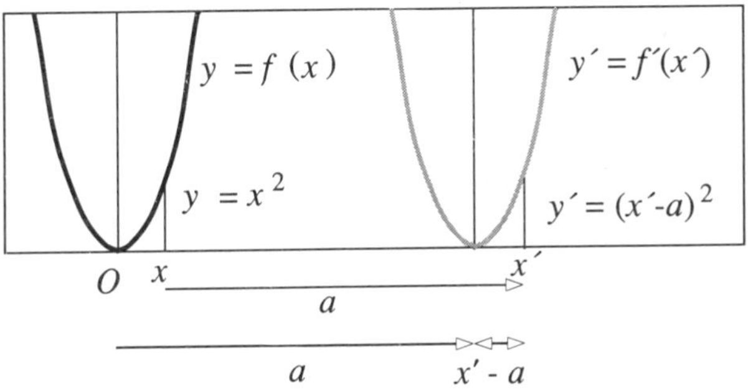

Fig. 2-2.2. Transformation of functions.

want to do is to obtain the new relation between y' and x'. It is clear from the figure that y' is not $(x')^2$ but, rather, $(x'-a)^2$. This means that, if the original functional relation was written as $y = f(x)$, it would be wrong to write it now as $y' = f(x')$, the right-hand side here being merely a shorthand for $(x')^2$. We must therefore write the new functional relation between x' and y' as $y' = f'(x')$. This is most important: it means that while a point of the configuration space changes from x into x', the function f also changes into the function f'. We can now see why we are interested in configuration space transformations: they induce corresponding *functional transformations.*

The rule to define these functional transformations is obtained as follows. We notice that, as the value of the function at a given point is the same as the value of the function at the transformed point we can write

$$f'(x') = f(x) \qquad \text{or} \qquad f'(\mathbf{r}') = f(\mathbf{r}), \tag{5}$$

the latter being the more general, three-dimensional, expression. We shall now refine our notation. We have seen that, if we call g the configuration space operator, it is convenient to write $g\,\mathbf{r}$ instead of $\mathbf{r}'$. We must do the same thing for the transformed function f' in (5), in order to make it explicit that the functional transformation originates from the transformation g. We cannot use this latter operator, though, since this is an operator which transforms not functions but points of space. We therefore introduce a new *function space operator* $\mathring{g}$, so that we write $\mathring{g}f$ for f':

5
$$\mathring{g}f(g\mathbf{r}) = f(\mathbf{r}). \tag{6}$$

Since (6) must be valid for all points $\mathbf{r}$, and in particular for those given in the form $g^{-1}\mathbf{r}$, then,

6
$$\mathring{g}f(\mathbf{r}) = f(g^{-1}\mathbf{r}). \tag{7}$$

Equation (7) will be our fundamental formula for transforming functions. The reader who is somewhat worried by it might find it useful to try Problem 1.

Notice that what we have done is this. We started with a group G of configuration space operators g, and we have found that the existence of this group entails the existence of a parallel group $\mathring{G}$ of function space operators $\mathring{g}$. Obviously, both groups are precisely of the same order and it can be proved that the multiplication rules are preserved in going from one to the other, which is most important since chaos would otherwise ensue:

$$g_i g_j = g_k \qquad \Rightarrow \qquad \mathring{g}_i \mathring{g}_j = \mathring{g}_k. \tag{8}$$

This result is formally expressed by saying that the groups G and $\mathring{G}$ are *isomorphic*.

A point about notation. Whenever an operator acts on a function it must be a function space operator and thus carry the circlet. It is possible in such a case, however, to dispense with the latter and identify the nature of the operator in question from the context, which will in fact be done after this chapter in the rest of the book.

Transformation of operators

We have points in configuration space, functions defined at these points, and operators (including of course quantum mechanical operators) acting on these functions. As we transform the points, the functions are transformed and we must expect the operators also to transform, as we shall now prove. Consider the eigenvalue equation of an operator $\mathbb{A}$, (this could be an operator like the momentum or the Hamiltonian),

$$\mathbb{A}\varphi = a\varphi. \tag{9}$$

We now act with g on the configuration space whereupon φ changes into $\mathring{g}\varphi$, which we call ψ:

$$\psi = \mathring{g}\varphi \qquad \Rightarrow \qquad \varphi = \mathring{g}^{-1}\psi. \tag{10}$$

Replace φ in (9) by its value as given on the right of (10):

$$\mathbb{A}\mathring{g}^{-1}\psi = a\mathring{g}^{-1}\psi, \tag{11}$$

11 $$\mathring{g}\mathbb{A}\mathring{g}^{-1}\psi = a\mathring{g}\mathring{g}^{-1}\psi = a\psi. \tag{12}$$

Equation (12) is the eigenvalue equation of the operator $\mathring{g}\mathbb{A}\mathring{g}^{-1}$ for the transformed eigenfunctions ψ. We summarize our results:

$$\varphi \mapsto \psi = \mathring{g}\varphi \qquad \Rightarrow \qquad \mathbb{A} \mapsto \mathring{g}\mathbb{A}\mathring{g}^{-1}. \tag{13}$$

3 The Schrödinger group

It follows at once from (2.13) that, when φ changes into $\mathring{g}\varphi$, the Hamiltonian $\mathbb{H}$ then changes into $\mathring{g}\mathbb{H}\mathring{g}^{-1}$. We define the *Schrödinger group S* as the group of all operations $\mathring{g}$ that leave the Hamiltonian invariant:

$$\mathring{g} \in S \qquad \Rightarrow \qquad \mathring{g}\mathbb{H}\mathring{g}^{-1} = \mathbb{H}. \tag{1}$$

It should be pretty obvious why we are interested in this group: its operations must leave the energy of a system invariant. Let us see how this idea develops. It follows at once from the right of (1) that all operations of the Schrödinger group commute with the Hamiltonian:

$$\mathring{g}\mathbb{H} = \mathbb{H}\mathring{g}. \tag{2}$$

This important result will allow us to discover the physical significance of the operations $\mathring{g}$ of the Schrödinger group by examining their effect on the eigenfunctions φ of the Hamiltonian,

$$\mathbb{H}\varphi = E\varphi, \tag{3}$$

3 $$\mathring{g}\mathbb{H}\varphi = E\mathring{g}\varphi, \tag{4}$$

2|4 $$\mathbb{H}\mathring{g}\varphi = E\mathring{g}\varphi. \tag{5}$$

Comparison of (3) with (5) shows at once that $\mathring{g}\varphi$ and φ are *degenerate* whenever $\mathring{g}$ belongs to S. This, as expected, agrees with our previous result (see the sentence following eqn 1) to the effect that the operations of the Schrödinger group should keep the energy invariant. All this, of course, applies perfectly well to the symmetry operations which we have been considering, since by their definition they cannot affect energy eigenvalues. Before we go any further it might be useful to revise quickly the major features of degeneracy.

Degeneracy

We must first make sure that we know correctly the definition of degeneracy. Consider the case when we can find n eigenfunctions for an eigenvalue a of an operator $\mathbb{A}$:

$$\mathbb{A}\varphi_i = a\varphi_i, \qquad i = 1, 2, \ldots, n. \tag{6}$$

We say that these eigenfunctions are n-fold degenerate if: (i) they all belong to the same eigenvalue, as shown; (ii) they are all *linearly independent*. This latter condition means that it is not possible to write down one of the functions as a linear combination of the rest, say,

$$\varphi_1 \neq \sum_{i=2}^{n} \varphi_i C_i. \tag{7}$$

This expression can first be rewritten by multiplying both sides by an arbitrary coefficient c_1, and by rewriting $c_1 C_i$ as c_i:

$$c_1\varphi_1 \neq \sum_{i=2}^{n} \varphi_i c_i \qquad \Rightarrow \qquad \sum_{i=1}^{n} \varphi_i c_i \neq 0. \tag{8}$$

This last condition is the most compact definition of linear independence, but one must require that the arbitrary coefficients c_i be not all equal to zero simultaneously. We have not yet finished our definition since a third condition is required to ensure that the degeneracy of (6) is precisely n and not higher: (iii) no function must exist which satisfies (6) and which is linearly independent of the n functions φ_i.

There are two fundamental properties of degeneracy which will be very important in our work. The first is that any linear combination of the functions φ_i in (6),

$$\psi = \sum_{i=1}^{n} \varphi_i c_i, \tag{9}$$

is also an eigenfunction of the same operator $\mathbb{A}$ for the same eigenvalue a. We leave the verification of this result to the reader.

The second result is very important. It says that, given that the equation (6) is precisely n-fold degenerate, then any eigenfunction Ψ of (6) with the same eigenvalue a and distinct from all the φ_i must be of the form (9):

$$\Psi = \sum_{i=1}^{n} \varphi_i c_i. \tag{10}$$

The proof is immediate: if (10) were not valid then Ψ would be linearly independent of the n degenerate eigenfunctions and the degeneracy would have to be of order $n + 1$, against the assumption.

The conjugator

We have seen that covering operations in the configuration space give us operations of the Schrödinger group. It must not be believed that this is the only way in which operations of this group arise. The quickest way to find operations of the Schrödinger group is by the commutation property (2). (It is very easy to prove that the product of two operators which commute with the Hamiltonian also commutes with the Hamiltonian, thus verifying closure.) We shall now introduce an operator which will immediately be seen to commute with the Hamiltonian under certain conditions, and which will therefore be a symmetry operator, although it has no geometrical origin. We first need a definition. Consider a complex number ω and a complex function $f(z)$ of a complex variable z. (An example of a complex function of a complex variable is $f(z)$ equal to $\exp(iz)$, the functional relation here involving the imaginary unit.) The *conjugator operator* $\mathbb{j}$ is defined as the operator that effects the complex conjugation of everything on its right, in the following way:

$$\mathbb{j}\omega f(z) = \omega^* \mathbb{j} f(z) = \omega^* f^*(z^*), \tag{11}$$

where, for example, if z is $a + ib$ then its conjugate z^* is $a - ib$. Thus, the conjugator $\mathbb{j}$ acting on $\omega \exp(iz)$ transforms it into $\omega^* \exp(-iz^*)$, the change of sign of the imaginary unit corresponding to the asterisk in the functional relation. Notice that all our previous operators were *linear*, which means that when $\mathring{g}$ acts, for example, on (R3), it leaves the constant E unchanged and acts only on the function, as shown in (R4). As we can see in the second term of (11) this is not so for $\mathbb{j}$: the conjugator is a *non-linear operator*.

We have now to show the credentials of this new operator, which we must do for the case when the Hamiltonian is a real operator. (It is well known in quantum mechanics that, unless spin is considered, the Hamiltonian can always be taken to be real.) Therefore, in the same manner as in (11), we have

$$\mathbb{j}\mathbb{H} f(z) = \mathbb{H}\mathbb{j} f(z), \tag{12}$$

which shows at once that $\mathbb{j}$ commutes with $\mathbb{H}$ for real Hamiltonians and thus belongs to their Schrödinger group. (The reality condition will not be a severe limitation for us since in this book we shall not be concerned with spin eigenfunctions: spin will only appear as doubling the orbital degeneracy. If spin, and thus complex Hamiltonians are treated explicitly, a lot more work has to be done to obtain a good new operation of the Schrödinger group, which is called the *time-reversal operator*.)

Because these things are so important we shall prove independently that when $\mathbb{j}$ acts on a wave function it always creates another wave function degenerate with the original one. (Remember that we always assume the Hamiltonian to be real.) Act with $\mathbb{j}$ on both sides of (3):

$$\mathbb{j}\mathbb{H}\varphi = \mathbb{j}E\varphi \quad \Rightarrow \quad \mathbb{H}\mathbb{j}\varphi = E\mathbb{j}\varphi \quad \Rightarrow \quad \mathbb{H}\varphi^* = E\varphi^*. \tag{13}$$

On comparing (3) and (13) φ and φ^* belong to the same eigenvalue. Also, (as long as they are not real), they are linearly independent (φ^* can never be written as a constant times φ!). Therefore, they are degenerate. We thus verify that $\mathbb{j}$ acts precisely as a symmetry operator $\mathring{g}$, in the sense that, for complex functions, φ and $\mathbb{j}\varphi$ are degenerate, like φ and $\mathring{g}\varphi$ are.

An important property of the conjugator is that it commutes with all symmetry operations. Consider an operator $\mathring{g}$ such that

2.7 $$\mathring{g} f(\mathbf{r}) = f(g^{-1}\mathbf{r}) \tag{14}$$

14 $$\mathbb{j}\mathring{g} f(\mathbf{r}) = \mathbb{j} f(g^{-1}\mathbf{r}) = f^*(g^{-1}\mathbf{r}). \tag{15}$$

In the last step here we recognize that $\mathbf{r}$, and therefore $g^{-1}\mathbf{r}$, is real. On the other hand, on acting directly with $\mathring{g}$ on $\mathbb{j} f(\mathbf{r})$, we get:

$$\mathring{g}\mathbb{j} f(\mathbf{r}) \equiv \mathring{g} f^*(\mathbf{r}) = f^*(g^{-1}\mathbf{r}). \tag{16}$$

We use here the definition (11) of the conjugator, bearing in mind the reality of $\mathbf{r}$, and, in the last step, eqn (14). From (16) and (15),

$$\mathbb{j}\mathring{g} = \mathring{g}\mathbb{j}. \tag{17}$$

4 Representations

We have already seen that group theory, from our point of view, will be mainly a tool to count and classify the degeneracies of the electron states in a crystal and this is mainly done through the concept of the group representation which we shall now discuss. Consider an n-fold degenerate energy eigenvalue,

$$\mathbb{H}\varphi_i = E\varphi_i, \qquad i = 1, 2, \ldots, n. \tag{1}$$

It follows from eqn (3.10) that any other eigenfunction belonging to the same eigenvalue must be a linear combination of the φ_i:

$$\mathbb{H}\sum_i \varphi_i c_i = \sum_i \mathbb{H}\varphi_i c_i = \sum_i E\varphi_i c_i = E\sum_i \varphi_i c_i. \tag{2}$$

It can be seen from (2), in fact, that the summation satisfies eqn (1). (The reader may have puzzled why have we put the constants c_i on the right, rather than on the left of the functions, as it is most often done. Of course, their position here is entirely irrelevant but our notation will soon be seen to be convenient.) We know from § 3 that if $\mathring{g}$ belongs to the Schrödinger group, now to be denoted with G, then $\mathring{g}\varphi_j$, say, must be degenerate with φ_j and thus with all the φ_i. From (2), $\mathring{g}\varphi_j$ must therefore be a linear combination of all the φ_i but, clearly, the coefficients c_i must depend on the particular φ_j which is being transformed so that we shall label them as c_{ij}:

$$\mathring{g}\varphi_j = \sum_i \varphi_i c_{ij}. \tag{3}$$

The coefficients c_{ij} form a matrix, since j must range also over the same n values of i shown in (1). This matrix clearly corresponds to the operation $\mathring{g}$ of the group G, for which reason it will be designated with the symbol $\hat{G}(\mathring{g})$. Notice that, for reasons to be revealed later, we have written this matrix in (3) so that it operates on the *right* of the functions. Thus, if we collect all equations of the form (3) for j ranging from 1 to n, then the n functions $\varphi_1, \varphi_2, \ldots, \varphi_n$ have to be disposed on a *row* (and not a column) vector, for which we shall use the symbol $\langle \varphi_1, \varphi_2, \ldots, \varphi_n|$. We therefore write

$$\langle \mathring{g}\varphi_1, \mathring{g}\varphi_2, \ldots, \mathring{g}\varphi_n| = \langle \varphi_1, \varphi_2, \ldots, \varphi_n|\,\hat{G}(\mathring{g}), \tag{4}$$

or, more compactly,

$$\mathring{g}\langle \varphi_1, \varphi_2, \ldots, \varphi_n| = \langle \varphi_1, \varphi_2, \ldots, \varphi_n|\,\hat{G}(\mathring{g}). \tag{5}$$

Clearly, one such matrix can be given for each $\mathring{g}$ in G and they have been defined so that they sensibly preserve the multiplication rules of G:

$$\mathring{g}_i\mathring{g}_j = \mathring{g}_k \qquad \Rightarrow \qquad \hat{G}(\mathring{g}_i)\,\hat{G}(\mathring{g}_j) = \hat{G}(\mathring{g}_k). \tag{6}$$

Although we shall not prove this statement, it is precisely in order that (6) be verified that we have placed the matrices on the right of the functions in (4) and thus that we have worked with row rather than column vectors. This also explains the positioning of the coefficients in (3) and (2). The set of matrices $\hat{G}(\mathring{g})$ for all $\mathring{g}$ in G is called a *representation* of the group G, and the set of functions $\langle \varphi_1, \varphi_2, \ldots, \varphi_n|$, to be abbreviated as $\langle \varphi|$, is called the *basis of the representation.* Each set of degenerate eigenfunctions of $\mathbb{H}$, $\langle \varphi|$, $\langle \psi|$, and so on, will form a representation, the dimension of which equals the degeneracy of the set. It is thus clear that the symbol $\hat{G}(\mathring{g})$ requires a further label, since it must depend on the basis chosen. We therefore rewrite (5) as follows:

$$\mathring{g}\langle \varphi| = \langle \varphi|\hat{G}_\varphi(\mathring{g}). \tag{7}$$

So far, we have assumed the basis to be a degenerate set of eigenfunctions of some Hamiltonian $\mathbb{H}$. We shall now relax this condition: all that we shall require is that the transform of each function of the basis, as expressed on (L7), be always given as a linear combination of the functions of the same basis. Thus, a matrix as on the right of (7) can always be defined. If, furthermore, these matrices satisfy the conservation of the multiplication rules required by (6) we shall agree that they form a *representation* of the group. The reader can see how this works from some of the problems at the end of this chapter.

Direct sum. Reduction of the representations

Suppose that $\langle \varphi|$ is a singly-degenerate basis (that is one-dimensional), which is thus fully defined as $\langle \varphi_1|$, where φ_1 could be an s type function, say. Suppose also that $\langle \psi|$ is a doubly-degenerate (two-dimensional) basis, $\langle \psi_1 \psi_2|$, like the pair of p_x, p_y functions, say. We therefore have now two representations of G, which we give with full matrices:

$$\mathring{g}\langle \varphi_1| = \langle \varphi_1|\hat{G}_\varphi(\mathring{g})_{11}, \tag{8}$$

$$\mathring{g}\langle \psi_1 \psi_2| = \langle \psi_1 \psi_2|\begin{bmatrix} \hat{G}_\psi(\mathring{g})_{11} & \hat{G}_\psi(\mathring{g})_{12} \\ \hat{G}_\psi(\mathring{g})_{21} & \hat{G}_\psi(\mathring{g})_{22} \end{bmatrix}. \tag{9}$$

We can now join the two bases above into a single, three-dimensional, basis $\langle \varphi_1 \psi_1 \psi_2|$, so that we shall have a new three-dimensional representation,

$$\mathring{g}\langle \varphi_1 \psi_1 \psi_2| = \langle \varphi_1 \psi_1 \psi_2|\begin{bmatrix} \hat{G}_\varphi(\mathring{g})_{11} & 0 & 0 \\ 0 & \hat{G}_\psi(\mathring{g})_{11} & \hat{G}_\psi(\mathring{g})_{12} \\ 0 & \hat{G}_\psi(\mathring{g})_{21} & \hat{G}_\psi(\mathring{g})_{22} \end{bmatrix}. \tag{10}$$

It is clear that for all $\mathring{g}$ in G the matrices must all have precisely the same form as shown in (10), often called a *block-diagonal form.* More precisely, it is called the *direct sum* of the matrices strung along its diagonal, that is, of the matrices in (8) and (9). Likewise, the basis in (10) is called the *direct sum of the bases* in (8) and (9).

It must be appreciated that no one in their senses would do what we have done. However, if we inadvertently join together degenerate bases and obtain a representation that has a block-diagonal form, what we have done shows that we can chop up the basis into smaller degenerate parts and consequently obtain corresponding representations of lower dimension that the original one. We say in this case that we have *reduced* the representation, which is obviously a worthwhile thing to do, since there is no point whatsoever in keeping together, as in (10), bases which do not 'mix'.

Similarity. Irreducibility

Given a basis $\langle\varphi|$, and corresponding representation $\hat{G}_\varphi(\mathring{g})$, another basis $\langle\psi|$ formed by taking some linear combinations of the φ's, that is formed by multiplying $\langle\varphi|$ with some given matrix, will have the same dimension as the original basis and will form a new representation $\hat{G}_\psi(\mathring{g})$. The new important result will be that, in this case, the two representations $\hat{G}_\varphi(\mathring{g})$ and $\hat{G}_\psi(\mathring{g})$ are very simply related:

7
$$\mathring{g}\langle\varphi| = \langle\varphi|\hat{G}_\varphi(\mathring{g}). \qquad (11)$$

$$\langle\varphi| = \langle\psi|U. \qquad (12)$$

12|11
$$\mathring{g}\langle\psi|U = \langle\psi|U\hat{G}_\varphi(\mathring{g}). \qquad (13)$$

13
$$\mathring{g}\langle\psi| = \langle\psi|U\hat{G}_\varphi(\mathring{g})U^{-1}. \qquad (14)$$

7′
$$\mathring{g}\langle\psi| = \langle\psi|\hat{G}_\psi(\mathring{g}). \qquad (15)$$

14, 15
$$\hat{G}_\psi(\mathring{g}) = U\hat{G}_\varphi(\mathring{g})U^{-1}. \qquad (16)$$

The relation (16) between the two matrix representations is called a *similarity*. We shall consider two consequences of this important concept. First, take a doubly-degenerate basis such as $\langle p_x p_y|$. On rotation of the x, y axes, these functions will become $\langle p'_x p'_y|$, say, and will be obtained from $\langle p_x p_y|$ by multiplication of this basis on the right by some matrix U as in (12). Correspondingly, the representation formed, (or *spanned*, as it is said) by the basis $\langle p_x p_y|$ will change by a similarity (16), under this change of axis, into the new representation spanned by $\langle p'_x p'_y|$. The important thing to remember is that the representation spanned by any basis (of dimension higher than one, otherwise the matrices on (R16) commute!) is not uniquely determined but depends on the choice of axes under a similarity.

The second important consequence of similarity is this. Imagine we apply a linear transformation with a matrix U on the basis of the reduced representation (10). Because the new matrix will take the form (16), it is most likely that the beautiful block-diagonal form may be lost. This shows that the reduction of a representation is not such a trivial operation as it appeared to be so far. It could in fact be very difficult, because, given a presumptive reducible representation, one has to hit upon the precise similarity transformation that will take it to its block-diagonal or reduced form. One of the major triumphs of group theory is that it provides prescriptions for finding such similarities or, when such a task is impossible, for asserting that this is so, in which case the basis is said to be *irreducible*. As it follows from our discussion, irreducible bases are in general formed by sets of degenerate eigenfunctions of the Hamiltonian (barring cases of the so-called *accidental degeneracies*).

The last result is one of extraordinary importance for us, for the following reason. Given a system, be it a molecule or a crystal, it is not difficult to find its symmetry group. Group theory, by various fairly routine procedures, with which we need not concern ourselves, allows us to obtain all the irreducible representations of this group. (In any case, tables of such irreducible representations exist for most important groups.) The result is that we have an immediate classification of the energy levels and their degeneracies. Clearly, this will be quite easy for a molecule, since its symmetry group is small and thus will have very few representations. (See § 6 for an example.) This is not so for a crystal, where the detailed tabulation of each of the irreducible representations is well-nigh impossible. In solid-state theory, however, one invents a geometrical object, the *Brillouin zone* (see § **5**-3) which permits the identification of the irreducible representations of the crystal.

Characters

We have seen that the matrix $\hat{G}_\varphi(\mathring{g})$ of a representation is not unique but only determined within a similarity. It can be proved, however, that matrices have an *invariant* under any similarity transformation, which is given by the sum of their diagonal elements, called the *trace*. (We merely mean here that the trace of a matrix is invariant when the matrix is subject to a similarity transformation as in (16).) The trace of $\hat{G}_\varphi(\mathring{g})$ is called the *character* of $\mathring{g}$ in the representation $\hat{G}$ (the basis need now not be given), and we shall denote it with the symbol $\chi(\mathring{g}|\hat{G})$:

$$\chi(\mathring{g}|\hat{G}) = \sum_i \hat{G}_\varphi(\mathring{g})_{ii}. \tag{17}$$

An important property of the characters is that they are the same for all elements of the same class. This property stems from the fact that the conjugation relation which must obtain between any two elements of a class acts somewhat like a similarity and thus leaves the character invariant.

5 Irreducible representations

A very important property of a group G of order n is that the number of its irreducible representations can be very easily obtained, since it can be proved that it equals the number of its classes. Once the number of irreducible representations is obtained in this way, their dimensions can often be easily guessed by means of a second simple rule, which states that the sum of the squares of the dimensions of the representations must equal the order of the group. The reader must appreciate the enormous power of these results: we now know precisely the types of energy levels, with their corresponding degeneracies, which are possible for any system with G as its Schrödinger group.

Given a representation $\hat{G}$ of matrices $\hat{G}(\mathring{g})$, an important problem is to be able to recognize whether it is irreducible or not. This is easily done by means of the following test, which involves the characters $\chi(\mathring{g}|\hat{G})$:

$$\sum_{\mathring{g}} \chi(\mathring{g}|\hat{G})^* \chi(\mathring{g}|\hat{G}) = n. \tag{1}$$

If the result of the sum in (L1) is not n, the order of the group, then the representation is not irreducible.

Given a reducible representation $\hat{G}$, it is important to know which are the irreducible representations which will appear on reduction, as in (4.10). Because the characters must be the same before and after reduction, $\chi(\mathring{g}|\hat{G})$ must be, for each $\mathring{g}$, a sum over the irreducible characters of the group representations (some of which may appear more than once or not at all), and it is usually quite easy to guess how this partition goes.

An important property of the characters of irreducible representations is their *orthogonality*, a relation which is expressed as follows. Let us call ${}^i\hat{G}$ and ${}^j\hat{G}$ two irreducible representations of the group G of order n. Then:

$$\sum_{\mathring{g}} \chi(\mathring{g}|{}^i\hat{G})^* \chi(\mathring{g}|{}^j\hat{G}) = n\,\delta_{ij}, \tag{2}$$

where δ_{ij} is *Kronecker's delta*, equal to unity for i equal to j and to zero otherwise. It can be seen that eqn (2) coincides with (1) for i and j equal. It is useful to keep in mind a second orthogonality relation:

$$\sum_{i} \chi(\mathring{g}_r|{}^i\hat{G})^* \chi(\mathring{g}_s|{}^i\hat{G}) = n(c_r)^{-1}\delta_{rs}, \tag{3}$$

where the summation is over all the irreducible representations of the group and c_r is the number of elements in the class of g_r.

Finally, it is important to be able to construct a basis for an irreducible representation ${}^i\hat{G}$, say. This is done by using the so-called *projection operator*, which works as follows. For any arbitrary function ϕ:

$$\sum_{\mathring{g}} \chi(\mathring{g}|{}^i\hat{G})^* \mathring{g}_s\phi = \text{function of the basis of } {}^i\hat{G}. \tag{4}$$

Representations of the direct product

We defined the direct product in (1.7) and (1.8):

$$G = \{l_r m_t\} = L \otimes M, \qquad lm = ml, \qquad \forall\, l \in L,\ \ \forall\, m \in M. \tag{5}$$

The multiplication rule for any two operations of G, therefore, is as follows:

$$l_r m_t \, l_s m_u = l_r l_s \, m_t m_u = l_v m_w. \tag{6}$$

From the general definition of a representation (see eqn 4.6), any representation of G must be given by matrices $\hat{G}$ which satisfy the multiplication rule (6):

$$\hat{G}(l_r m_t)\,\hat{G}(l_s m_u) = \hat{G}(l_v m_w). \tag{7}$$

We cannot solve here the problem of finding these matrices in the most general case, but we shall be able nevertheless to discuss the particular case, which will be the only one needed in this book, when the irreducible representations of L and M are one-dimensional. They are therefore merely numbers rather than matrices, and always commute. On comparing with the multiplication rules implicitly used in (6), we must have:

$$\hat{L}(l_r)\hat{L}(l_s) = \hat{L}(l_v), \qquad \hat{M}(m_t)\hat{M}(m_u) = \hat{M}(m_w). \tag{8}$$

We assert that

$$\hat{G}(lm) = \hat{L}(l)\,\hat{M}(m), \tag{9}$$

for all l and m, satisfies (7) and is therefore a representation. (Like $\hat{L}$ and $\hat{M}$, it must be one-dimensional and therefore irreducible.) It is very easy to verify this statement, remembering that our one-dimensional matrices commute:

$$\hat{G}(l_r m_t)\,\hat{G}(l_s m_u) = \hat{L}(l_r)\,\hat{M}(m_t)\,\hat{L}(l_s)\,\hat{M}(m_u) \tag{10}$$

$$= \hat{L}(l_r)\hat{L}(l_s)\,\hat{M}(m_t)\,\hat{M}(m_u) \tag{11}$$

8,9 $$= \hat{L}(l_v)\hat{M}(m_w) = \hat{G}(l_v m_w). \tag{12}$$

Clearly, eqn (9) can immediately be extended for triple direct products of the form (1.10). A very simple example of this result will be found § **5**-2 and many readers will probably prefer to return to a more serious assessment of the present section after they see how that example works.

6 An example: benzene

We depict the benzene molecule in Fig. 1, where the functions $\phi_1, \phi_2, \ldots, \phi_6$, represent the π orbitals of the carbon atoms, which are p-type orbitals with their axes normal to the plane of the molecule: they are of course necessary in order to agree with the valence of four of the carbon atoms. We recognize at once that there is a six-fold rotation axis, (represented with the black hexagon at the centre) to which there correspond rotations by $2\pi/6$, ($C_6^{\pm}$), $2\pi/3$, ($C_3^{\pm}$), and π, (C_2). We also find a set of three reflection planes $\sigma_{d1}, \sigma_{d2}, \sigma_{d3}$ and another set $\sigma_{v1}, \sigma_{v2}, \sigma_{v3}$. Each of these sets form one class, the elements of which are conjugated, for example, by a rotation by

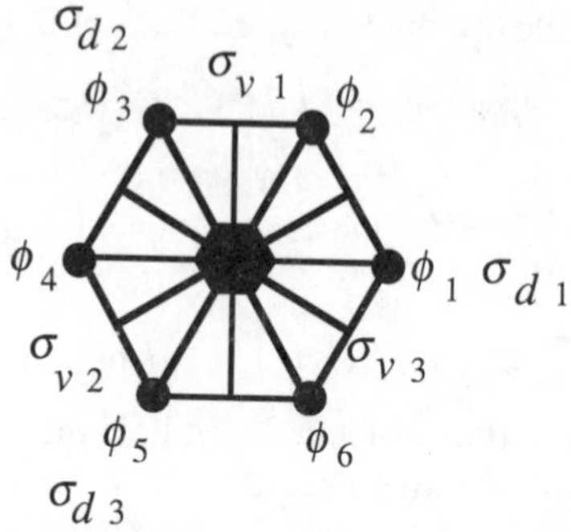

Fig. 2-6.1. Benzene and its symmetry operations. (Group $\mathbf{C}_{6v}$.) The hydrogen atoms are not shown.

$2\pi/6$. There is, however, no operation which conjugates one σ_d into some other σ_v, (a rotation by $2\pi/12$ would be required, which does not belong to the group) and this is the reason why we have two separate classes.

Strictly speaking, the plane of the molecule is a symmetry plane, so that this operation, as well as others which appear as products of this reflection with any of the operations previously listed, also belong to the group. For simplicity of the example, however, we shall only consider the twelve operations listed, which form a group called $\mathbf{C}_{6v}$. This is the group of operations of an hexagonal pyramid, the base of which is not, of course, a symmetry plane of the solid, whereas the complete group of benzene, $\mathbf{D}_{6h}$, is the group of the hexagonal prism. Although some details have to be refined, it can be proved that on using $\mathbf{C}_{6v}$ the correct classification of degeneracies is obtained.

It is a simple, although tedious task (unless clever shortcuts are used), to verify that the twelve operations of $\mathbf{C}_{6v}$ separate out in six classes: E, C_2, $2C_3$, $2C_6$, $3\sigma_d$, $3\sigma_v$. (The numerals here give the number of elements in each class.) We can now find the number and the dimensions of the representations by the method explained right at the beginning of §5. Because there are six classes, there must be six irreducible representations. It can be guessed that their dimensions must be 1, 1, 1, 1, 2, 2, since the sum of the squares of these numbers correctly gives 12, the order of the group. Group theory gives methods for obtaining irreducible representations of any group and, in any case, tables are easily available. We show a table of the characters of the irreducible representations of $\mathbf{C}_{6v}$ in Table 1, the first column of which provides conventional symbols for the various irreducible representations. Notice that the characters of the identity E in the second column agree, as they must, with the dimensions of the representations, so that the vital information about the number and type of the permitted degeneracies of the group can immediately be read from such a table.

Let us now try to understand in more detail the nature of the energy levels associated with the π orbitals of benzene. We want to know, for instance,

Table 2-6.1. Characters of the irreducible representations of $\mathbf{C}_{6v}$.

	E	C_2	$2C_3$	$2C_6$	$3\sigma_d$	$3\sigma_v$	
A_1	1	1	1	1	1	1	1
A_2	1	1	1	1	-1	-1	0
B_1	1	-1	1	-1	1	-1	1
B_2	1	-1	1	-1	-1	1	0
E_1	2	-2	-1	1	0	0	1
E_2	2	2	-1	-1	0	0	1
χ	6	0	0	0	2	0	

whether the six π orbitals $\phi_1, \phi_2, \ldots, \phi_6$ could form a degenerate set and, if they are not degenerate, which are the irreducible representations into which they reduce. We must form for this purpose the basis $\langle \phi_1, \phi_2, \ldots, \phi_6 |$ and determine the corresponding characters. Consider, for example, σ_{d1}. It follows at once from Fig. 1 that this operation leaves ϕ_1 and ϕ_4 invariant while exchanging the other orbitals. A little thought (see Problem 7.7) will show that the corresponding 6×6 matrix will have unity on the diagonal in the first and fourth rows, and zeros in all the other diagonal elements. Therefore, its character must be 2. In general, it is easy to satisfy oneself that the character of an operation on the basis $\langle \phi_1, \phi_2, \ldots, \phi_6 |$ equals the number of functions left invariant by the operation in question. The characters thus obtained for this basis are given in the last row of the table. We immediately conclude that this basis must be reducible, since the sum of the square of the characters (see eqn 5.1) equals not 12, the order of the group but rather 48. (If you work it out to be 40 you are making a mistake: remember to add up over all the *operations*, not just over the classes.) In order to see how the basis reduces all that we need is to find linear combinations of the irreducible characters which add up to the characters of the reducible representation. The corresponding coefficients are listed in the last column of the table. The result is most important, since it means that the π-electron system will produce four energy levels, two singly and two doubly degenerate. We illustrate this in Fig. 2, where the order of the levels, which does not follow from group theory, can be guessed with a bit of experience. Since there are six electrons, only the two lower levels are occupied, as shown. The great power of group theory should be evident from this example, which in fact is more relevant to solid-state theory than it appears, since the π electron system has very much the characteristics of a free-electron gas and the six levels shown in the figure can be considered the band that arises from the π atomic state of the carbon atom. The energy of the top-occupied level in Fig. 2 corresponds indeed to the Fermi energy in a solid.

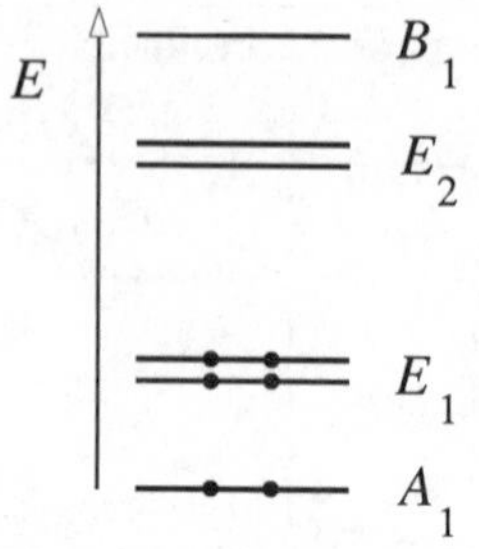

Fig. 2-6.2. The energy levels in benzene.

7 Problems

1. Imagine that the black circles in Fig. 2.1 represent a set of identical atomic orbitals centred on the respective atoms sites. It should be clear that, if $\phi_i(\mathbf{r})$ is such an orbital centred on the grid site with $\mathbf{r}_i$ as position vector, then $\phi_i(\mathbf{r})$ must be of the form $\phi(\mathbf{r} - \mathbf{r}_i)$, where the function ϕ determines the nature (say 1*s*) of the atomic orbital considered. Verify from the figure that $\mathring{t}$ acting on ϕ_1 gives ϕ_6. Use eqn (2.7) with (2.3) to obtain $\mathring{t}\phi_1$, that is $\mathring{t}\phi(\mathbf{r} - \mathbf{r}_1)$. Verify that unless you *subtract t* from the position vectors, as required by the formulae, you will not get the result obtained geometrically.
2. Prove that in an *abelian group* (a group in which all operators commute) each operation is its own class. Thus prove that the number of irreducible representations in such a group equals its order and therefore that all the irreducible representations are one-dimensional.
3. Prove that, given a rotation C_n by $2\pi/n$, the set

$$C_n, (C_n)^2, (C_n)^3, \ldots, (C_n)^n \equiv E, \qquad (1)$$

where $(C_n)^m$ is the rotation in which C_n is repeated m times, forms a group. (Such a set, in which all elements are given as powers of a single element, is called a *cyclic group*.) Prove that this group is abelian.
4. Consider the cyclic group $\mathbf{C}_3$, depicted in Fig. 1, with irreducible representations in Table 1. Prove that the representations listed in the table are

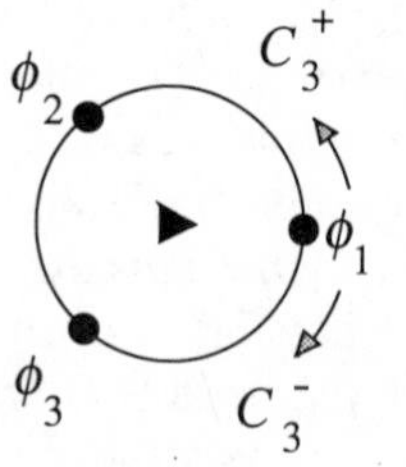

Fig. 2-7.1. The group $\mathbf{C}_3$.

Table 2-7.1. The irreducible representations of $\mathbf{C}_3$.

$\omega = \exp(\frac{2}{3}\pi i)$, $\omega^* = \exp(-\frac{2}{3}\pi i)$.

	E	C_3^+	C_3^-
A	1	1	1
^1E	1	ω	ω^*
^2E	1	ω^*	ω

irreducible. Prove, on considering the various products involving C_3^+ and C_3^-, that their multiplication rules are preserved by the (one-dimensional) matrices shown in Table 1.

5. Verify that the projection operator (5.4), applied, for the representation ^1E of the group $\mathbf{C}_3$, on the function ϕ_1 of Fig. 1 forms the function

$$\psi^{^1E} = \phi_1 + \omega^* \mathring{C}_3^+ \phi_1 + \omega \mathring{C}_3^- \phi_1 = \phi_1 + \omega^* \phi_2 + \omega \phi_3. \tag{2}$$

Verify that $\psi^{^1E}$ is actually a basis of ^1E, that is that $\mathring{g}\psi^{^1E}$, for all $\mathring{g}$, transforms in accordance to Table 1, as shown below as an example for $\mathring{C}_3^+$:

$$\mathring{C}_3^+ \psi^{^1E} = \phi_2 + \omega^* \phi_3 + \omega \phi_1 = \omega(\phi_1 + \omega^* \phi_2 + \omega \phi_3) = \omega \psi^{^1E}. \tag{3}$$

Form $\psi^{^2E}$ in the same manner and verify its transformation properties.

6. Show that, if $\mathbb{H}$ is real and the orbitals in Fig. 1 are also real, then the functions $\psi^{^1E}$ and $\psi^{^2E}$ of Problem 5 are degenerate. (Hint: use the conjugator operator $\mathbb{j}$. Notice that, in this case, the conjugator introduces degeneracies over and above those given by the group of covering operations.)

7. Show that for the orbitals and symmetry operations defined in Fig. 6.1,

$$\mathring{\sigma}_{d1} \langle \phi_1 \phi_2 \phi_3 \phi_4 \phi_5 \phi_6 | = \langle \phi_1 \phi_6 \phi_5 \phi_4 \phi_3 \phi_2 | \tag{4}$$

$$= \langle \phi_1 \phi_2 \phi_3 \phi_4 \phi_5 \phi_6 | \begin{bmatrix} 1 & 0 & 0 & 0 & 0 & 0 \\ 0 & 0 & 0 & 0 & 0 & 1 \\ 0 & 0 & 0 & 0 & 1 & 0 \\ 0 & 0 & 0 & 1 & 0 & 0 \\ 0 & 0 & 1 & 0 & 0 & 0 \\ 0 & 1 & 0 & 0 & 0 & 0 \end{bmatrix} \tag{5}$$

Hence verify that the character of this operation equals 2.

8. Consider a molecule with symmetry $\mathbf{C}_{3v}$ (Fig. 2), with three identical orbitals ϕ_1, ϕ_2, ϕ_3. (They could be, for example, the hydrogen atomic orbitals in the ammonia molecule.) The character table of this group is

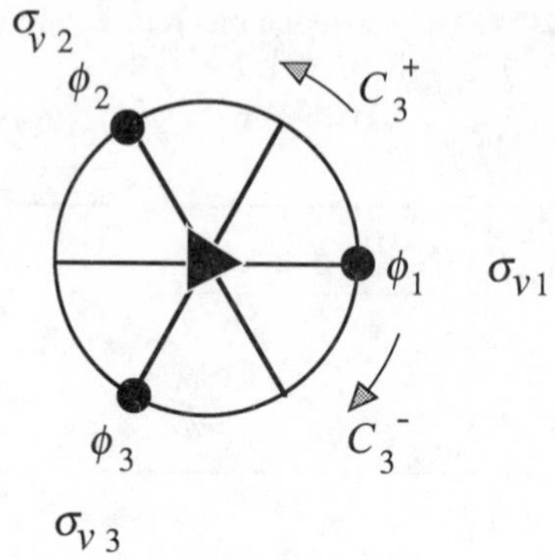

Fig. 2-7.2. The group $\mathbf{C}_{3v}$.

Table 2-7.2. The irreducible representations of $\mathbf{C}_{3v}$.

	E	$2C_3$	$3\sigma_v$
A	1	1	1
A_2	1	1	-1
E	2	-1	0

given in Table 2. Verify the following results:

$$\phi_1 + \phi_2 + \phi_3 \qquad \text{spans } A_1. \tag{6}$$

$$\left.\begin{matrix} 2\phi_1 - \phi_2 - \phi_3 \\ \phi_2 - \phi_3 \end{matrix}\right\} \quad \text{span } E. \tag{7}$$

9. With the notation of Problem 8, prove that the symmetrized functions

$$\psi_1 = \phi_1 + \phi_2 + \phi_3, \qquad \psi_2 = \phi_2 - \phi_3. \tag{8}$$

are orthogonal. (This is an example of a general result whereby functions which belong to different irreducible representations are orthogonal.)

Further reading

The most satisfactory general introduction to all the topics treated in this chapter is provided by the first 90 pages or so of Tinkham (1964). Falicov (1966) is a very concise and useful little book. McWeeny (1963) is a useful general introduction, as well as, at much greater length, Heine (1960) and Lax (1974). For density and precision of information per unit length, however, the chapter on group theory (Chapter 12) in the book on quantum mechanics by

Landau and Lifshitz (1977) cannot be beaten. More chemical approaches may be found in Cotton (1971) and in Atkins (1983). Although they were among the first books ever published on group theory in physics, both Wigner (1959) and van der Waerden (1974) are still very good reading, although well above the level at which we aim, as is the case with Hamermesh (1962), Jansen and Boon (1967), and Cornwell (1984). Proofs of all the results quoted in the present chapter may be found in Altmann (1977).

The definition (2.7) of the function space operator has not always been accepted. Slater (1965) used a variant form which can now be ruled out. Care must be exercised in using results obtained with this variant definition, since errors may appear.

GOSHEN COLLEGE LIBRARY
GOSHEN, INDIANA

3

Space groups

We have learnt in Chapter **2** how group theory allows us to classify the degeneracies of a physical system, as long as we know its symmetry group. Symmetry groups of crystals are substantially more involved than those of molecules and they are called *space groups.* We must learn some of their properties in order to be able to study those of the energy eigenvalues in crystals.

The first fact which must be considered is that an actual crystal may contain an extremely complex arrangement of atoms because centred at each crystal site one might have a whole molecule like, say, anthracene. We call this arrangement of all atoms in the crystal the *crystal structure* and, because it is in general so complicated, we shall replace it by a set of geometrical and not of material points which possesses the same symmetry as the crystal structure and which we shall call the *crystal pattern.* This is defined as the minimal set of points which is left invariant by all the symmetry operations of the crystal structure. We must explain. First, 'left invariant' here means that a symmetry operation of the crystal is a covering operation of the pattern. Secondly, we say 'minimal' because, obviously, given a pattern, one can always add any number of points to it, as long as they are carefully chosen so as to maintain symmetry. (This is, in fact, often done in order to make symmetries more evident, by using *ornaments* instead of points. An L-shaped ornament, for example, superimposed on a line, shows at once that this line is not a reflection plane.) It should be appreciated that, when drawings are made, a pattern may look like a structure and vice versa: the correct appellation is necessary for the interpretation of any figure of this type. In practice, unless we are dealing with a definite material, a silicon crystal, say, it can always be safely assumed in this book that the illustrations refer to crystal patterns and not to crystal structures. This makes the work not only simpler but vastly more general, since one crystal pattern may describe the symmetry operations of a very large number of entirely different crystal structures.

It should be appreciated that the above definition of a crystal pattern does not, of course, provide a method for its construction starting from the crystal structure. It is only after enough experience is acquired in recognizing and listing *all* the symmetry operations of a crystal structure that such a construction is possible.

The set of all symmetry operations of a crystal pattern forms a group which, in accordance to what we have said, is the *space group* of the crystal. The most distinctive feature of a crystal pattern is its *periodicity*, a property which is manifested by the fact that some specific translations are covering operations of the pattern. Thus, a space group must always contain some translations as symmetry operations.

1 Translations and unit cells

We shall always assume that our crystal patterns are infinite since, if this were not so, translations, as discussed in § **2**-2, would not be covering operations, a point particularly important in this section since we want to discuss translations as symmetry operations. It must be stressed that, of course, when we refer to translations (or for that matter to any other symmetry operations, to be considered not here but in later sections) we always have in mind the precise definition described in § **2**-2, so that a word such as 'translation' is a mere shorthand for the *covering* operation defined there in detail. It is pretty clear that the pattern of Fig. 1 admits of two translation operations by the vectors **a** and **b**, and of an infinite number of translations by any integral linear combinations of these vectors. We shall loosely refer to all this as 'repeated translation by the vectors **a** and **b**'. Such vectors, which by repeated translation generate the translations of a crystal pattern (not necessarily all, as we shall see), are called *unit vectors.* (It is important to remember that, despite their name, crystallographic unit vectors are not generally vectors of unit length. Their length is usually given in nanometers.) In three dimensions, of course, we shall have three unit vectors. The unit vectors span a parallelogram (parallelepiped in three dimensions) which is called the *unit cell*, repeated translation of which *covers* the whole of the crystal pattern.

It is very important to realize that the choice of unit vectors, and thus of unit cells, is not unique. In principle, in Fig. 1, we could have taken 2**a** and

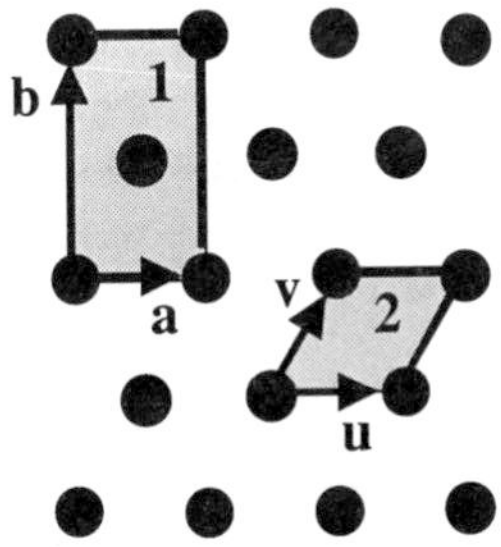

Fig. 3-1.1. A crystal pattern in which a unit cell, (1), and a primitive cell, (2), are shown.

b as unit vectors, although this would have been an eccentric choice. Conventionally, once the direction of a unit translation is selected, the unit vector is chosen as the shortest translation vector along that direction. We thus ensure that, at least, we do not miss any translation operations along that direction. (With the choice just stated, translations by **a** would of course be missed.) On the other hand, the unit vectors **u** and **v** shown in the figure also generate by repeated translation a set of translation operations of the pattern and it is clear that this set contains all the translations spanned by **a** and **b** (because **b** is given $-\mathbf{u} + 2\mathbf{v}$) plus a new infinite set of translations (of the form $n\mathbf{v}$, for any integer n). It is evident, in fact, in this simple example, that **u** and **v** span all the possible translations of the crystal pattern (because all pattern points are *equivalent* under **u** and **v**, by which we mean that any pattern point can be reached by repeated translations by **u**, **v**, and their *integral* linear combinations, whereas this is not the case when **a** and **b** are used).

We shall now summarize and refine the definition of our objects. We consider three independent directions in the crystal pattern and take the shortest translation vectors along these directions. They are the *unit vectors* and they span a *unit cell*. If the vectors chosen are such that their integral linear combinations provide by repetition *all* the translations of the crystal pattern, they are then called *primitive vectors* and the cell which they span is called a *primitive cell*. (It should be clear that whereas all primitive cells are also unit cells, not all unit cells are primitive.)

For the definition given to work, one must make sure that *all* translations are included, which might require a bit more thought than in the very simple example of Fig. 1 (see below). It is evident that we shall be primarily interested in primitive vectors and primitive cells, because our job is to enumerate all the symmetry operations of the crystal pattern. In crystallography, however, unit cells which are not primitive are much used because, as in Fig. 1, they often reflect more quickly than primitive cells the existence of important, non-translational, symmetry elements of the crystal (two perpendicular symmetry planes in the case of Fig. 1).

It is important for us to establish a fundamental distinction between primitive and non-primitive cells. The unit cell 1 of Fig. 1 contains pairs of points which are translationally equivalent, that is that are related by some translation of the crystal pattern. Two points anywhere inside the cell which differ by **v** are, in fact, translationally equivalent. A primitive cell, instead, cannot contain any pair of internal points which are translationally equivalent. If we add **u**, for example, to the position vector of any internal point, we take it to a point outside the primitive cell. A little care has to be taken with points on the surface (boundary in two dimensions) of the cell. Each point along the vector **v**, is translationally equivalent, under a translation by **u**, to a point on the opposite side (face in three dimensions) of the cell. In order to ensure, however, that each point of the cell, *including its surface*, belongs to one and only one cell, we shall agree that only one of each pair of parallel faces (sides) of the primitive cell belongs to the cell. (Such a surface, or boundary is called *open*. The sides of the cell 2 of Fig. 1, for example, which are not marked with vectors, are *outside* the cell.) Once this is done, no two points of the boundary of the cell can be translationally equivalent, since the corresponding face for one of these two points must now be *outside* the cell. We can now re-define a primitive cell of a crystal pattern in a way which will prove very useful for our purposes: it is an open polyhedron which contains a *maximal* set of translationally inequivalent points. The condition that this set be *maximal* is required because any region of the crystal pattern which is smaller than the primitive cell contains only inequivalent points and any region which is larger contains some points which are equivalent to points of the primitive cell.

We have chosen the primitive cell in Fig. 1 in such a way that all the points of the crystal pattern are equivalent under the primitive vectors chosen (it is this property that ensures that all translations are enumerated). This, however, is not always possible and it must be appreciated that there are crystal patterns such that not all their points are equivalent under translations. If the crystal pattern is such that all its points are equivalent under translations, then it is called a *lattice*. If this property is not valid then the crystal pattern is called a *lattice with basis*. It must be understood, though, that a lattice with basis is not a lattice.

An example of a lattice with basis is shown in Fig. 2, in which the pattern is obtained from that in Fig. 1 by moving away from the centre the pattern point in the middle of the unit cell 1 of that figure. It is clearly no longer possible to choose the vector from the origin to the internal pattern point of the unit cell (**v** in Fig. 1) as a translational vector. The cell shown in Fig. 2*a* is a primitive cell, because there is no possible choice of translational vectors which would make its internal pattern point translationally equivalent to one

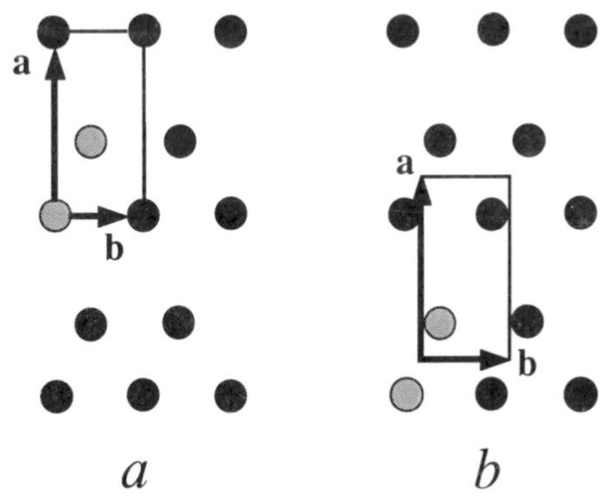

Fig. 3-1.2. A lattice with basis. The two pattern points of the basis are shown in grey.

of the vertices of the cell. Because this cell has an internal pattern point in it, it is called a *primitive cell with a basis.* The number of pattern points which belong to the cell is worked out as follows. The cell has a pattern point at each of its four vertices, each of which is shared by four cells, thus contributing a total of one point to the cell. Add to this the internal pattern point, so that we have two pattern points per primitive cell. These two points can be collected together to form a *basis*, the translations of which generate the whole of the crystal pattern. This property is emphasized by the choice of primitive cell displayed in Fig. 2*b*. (Notice that, as shown here, neither unit nor primitive vectors need be determined by pairs of pattern points.)

One final remark about translations. It should be pretty clear at this stage that the product of two translations is always a translation, so that the set of all translations of a space group *G* forms a subgroup of *G*, called the *translation subgroup*. (See § 8 for a formal proof of this result.)

2 Centred primitive cells

We shall now construct primitive cells anew, from their definition as maximal sets of translationally inequivalent points. Given a translation vector **a** (not necessarily primitive), as in Fig. 1, it is possible to give a very simple prescription in order to construct a set of points which are all translationally inequivalent with respect to **a**. The vector − **a** must necessarily be a translation vector along the same axis as **a**. Form the normal bisector planes of **a** and − **a**. Between them they limit an infinite slab. Let us agree, for the reasons explained in § 1, that of the two parallel surfaces of the slab the one along the positive direction belongs to the slab, whereas the other is outside it. It is now quite clear that there are no two points both of which belong to the slab and which are equivalent under a translation by **a**. Also, that the slab is a maximal set of points with this property, any point outside the slab being equivalent to some internal point. If we now take a crystal pattern point *O* in three

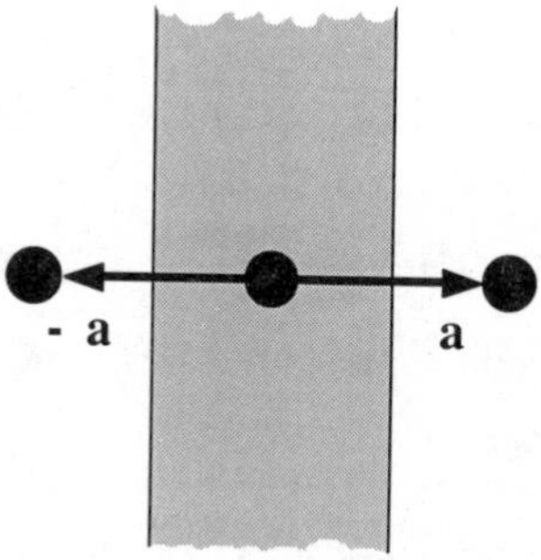

Fig. 3-2.1. Construction of translationally inequivalent sets of points.

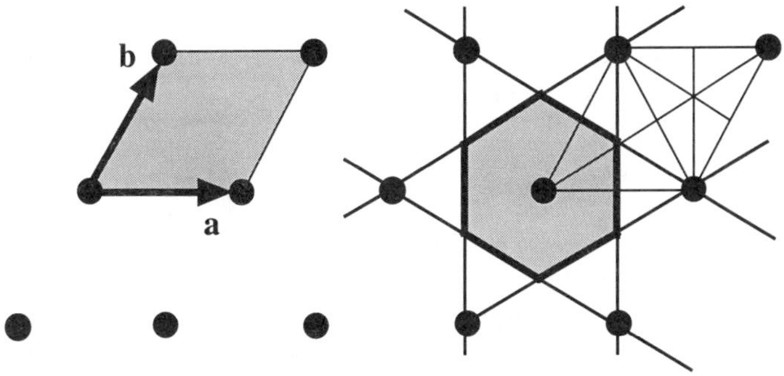

Fig. 3-2.2. Construction of the centred primitive cell of a two-dimensional hexagonal close-packed lattice. The ordinary crystallographic primitive cell spanned by the primitive translation vectors **a** and **b** is shown on the left of the figure.

dimensions, do the work described for the three primitive vectors **a**, **b**, **c** from *O*, and we then take the volume common to all the slabs thus formed, we shall have a maximal set of translationally inequivalent points, that is, a primitive cell. This cell is called a *centred primitive cell* because, whereas the origin *O* of the crystal pattern is at a vertex of an ordinary cell, it is at the centre of the new cell, as seen in Fig. 2 for a two-dimensional example. Such cells are often called also *Wigner–Seitz cells*, after the authors who first used them. It should be remembered that, as all primitive cells, they are open, only three of the six sides (which must be conventionally chosen) belonging to the cell illustrated in the figure.

Because the standard primitive cell on the left of Fig. 2, and the centred cell on its right are both maximal sets of translationally inequivalent points, we must expect that their extensions be identical, which we shall now verify. The centred cell contains twelve identical triangles, (of which only two are shown), precisely like the standard unit cell. Thus, both cells have the same area. It is also clear from the figure that the centred cell reflects very directly the hexagonal symmetry of the crystal pattern, which is not the case for the standard cell. This is the reason why centred primitive cells are most useful in solid-state theory, whereas they are rarely used in crystallography.

3 The Bravais lattice

We have already seen that not all crystal patterns are such that all their points are translationally equivalent. It is therefore natural, in order to display as clearly as possible the translational symmetry of a crystal pattern, to abstract from it another pattern which contains only translationally equivalent points and which will be called the *Bravais lattice*. Given a crystal pattern, its Bravais lattice is formed by taking a set of points in the original pattern which are all translationally equivalent amongst themselves, and such that there are no other points in the crystal pattern which are translationally equivalent to the points of this set. We show in Fig. 1 how the Bravais lattice of the lattice with basis of Fig. 1.2 is formed. Notice that a Bravais lattice is itself a crystal pattern of the type called a lattice in §1. For all uses and purposes the expressions Bravais lattice, translational lattice, and lattice are all synonymous. Notice, however, that a lattice with basis *is not a lattice* but a pattern.

A three-dimensional example of a Bravais lattice is given in Fig. 2. This picture depicts the unit cell of the so-called body-centred cubic lattice, translation of which by the vectors **a**, **b**, **c** shown generates the whole pattern. Although the cell shown is a unit and not a primitive cell, it is clear that all the points of this crystal pattern are translationally equivalent: the vector that goes from the origin to the central pattern point is a primitive translation vector, the repetition of which reaches the point at the opposite end of the cube diagonal. Thus, the Bravais lattice of the pattern shown coincides with it.

Although a picture of the Bravais lattice looks like that of a crystal pattern (with which it can formally coincide) it is nothing more than a graphical

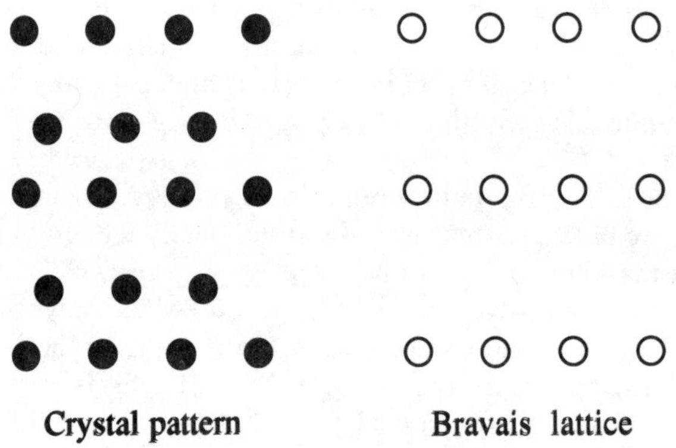

Fig. 3-3.1. A lattice with basis and its Bravais lattice.

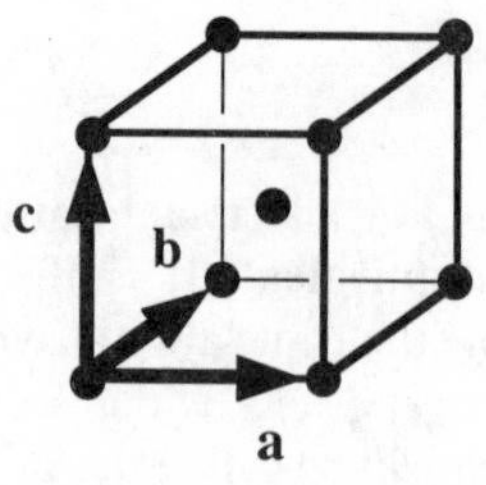

Fig. 3-3.2. The body-centred cubic lattice (a Bravais lattice).

depiction of the translation vectors of the pattern, *so that it can always be slid parallel to itself in any manner whatsoever.* The empty points in Fig. 1 do not bear, in fact, any necessary relation to the full pattern points of the figure on their left, except as regards their orientation.

If we consider the crystal pattern of Fig. 1.1, it is easy to see that it coincides with its Bravais lattice. It thus follows that each unit cell of the type (1) therein depicted contains two points of the Bravais lattice, whereas the primitive cell (2) contains only one such point. It should be clear that primitive cells cannot contain more than one Bravais lattice point per cell. (The primitive cell of a lattice with basis contains several pattern points but only one lattice point!)

It should be clear that all Bravais lattices must be lattices without bases, that is, a primitive cell can always be constructed in them, such that all the points of the Bravais lattice are translationally equivalent under the primitive vectors which span the cell. It is proved in crystallography that only 14 types of Bravais lattices are possible in three dimensions.

4 Space groups and their symmetry operations

The *space group* is the set of all the symmetry operations that leave the crystal pattern invariant, by which we mean that, if we make a copy of the crystal pattern and transform this copy under an operation, then this transformed copy covers the original. We give in Fig. 1 an example of a *strictly two-dimensional* space group, also called a *plane group*, (that is, a space group in which the third dimension is supposed not to exist). We must now discuss all the symmetry operations of such a system, which as always we assume to extend to infinity in all directions. First of all, of course, we have translations by $\mathbf{a}'$ and $\mathbf{b}'$ and by all possible integral linear combinations of them. We next consider reflections σ. Whereas σ'_y, σ_2, and σ_4 are clearly symmetry operations, σ_1 is not on its own a covering operation, since it takes the point 3 half way along $\mathbf{a}'$, to a position where there is no pattern point. If we follow σ_1, however, with the translation v by $\mathbf{v}$, then the compound operation $v\sigma_1$ is a covering operation. (Remember to read correctly *from right to left* the product $v\sigma_1$: σ_1 here is the *first* operation. This is the way in which all operator products, such as log sin, say, are read.) As always, this assertion should really be checked by transforming a tracing of the original system. The same result arises for all the vertical reflection planes shown in broken lines and, likewise, σ'_x must be followed by v, forming the operation $v\sigma'_x$ which is a covering operation. We now consider the *binary rotations* (rotations by π). The binary axis C'_2 (remember that C_n is a rotation by $2\pi/n$) is a symmetry operation on its own right, whereas g_2 is a binary rotation which must be followed by v.

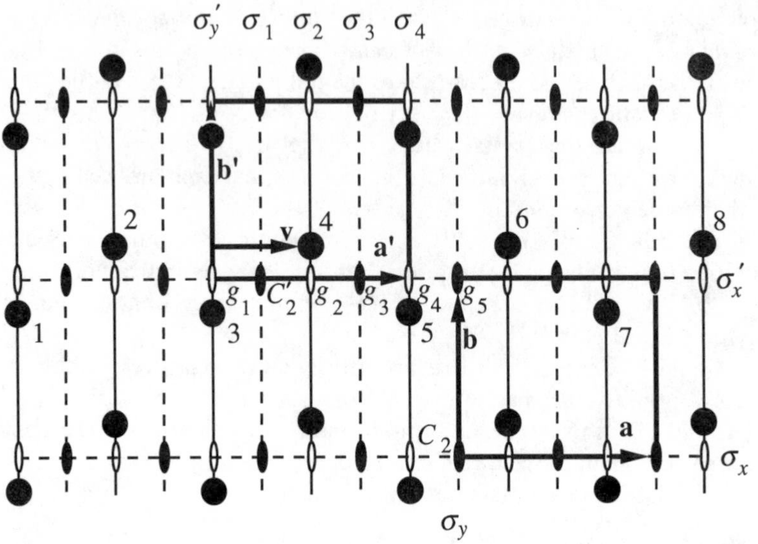

Fig. 3-4.1. A space group in two dimensions. Full lines are reflection planes. Broken lines are symmetry planes that must be associated with the fractional translation vector **v** shown. Digons (ellipse-like two-sided figures) stand for binary axes C_2 (full digons) or binary axes associated with a translation by **v** (open digons).

It appears from the above description of the symmetry operations of the space group that their enumeration is an almost impossible task, since there is an infinite number of them. We can, however, considerably simplify this task by factoring out the translation subgroup, a procedure which we shall describe by means of an example. Consider the operation σ_2: its effect is to leave the point 4 invariant and to exchange 2 and 6, which can identically be obtained by the compound operation $a'\sigma_y'$ (reflection σ_y' followed by translation by **a**′). We are drastically simplifying here the effects of symmetry operations, which should be properly expressed in terms of all the pattern points. The description we give, however, is sufficient to recognize that σ_2 is the same operation as $a'\sigma_y'$. This means that, if σ_y' is listed in the group, and the whole of the translation group is also listed, then $a'\sigma_y'$ will necessarily appear in the set, as the product of σ_y' with one of the operations of the translation subgroup, and therefore need not be listed separately. Once we understand this principle, we realize that all operations outside the primitive cell spanned by **a**′ and **b**′ can be given as products of operations defined inside the cell multiplied with translations. We shall consider an example. The operation g_5 exchanges points 4 and 7, which is the same effect as that of $(2a')C_2'$.

The net result so far is that we need consider only operations inside the primitive cell but, even then, not all of them are necessary since we have already seen that, for example, σ_2 is $a'\sigma'_y$. When all operations such as this one are discarded, all that remains is the set of four operations

$$E, C'_2, v\sigma'_x, \sigma'_y, \tag{1}$$

so that the *whole* of the space group is obtained by multiplying one of these four operations with some translation of the translation group (the geometrical realization of which in the Bravais lattice is obtained by infinite repetition of **a**′, **b**′, and their integral linear combinations).

Two points must be carefully understood here. One is that, having chosen the unit cell and having listed the symmetry operations as in (1), we must stick precisely to this list: C'_2, for example, must always be the operation so labelled and not any other of the binary axes, such as g_5, say. The second point is that the way in which the symmetry operations are presented in a list such as (1) crucially depends on the origin and disposition of the unit cell chosen. To illustrate this important question we display in Fig. 1 a second unit cell, spanned by the primitive vectors **a** and **b**. Once the translation subgroup is factorized out, the space group operations that remain are

$$E, C_2, v\sigma_x, v\sigma_y, \tag{2}$$

a list which, even if one were to forget the primes, is manifestly different from (1).

It is important to notice that we have required in the above work the use of a vector **v** which is not a translation vector but rather an *integral fraction* of one:

$$\mathbf{v} = \tfrac{1}{2}\mathbf{a}. \tag{3}$$

A translation by a vector such as **v** is not a covering operation of the crystal pattern. Thus, it does not belong to the space group but, because it is required when associated with other operations, it will be called a *fractional translation.*

The compound operation $v\sigma'_x$ in (1) is called in crystallography a *glide reflection.* This is a reflection followed by a fractional translation *parallel* to the reflection plane. In three dimensions, composite operations also appear in which rotations are associated with fractional translations. A *screw rotation* is a rotation followed by a fractional translation *parallel* to the rotation axis. Notice that, as we have seen, g_2 is vC'_2, and therefore it is not a screw rotation, because the vector **v** is normal and not parallel to the rotation axis. It should be noted that glide reflections and screw rotations as just defined are the only compound operations admitted in crystallography.

We now return briefly to the factorizations of the space group entailed in (1) and (2). It can be seen that the four operations in (1) are all standard crystallographic operations and because of this, this is the setting used in crystallography. In solid-state theory, however, such a setting is very inconvenient for the following reason. The three operations E, σ'_x, and σ'_y in the set

(1) are all referred to the origin of the unit cell, whereas this is not the case for the remaining operation C'_2, (see Fig. 1). Therefore, whereas the first three operations mentioned leave invariant the point at the origin, this is not so for the last one. In the setting used in (2), instead, not all operations are crystallographically standard, $v\sigma_y$ not being a correct glide. This setting, however, has an overwhelming advantage in solid-state theory, namely that all the four non-translational operations in it, E, C_2, σ_x, σ_y, are referred to the origin of the unit cell, that is, they all leave invariant exactly the same point of the crystal pattern (its origin). We shall see in § 5 that this is most important in our work and for this reason the setting (1) must not be used. We shall always assume that a setting is chosen so that all the non-translational operations required leave the same point invariant, normally chosen to be the origin of the unit cell. We shall not worry in the least that operations which are not standard in crystallography are used in order to achieve this result.

5 The point group and the Bravais lattice

It will be convenient, first of all, to depict the crystal pattern of Fig. 4.1 in a clearer relation to its Bravais lattice, with we do in Fig. 1. Of course, there are infinitely many ways in which this could be done, but we have chosen to do so in terms of the unit cell shown in the lower part of Fig. 4.1, since it is this setting which has led us to express the space group G of the crystal as products of translations with one of the four operations

4.2 $$E,\ C_2,\ v\sigma_x,\ v\sigma_y. \qquad (1)$$

We have said in § 4 that we strongly prefer this set to the set (4.1), and we shall now see more clearly why this is so. Because the four non-translational operations which appear in the above list,

$$E,\ C_2,\ \sigma_x,\ \sigma_y, \qquad (2)$$

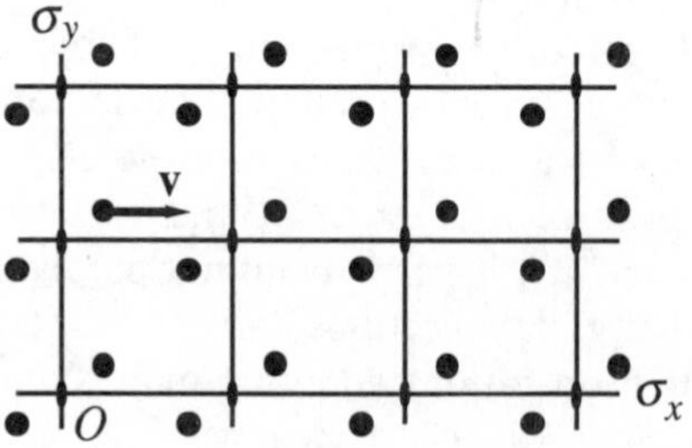

Fig. 3-5.1. The space group of Fig. 4.1 (in half the scale) and its Bravais lattice. The setting implied in eqn (4.2) has been used, that is, the origin O of the Bravais lattice (shown in thick full lines) has been taken to coincide with the binary axis C_2 of Fig. 4.1.

all leave identically the same point (the origin of coordinates) invariant (see §4), it follows that the product of any two operations in this set will also have the same property, so that they may form a group, which it is quite easy to verify is actually the case. Such a group is called a *point group*, to emphasize that all operations in it leave one point invariant. The group in (2) is actually the well-known point group $\mathbf{C}_{2v}$.

Given a space group G, with its operations listed in an appropriate setting as products of non-translational operations, such as (1), times the translation group, then the *point group* P of the space group G is the set of *all point operations* which appear in the space group listing, either on their own or associated with fractional translations. It cannot be emphasized too strongly that the point group P is *not* necessarily a subgroup of its space group. It should be clear that, whereas the operations in (1) are operations of the space group, those in (2) are not. It seems somewhat strange, that, given a space group G we take the trouble to define another group P which is not a subgroup of it. The following discussion will show, however, that this is a sensible thing to do.

We go from G to P, that is from (1) to (2), by ignoring all translational operations and by taking v as the identity in (1). The fractional translation v, however, exists in G because we have a lattice with a basis. Taking v as the identity is equivalent to collapsing the basis into a single point, say at the corners of the rectangles in Fig. 1. Therefore, the operations in (2), although they are not covering operations of G, are covering operations of the rectangular crystal pattern in Fig. 1, that is, they are covering operations of the Bravais lattice. It can be seen at once, in fact, that $\mathbf{C}_{2v}$ as defined in (2) has this property. Thus, given a space group G, its point group P is a covering group of its Bravais lattice. We shall often say that P leaves the Bravais lattice *invariant*, in the sense that it transforms the position vector of one of the lattice points of the Bravais lattice into the position vector of another point of the same lattice. It should be appreciated, though, that the Bravais lattice $\mathscr{B}$ may have a point group $P_{\mathscr{B}}$ larger than P. This point group $P_{\mathscr{B}}$ is, of course, the set of all operations that leave the Bravais lattice and one of its points (normally taken to be the origin) invariant. We illustrate this in Fig. 2. The two crystal patterns a and b in the picture have both precisely the same Bravais lattice, shown in c, although in the pattern in b no fractional translations can be associated either with σ_x or σ_y in order to construct any possible symmetry operations. The corresponding point groups are:

$$a\!:\!P = E, C_2, \sigma_x, \sigma_y; \qquad b\!:\!P = E, C_2; \qquad c\!:\!P_{\mathscr{B}} = E, C_2, \sigma_x, \sigma_y. \tag{3}$$

This means that P will either be a subgroup of $P_{\mathscr{B}}$ or identical to it:

$$P \subseteq P_{\mathscr{B}}. \tag{4}$$

Whenever P coincides with $P_{\mathscr{B}}$ we say that the point group P is *holohedric*, that is, that it has as much point group symmetry as it is compatible with its

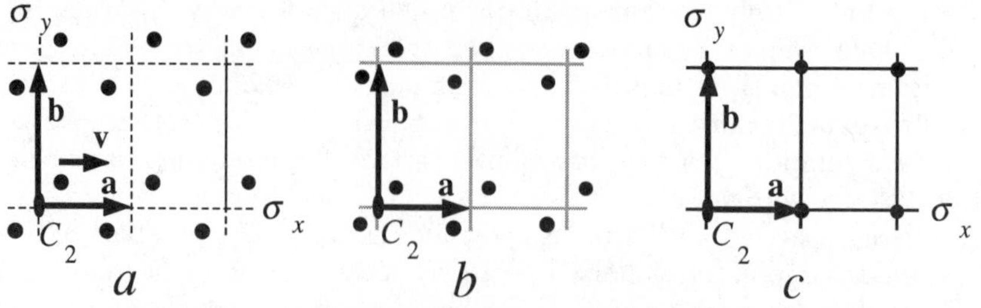

Fig. 3-5.2. Space groups with the same Bravais lattice but with different point groups. The broken lines are symmetry planes which have to be associated with the translation vector **v**. The grey lines are not planes of symmetry of any type: they merely indicate the Bravais lattice. Full lines are true symmetry planes.

Bravais lattice. Notice that the non-holohedric case of Fig. 2*b* is obtained by ornamenting the Bravais lattice with a basis that destroys some of the point group symmetry.

The stated property that the point group P leaves the Bravais lattice $\mathscr{B}$ invariant is very important and we shall discuss it again in a more detailed fashion. We want to prove that

$$p \in P \qquad \Rightarrow \qquad p\mathscr{B} = \mathscr{B}. \tag{5}$$

On the other hand,

$$p \in P \qquad \Rightarrow \qquad vp \in G, \text{ (for some } v) \qquad \Rightarrow \qquad vp\mathscr{B} = \mathscr{B}, \tag{6}$$

since all operations of the space group must leave $\mathscr{B}$ invariant. Suppose now, in the last equation in (6), that

$$p\mathscr{B} = \mathscr{B}', \tag{7}$$

where $\mathscr{B}'$ differs from $\mathscr{B}$ in orientation. (Since the origins of P and $\mathscr{B}$ can always be taken as coinciding, if necessary, by a parallel translation of $\mathscr{B}$, $\mathscr{B}'$ and $\mathscr{B}$ must also *have the same origin.*) On substitution in (6), this relation would require $v\mathscr{B}'$ to coincide with $\mathscr{B}$, which is not possible, since the only operation that could transform $\mathscr{B}'$ into $\mathscr{B}$ is a rotation. Therefore, $p\mathscr{B}$ must always coincide with $\mathscr{B}$, (as stated in eqn 5), in order to satisfy (6), since $v\mathscr{B}$ and $\mathscr{B}$ are always identical. (Remember that a Bravais lattice can always be translated by an entirely arbitrary vector, as stated in the italicized sentence in § 3.)

We shall now briefly discuss what happens if we were to use the space group in the crystallographically preferred setting (4.1). If we try to extract from it the point group by using the above-given prescription, the point operations chosen would be

$$E, C'_2, \sigma'_x, \sigma'_y, \tag{8}$$

which do not even form a group, since they do not close, because C'_2 is not referred to the same centre as the other three operations so that it must introduce some additional translation whenever it acts. It can readily be seen, in fact, that the product $C'_2\sigma'_y$ equals $v\sigma'_x$, which is not in (8).

6 The Seitz operators

We have seen that in a space group we have operations which are products of translations t of the Bravais lattice with point group operations (as was the case with the operations $(2a')C'_2$ and $a'\sigma'_y$ discussed in §4), as well as operations in which point group operations are associated with a fractional translation v. Let us denote with p a point group operation and with $\mathbf{w}$ a vector that can be either a vector $\mathbf{t}$ of the Bravais lattice or some fractional translation vector $\mathbf{v}$. The most general space group operation, therefore, is of the form wp, that is, some point group operation *followed* by some translation. It is traditional in solid-state theory to denote such operations with a symbol introduced by F. Seitz, called the Seitz operator:

$$wp \equiv \{p|\mathbf{w}\}. \tag{1}$$

In order to define it fully, we must determine how this operator acts on a point $\mathbf{r}$ of the configuration space:

$$\{p|\mathbf{w}\}\mathbf{r} =_{\text{def}} wp\mathbf{r} \equiv w(p\mathbf{r}) = p\mathbf{r} + \mathbf{w}. \tag{2}$$

Here, as in §**2**-2, and illustrated in Fig. 1, $p\mathbf{r}$ means the position vector $\mathbf{r}'$ which is the transform of $\mathbf{r}$ under p. (Notice that, for unfortunate historical reasons, the Seitz symbol contains the operators in the wrong order. This, with a little practice, is no trouble, if eqn R2 is used directly.) We must obtain the *product* of two Seitz operators:

2 $$\{p|\mathbf{w}\}\,\{p'|\mathbf{w}'\}\mathbf{r} = \{p|\mathbf{w}\}\,(p'\mathbf{r} + \mathbf{w}') \tag{3}$$

2|3 $$= p(p'\mathbf{r} + \mathbf{w}') + \mathbf{w} = pp'\mathbf{r} + p\mathbf{w}' + \mathbf{w} \tag{4}$$

4|2 $$= \{pp'|p\mathbf{w}' + \mathbf{w}\}\mathbf{r}. \tag{5}$$

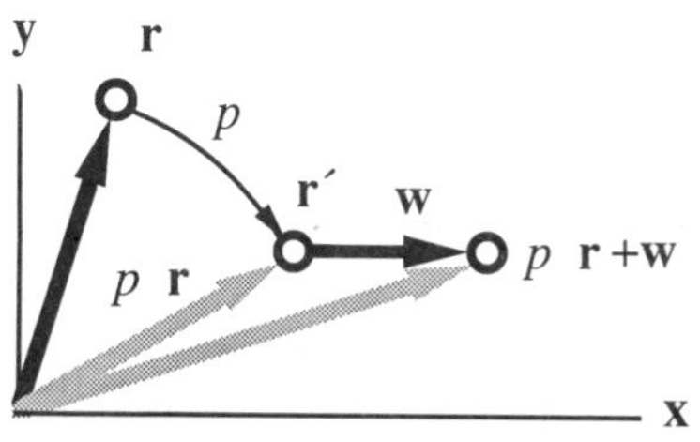

Fig. 3-6.1. Definition of the Seitz space group operator.

Therefore,

5 $$\{p|\mathbf{w}\}\,\{p'|\mathbf{w}'\} = \{pp'|p\mathbf{w}' + \mathbf{w}\}. \tag{6}$$

We also need the *inverse* $\{p|\mathbf{w}\}^{-1}$ of a Seitz operator, for which we assert:

$$\{p|\mathbf{w}\}^{-1} = \{p^{-1}| -p^{-1}\mathbf{w}\}. \tag{7}$$

The proof is as follows.

6 $$\{p|\mathbf{w}\}\,\{p^{-1}| -p^{-1}\mathbf{w}\} = \{pp^{-1}| -pp^{-1}\mathbf{w} + \mathbf{w}\} \tag{8}$$

$$= \{E| -\mathbf{w} + \mathbf{w}\} = \{E|\mathbf{0}\}. \tag{9}$$

The operator $\{E|\mathbf{0}\}$ here is, of course, the identity of the space group G, since it does not alter anything at all.

One final point. In principle, we should make a notational distinction between the operator $\{p|\mathbf{w}\}$ in configuration space, say g, and its corresponding function space operator, say $\mathring{g}$. We shall not do this, since it is not difficult to guess from the context which type of operator one is using.

7 The space group and its point group

In this Seitz notation, the space group G is the set

$$G = \{\{p|\mathbf{w}\}\}, \tag{1}$$

from which the point group P can be formed at once:

$$P = \{\{p|\mathbf{0}\}\}. \tag{2}$$

Given that G is a group, it follows at once, in fact, that P is a group:

6.6 $$\{p|\mathbf{0}\}\,\{p'|\mathbf{0}\} = \{pp'|\mathbf{0}\}. \tag{3}$$

The product pp' in (3) must appear in G from its closure and eqn (6.6), whence, from (1) and (2), it must also appear in P.

8 The translation subgroup

Given a Bravais lattice, with primitive vectors **a, b, c**, along the respective directions x, y, z, a translation vector **t** which goes from the origin of the lattice to one of its lattice points must have integral components in **a**, **b**, and **c**:

$$\mathbf{t} = m\mathbf{a} + n\mathbf{b} + p\mathbf{c}; \qquad m, n, p \text{ integers}. \tag{1}$$

Such vectors are called *vectors of the lattice* (in contrast to vectors *in* the lattice, which have arbitrary real components and thus denote any points in the space of the lattice). Given a space group G as in (7.1), the translations are operations of the form $\{E|\mathbf{t}\}$ and they form a subgroup of G called the *translation subgroup* T:

6.6 $$\{E|\mathbf{t}\}\,\{E|\mathbf{t}'\} = \{E|\mathbf{t}+\mathbf{t}'\}. \qquad (2)$$

In fact, because $\mathbf{t}+\mathbf{t}'$ must also be of the form (1), closure is satisfied. Notice that as a difference with the point group P, T is a subgroup of G.

Commutation of translations

An important property of T is that all its operations commute and it is thus *abelian* (this being the generic name of groups with this property). The proof of this result is immediate:

6.6 $$\{E|\mathbf{t}\}\,\{E|\mathbf{t}'\} = \{E|\mathbf{t}+\mathbf{t}'\}, \qquad (3)$$

6.6 $$\{E|\mathbf{t}'\}\,\{E|\mathbf{t}\} = \{E|\mathbf{t}'+\mathbf{t}\} \equiv \{E|\mathbf{t}+\mathbf{t}'\}. \qquad (4)$$

A useful consequence of the commutation property is that T can always be written as a direct product.

Direct product form of T

Consider first translations along the x direction which, from (1), must have the form $\{E|m\,\mathbf{a}\}$. Clearly, the product of two such translations will still be a translation of the same form, that is along the x axis. Thus the set of all translations along the x axis forms a subgroup of T which we shall call T_x. In the same manner, we can define subgroups T_y and T_z. If we ignore for the time being in T the translations along the z axis, it is clear from (**2**-1.7) and (**2**-1.8) that T can be written as the direct product $T_x \otimes T_y$, because any operation of the first group here commutes with any operation of the second group. If we now do the same work between the operations of this direct product and those of T_z, (**2**-1.7) and (**2**-1.8) can be applied once more to give a new direct product and thus the final form for T:

$$T = T_x \otimes T_y \otimes T_z. \qquad (5)$$

Basically, all that this formula says is that any translation of T can be performed as a translation along the z axis followed by a translation along the y axis, followed by a translation along the x axis, and that the order in which these partial translations are performed is irrelevant.

Invariance

We shall now prove that T is an invariant subgroup of G,

$$T \lhd G, \qquad (6)$$

which means that the conjugate of any operation $\{E|\mathbf{t}\}$ of T must also belong to T,

§2-1 $$\{p|\mathbf{w}\}\ \{E|\mathbf{t}\}\ \{p|\mathbf{w}\}^{-1} \in T. \qquad (7)$$

This property of T is very important: we shall see that the bases of the irreducible representations of T will allow us to construct bases for the irreducible representations of G and that this remarkable result is due to the invariance property of the translation subgroup. (See §6-2.) The proof of the invariance property of T is as follows.

6.7|L7 $$\{p|\mathbf{w}\}\ \{E|\mathbf{t}\}\ \{p|\mathbf{w}\}^{-1} = \{p|\mathbf{w}\}\ \{E|\mathbf{t}\}\ \{p^{-1}| -p^{-1}\mathbf{w}\} \qquad (8)$$

6.6|8 $$= \{p|\mathbf{w}\}\ \{p^{-1}| -p^{-1}\mathbf{w} + \mathbf{t}\} \qquad (9)$$

6.6|9 $$= \{pp^{-1}| -pp^{-1}\mathbf{w} + p\mathbf{t} + \mathbf{w}\} \qquad (10)$$

10 $$= \{E|p\mathbf{t}\}. \qquad (11)$$

We have seen in (5.5) that the point group P leaves the Bravais lattice invariant which means that, for all $p \in P$, and all $\mathbf{t}$ belonging to the Bravais lattice, $p\mathbf{t}$ must also belong to the Bravais lattice. Thus (R11) is a translation, which proves our assertion.

9 Classification of space groups

There are two types of space groups, called *symmorphic* and *non-symmorphic* respectively. In *symmorphic space groups* no fractional translation vectors appear, so that the expression (7.1) of the space group takes the form

7.1 $$G = \{\{p|\mathbf{t}\}\}, \qquad \mathbf{t} \in T. \qquad (1)$$

Since

6.6 $$\{p|\mathbf{t}\} = \{E|\mathbf{t}\}\ \{p|\mathbf{0}\}, \qquad (2)$$

all operations of G are products of an operation of T times an operation of P. Another important feature of symmorphic groups is that for them, and only for them, the point group P is actually a subgroup of G. This follows from (1), since the null vector $\mathbf{0}$ belongs to T (where it is the identity). Hence $\{p|\mathbf{0}\}$, for all p, must belong to the set (1) so that every operation of the point group belong to G.

In a *non-symmorphic space group*, some point group operations p are associated with fractional translation vectors $\mathbf{v}$. These vectors do not belong to the translation lattice but when they are repeated by a specific integral number of times they give a vector of the lattice. Each point group operation will be associated with a particular vector $\mathbf{v}$ as follows:

6.6 $$\{p|\mathbf{v}\} = \{E|\mathbf{v}\}\ \{p|\mathbf{0}\}. \qquad (3)$$

Here, as a difference with (2), $\mathbf{v}$ cannot vanish for all p, since it must be a fractional vector for at least some p operations. Therefore, the point group operation $\{p|\mathbf{0}\}$ cannot belong to G for all p, that is, it cannot be made to coincide with (L3) except for the special operations for which $\mathbf{v}$ vanishes. It should be noted in (3) that, like $\{p|\mathbf{0}\}$, the operation $\{E|\mathbf{v}\}$ does not belong to G. It is only their product that belongs to the space group. Not all operations of P, therefore, belong to G.

As an example, the space group of Fig. 5.2*a* is non symmorphic, whereas that of Fig. 5.2*b* is symmorphic. Notice that both corresponding patterns, however, are lattices with bases.

10 Problems

1. Verify in Fig. 4.1 the following equalities:

$$g_1 = (-v)C'_2, \quad g_2 = vC'_2, \quad g_3 = a'C'_2, \quad g_4 = (3v)C'_2. \tag{1}$$

$$\sigma_1 = v\sigma'_y, \quad \sigma_2 = a'\sigma'_y, \quad \sigma_3 = (3v)\sigma'_y, \quad \sigma_4 = (2a')\sigma'_y. \tag{2}$$

2. On using Fig. 4.1, identify Seitz operators which have as point group operations the binary rotations g_1 to g_4 and the reflections σ_1 to σ_4 listed in (1) and (2) above. Hence, on using the results in (1) and (2), write these Seitz operators in the form $\{p|\mathbf{w}\}$, with p one of the point group operations listed in (4.1) and $\mathbf{w}$ either a translation of the lattice or a fractional translation.
3. Show that

$$\{p|\mathbf{w}\}\ (\mathbf{r} + \mathbf{r}') \neq \{p|\mathbf{w}\}\mathbf{r} + \{p|\mathbf{w}\}\mathbf{r}'. \tag{3}$$

4. Write all operations in (4.2) in the Seitz form and hence form their multiplication table. Show that this table does not close but that all products in it can be written as products of the same set of operators (4.2) times translations.
5. Prove that the only operations which commute with a translation by $\mathbf{t}$ are: (i) all translations; (ii) all rotations around $\mathbf{t}$ as axis of rotation; (iii) all reflections on any plane which contains $\mathbf{t}$.

Further reading

Good introductions to the basic properties of crystal structures can be found in Buerger (1956) and Kelly and Groves (1973), as well as in Glazer (1987), McKie and McKie (1974), and Ashcroft and Mermin (1976). A simple discussion of space group operators appears in Altmann (1977). Knox and Gold (1964) contains a good introduction to the subject as well as a very useful collection of reprints. Full discussions on space groups are given in Janssen (1973) and Streitwolf (1971). Burns and Glazer (1978) provide a very clear treatment and Bradley and Cracknell (1972) is an authoritative treatise on the subject.

4

The reciprocal lattice and the Fourier series

A crystal pattern is periodic in the translation lattice, that is, it is obtained in its (infinite) entirety by repeated translations of the unit cell, these translations thus providing periodicity along each of the corresponding directions. As a result, many functions of physical importance in the crystal will be *periodic functions* which, once defined inside the unit cell, are then known throughout the crystal by the periodicity condition. An important example of this situation is the crystal potential, but we shall see that various other periodic functions play a very important part in the study of bands in crystals.

It is important to recognize that the periodicity of a crystal pattern (or, for that matter, of a crystal structure) is fully determined by its translation lattice. Thus periodic functions in a crystal are periodic in the lattice. Likewise, the lattice provides a reference framework in order to specify *position vectors* in the crystal pattern, that is, vectors that denote points in the space in which the pattern is given, and thus such vectors are always referred to the lattice (rather than to the crystal pattern). For this reason, they will be called *vectors of the lattice* or *vectors in the lattice*, a distinction which will be explained in §§ 1 and 6. Finally, the translation lattice determines sets of planes which pass through the lattice points and which are very significant in discussing physical properties of the pattern. Such planes are, in accordance to our present description, entirely described in terms of the lattice rather than the pattern and can thus be called *lattice planes.*

We shall be largely concerned in this chapter with the construction of periodic functions in a *crystal lattice* (that is, the lattice of a crystal pattern), a problem which is far from trivial, so that we must do some serious work towards this goal. The main tool which we shall use for this purpose is the *reciprocal lattice* of a crystal lattice, which we shall define in § 2, after some simple classes of periodic functions are discussed in § 1. The *Fourier series*, to be defined in § 9, will provide us with the most complete answer to our problem, and it will allow us to expand any function which satisfies the periodicity of any arbitrary crystal in terms of carefully selected plane waves.

1 Periodic functions

Consider to start with a one-dimensional lattice along the x axis with periodic vector **a**. We shall prove that a plane wave $\exp(ikx)$, (see eqn **1**-8.1), will be a periodic function in a when

$$k = 2\pi h/a, \qquad h \text{ integral}; \tag{1}$$

(the letter h has nothing to do with Planck: it appears here because it shall later form part of a crystallographic symbol in which this letter, as well as k and l, are standard, but they are printed in sanserif to avoid confusion). To emphasize that the values defined in (1) are special values of k, they will be relabelled with the letter g instead, always to be reserved for k values of the form (1):

$$g = 2\pi h/a, \qquad h \text{ integral}. \tag{2}$$

The corresponding plane wave will then be written as $\exp(igx)$, and we shall now verify that it is indeed periodic in a:

$$\varphi(x) = \exp(igx), \tag{3}$$

,2
$$\varphi(x + a) = \exp(igx)\exp(iga) = \varphi(x)\exp(2\pi i h) = \varphi(x). \tag{4}$$

We shall want to extend this work to three-dimensional lattices, for which purpose we shall start with the simplest possible case, in which we have a simple cubic lattice the unit vectors of which are of unit length. (This is not a redundant statement: the unit vectors are normally given lengths in Å or nm):

$$\mathbf{a}\cdot\mathbf{b} = \mathbf{b}\cdot\mathbf{c} = \mathbf{c}\cdot\mathbf{a} = 0, \tag{5}$$

$$\mathbf{a}\cdot\mathbf{a} = \mathbf{b}\cdot\mathbf{b} = \mathbf{c}\cdot\mathbf{c} = 1. \tag{6}$$

Equation (5) states that **a**, **b**, **c** are *orthogonal*, and eqn (6) that they are *normalized*. Such vectors are said to be *orthonormal*. The vectors of the lattice are

3-8.1
$$\mathbf{t} = m\mathbf{a} + n\mathbf{b} + p\mathbf{c}; \qquad m, n, p\text{: integers}. \tag{7}$$

A periodic function in this lattice must satisfy the condition:

$$\varphi(\mathbf{r} + \mathbf{t}) = \varphi(\mathbf{r}). \tag{8}$$

It is reasonable to expect, as we shall prove in a moment, that $\varphi(\mathbf{r})$ be given by the product of three functions of the form (3):

$$\varphi(\mathbf{r}) = \exp(ig_x x)\exp(ig_y y)\exp(ig_z z), \tag{9}$$

where, as in (2), we must have, remembering that a, b, c are unity,

$$g_x = 2\pi h, \qquad g_y = 2\pi k, \qquad g_z = 2\pi l; \qquad h, k, l\text{: integers}. \tag{10}$$

On introducing (10) into (9), we have:

$$\varphi(\mathbf{r}) = \exp 2\pi i(hx + ky + lz). \tag{11}$$

In order to simplify this expression, we write **r** by components and we define a vector **g** with components along **a**, **b**, **c** given by (10):

$$\mathbf{r} = x\mathbf{a} + y\mathbf{b} + z\mathbf{c}, \tag{12}$$

$$\mathbf{g} = 2\pi(h\mathbf{a} + k\mathbf{b} + l\mathbf{c}), \tag{13}$$

so that

$$\mathbf{g}\cdot\mathbf{r} = 2\pi(hx + ky + lz), \tag{14}$$

whence,

14|11 $$\varphi(\mathbf{r}) = \exp(i\mathbf{g}\cdot\mathbf{r}). \tag{15}$$

We can now confirm our expectation that this function is periodic as desired:

$$\varphi(\mathbf{r} + \mathbf{t}) = \exp\{i\mathbf{g}\cdot(\mathbf{r} + \mathbf{t})\} = \varphi(\mathbf{r})\exp(i\mathbf{g}\cdot\mathbf{t}), \tag{16}$$

but,

7,13 $$\mathbf{g}\cdot\mathbf{t} = 2\pi(hm + kn + lp) = 2\pi v; \quad v\text{: integer.} \tag{17}$$

The exponential in (16) is therefore unity and (8) is thus verified.

Equation (15) is a very handy form for a periodic function, but it must be appreciated that (14) is not valid unless the unit vectors are orthonormal, that is, unless (5) and (6) are satisfied. The scalar product (14) would otherwise contain obnoxious cross terms, such as $\mathbf{a}\cdot\mathbf{b}$ and the like. The same is the case also for the vitally important product $\mathbf{g}\cdot\mathbf{t}$ in (17). In order to get rid of cross terms in the scalar product even when the unit vectors are not orthonormal, some new objects will be invented, called *reciprocal vectors*. They will allow us to obtain an expression of the same form as (15), which will be valid not just for the cubic lattice but for all lattices. Also, we shall be able to dispense with the impractical requirement that the unit vectors be of unit length.

Before we go into this work, it is useful to note that, as announced in the introduction to this chapter, the position vectors **r** in the crystal pattern are always expressed in terms of the crystal lattice, since this is fully determined by the vectors **a**, **b**, **c** used in (12). The components x, y, z in this expression are general real numbers. If they are integers, the position vector **r** designates a point of the crystal lattice itself, in which case the position vector will always be denoted with **t**, as in eqn (7), and called a *vector of the lattice*, whereas the general position vector **r** will be called a *vector in the lattice*.

2 Reciprocal vectors and general periodic functions

The work of the last section succeeded because, for the very special case of the cubic lattice with unit vectors of unit length, these vectors satisfy the orthonormality conditions

$$\mathbf{a}\cdot\mathbf{a} = \mathbf{b}\cdot\mathbf{b} = \mathbf{c}\cdot\mathbf{c} = 1, \qquad \mathbf{a}\cdot\mathbf{b} = \mathbf{b}\cdot\mathbf{c} = \mathbf{c}\cdot\mathbf{a} = 0, \tag{1}$$

which allowed us to write the exponent of the periodic functions

1.11 $$\varphi(\mathbf{r}) = \exp 2\pi \mathrm{i}(hx + ky + lz), \tag{2}$$

in the form (1.15) as a scalar product of the two vectors

1.13,1.12 $$\mathbf{g} = 2\pi(h\mathbf{a} + k\mathbf{b} + l\mathbf{c}), \qquad \mathbf{r} = x\mathbf{a} + y\mathbf{b} + z\mathbf{c}. \tag{3}$$

It was, in fact, this scalar product which allowed us to carry out the proof of periodicity in a very compact manner in (1.16). In a general crystal pattern, **a**, **b**, **c** (which we shall refer to as unit vectors, although in most cases they will be chosen to be primitive) are neither of unit length, nor orthogonal, so that the scalar product of two vectors such as (3) would never have the desired form in (2). The trick we shall introduce in order to circumvent this difficulty is based on the strange idea of describing a crystal pattern not by one but by *two* sets of unit vectors to be used *simultaneously*. The first of these two sets is formed by the three crystal unit vectors **a**, **b**, **c**, no assumptions being made about their length or direction. They will be called the *direct vectors*. The second set is given by three new vectors, $\mathbf{a}^{\#}, \mathbf{b}^{\#}, \mathbf{c}^{\#}$, called the *reciprocal vectors* and so ingeniously chosen that, with them, equations of precisely the same form as in (1) are valid, as long as in any expression used which involves two vectors, one of them is always taken as a direct and the other as a reciprocal vector:

$$\mathbf{a}^{\#}\cdot\mathbf{a} = \mathbf{b}^{\#}\cdot\mathbf{b} = \mathbf{c}^{\#}\cdot\mathbf{c} = 1, \tag{4}$$

$$\mathbf{a}^{\#}\cdot\mathbf{b} = \mathbf{b}^{\#}\cdot\mathbf{c} = \mathbf{c}^{\#}\cdot\mathbf{a} = \mathbf{a}\cdot\mathbf{b}^{\#} = \mathbf{b}\cdot\mathbf{c}^{\#} = \mathbf{c}\cdot\mathbf{a}^{\#} = 0. \tag{5}$$

(It will be useful, before we go any further, to have a name for the symbol $^{\#}$. This is called *hash*, so that $\mathbf{a}^{\#}$ is read '*a hash*'. It should be noted that the usual symbol for the reciprocal vectors is a *star* or *asterisk*, which has been changed in this book by the hash in order to avoid confusion with the standard mathematical notation for complex conjugation.) The first thing we must observe is that the new vectors $\mathbf{a}^{\#}, \mathbf{b}^{\#}, \mathbf{c}^{\#}$, strange as they might seem, can in principle easily be constructed. The vector $\mathbf{a}^{\#}$, for example, can be seen from (5) to be normal to **b** and **c**. It is therefore normal to the plane spanned by these vectors, which fixes its direction. Its length is fixed by (4), which requires $\mathbf{a}^{\#}\cdot\mathbf{a}$ to equal unity. (Remember that **a** is fully known.) Secondly, it is useful to know how to remember (4) and (5), for which the rules are simple: (i) all pairs of vectors must always be mixed, one direct, one reciprocal; (ii) if, ignoring the hashes, the two vectors in the same expression have the same names, the result is unity; if they have different names the result is zero.

So far, so good, but, unfortunately, we have to introduce a small and in principle totally unnecessary change, which is nevertheless *essential* in order to agree with the conventions always used in solid-state theory (which, beware, do not agree with the standard crystallographic conventions, as so far used.) The object of this change is to get rid of the 2π factor in (3), for which purpose, as we shall soon verify, it is sufficient to replace (R4) by 2π. Thus, on repeating (5) for completeness, the solid-state reciprocal vectors are defined as follows:

$$\mathbf{a}^{\#} \cdot \mathbf{a} = \mathbf{b}^{\#} \cdot \mathbf{b} = \mathbf{c}^{\#} \cdot \mathbf{c} = 2\pi, \tag{6}$$

$$\mathbf{a}^{\#} \cdot \mathbf{b} = \mathbf{b}^{\#} \cdot \mathbf{c} = \mathbf{c}^{\#} \cdot \mathbf{a} = \mathbf{a} \cdot \mathbf{b}^{\#} = \mathbf{b} \cdot \mathbf{c}^{\#} = \mathbf{c} \cdot \mathbf{a}^{\#} = 0. \tag{7}$$

The next problem is this: how are we going to juggle with two, rather than one, set of unit vectors at the same time? This is not so difficult. In any piece of work we shall find that there are position vectors like **r** in (1.12) and **t** in (1.7), which are firmly tied up to the direct lattice, because they either denote vectors *in*, or *of*, this lattice. (See §§1 and 6 for this distinction.) We shall agree that the components of such vectors be always given in terms of the direct unit vectors:

$$\mathbf{r} = x\mathbf{a} + y\mathbf{b} + z\mathbf{c}, \qquad x, y, z : \text{reals} \tag{8}$$

$$\mathbf{t} = m\mathbf{a} + n\mathbf{b} + p\mathbf{c}, \qquad m, n, p : \text{integers} \tag{9}$$

just as before. On the other hand, the vector **g** in (3) was invented because we needed a vector with components $2\pi h$, $2\pi k$, $2\pi l$, in order to express the exponent in (2) as a scalar product. Clearly, this vector is not primarily related to the vectors **a**, **b**, **c** and we can agree that those components, (deprived now of the bothersome 2π factor), should now be reckoned along the reciprocal vectors, thus defining the reciprocal components of some vector **g**:

$$\mathbf{g} = h\mathbf{a}^{\#} + k\mathbf{b}^{\#} + l\mathbf{c}^{\#}, \qquad h, k, l : \text{integers}. \tag{10}$$

It must be clearly appreciated that at this stage the vector **g** is nothing more than an artificial construct, designed so as to carry the numbers h, k, l, in the most economical way possible. Like most things useful, however, it will be found later on to have a clear physical meaning. In any case, it is evident that the dot product of (10) with (8) gives

$$\mathbf{g} \cdot \mathbf{r} = 2\pi(hx + ky + lz), \tag{11}$$

once relations (6) and (7) are used. This is precisely that part of the exponent on (R2) which we wanted to write as a scalar product:

$$\varphi(\mathbf{r}) = \exp(i\mathbf{g} \cdot \mathbf{r}). \tag{12}$$

Notice that although this expression is formally identical to (1.15), the vector **g** in it has an entirely new meaning, being given by (10) and not by (1.13).

At long last, we can reap the fruit of our labours and prove that $\varphi(\mathbf{r})$ from (12) is periodic *in any lattice*:

12 $$\varphi(\mathbf{r} + \mathbf{t}) = \exp(\mathrm{i}\mathbf{g}\cdot\mathbf{r})\exp(\mathrm{i}\mathbf{g}\cdot\mathbf{t}). \tag{13}$$

The dot product of $\mathbf{g}$ from (10) and $\mathbf{t}$ from (9) is done at once on using (6) and (7): only the dot products of vectors with the same 'name', as $\mathbf{a}$ and $\mathbf{a}^{\#}$ remain, and they equal 2π:

$$\mathbf{g}\cdot\mathbf{t} = 2\pi(hm + kn + lp) = 2\pi v, \tag{14}$$

for some integer v, since h, k, l, m, n, p, are all integers. Therefore

14|13 $$\varphi(\mathbf{r} + \mathbf{t}) = \exp(\mathrm{i}\mathbf{g}\cdot\mathbf{r})\exp(2\pi\mathrm{i}v) = \varphi(\mathbf{r}), \tag{15}$$

as we had to prove. It can thus be seen that the work of learning reciprocal vectors provides handsome rewards, and this is only the beginning.

3 Construction of the reciprocal vectors

We have already remarked that, from (2.5), $\mathbf{a}^{\#}$ is orthogonal to $\mathbf{b}$ and $\mathbf{c}$, whence, for some constant λ, it must have the form

$$\mathbf{a}^{\#} = \lambda\mathbf{b} \times \mathbf{c}. \tag{1}$$

The constant here is obtained from the condition of eqn (2.6) that $\mathbf{a}\cdot\mathbf{a}^{\#}$ be equal to 2π.

1 $$\mathbf{a}\cdot\mathbf{a}^{\#} = \lambda\mathbf{a}\cdot\mathbf{b} \times \mathbf{c} = 2\pi \quad \Rightarrow \quad \lambda = 2\pi(\mathbf{a}\cdot\mathbf{b} \times \mathbf{c})^{-1}. \tag{2}$$

2|1 $$\mathbf{a}^{\#} = 2\pi(\mathbf{b} \times \mathbf{c})(\mathbf{a}\cdot\mathbf{b} \times \mathbf{c})^{-1}. \tag{3}$$

On cycling

$$\mathbf{a}^{\#} \mapsto \mathbf{b}^{\#} \mapsto \mathbf{c}^{\#}, \qquad \mathbf{a} \mapsto \mathbf{b} \mapsto \mathbf{c}, \tag{4}$$

and remembering that the triple product $\mathbf{a}\cdot\mathbf{b} \times \mathbf{c}$ is invariant with respect to such an operation, we obtain the other reciprocal vectors:

$$\mathbf{b}^{\#} = 2\pi(\mathbf{c} \times \mathbf{a})(\mathbf{a}\cdot\mathbf{b} \times \mathbf{c})^{-1}, \tag{5}$$

$$\mathbf{c}^{\#} = 2\pi(\mathbf{a} \times \mathbf{b})(\mathbf{a}\cdot\mathbf{b} \times \mathbf{c})^{-1}. \tag{6}$$

It is useful to remember that, from the geometrical meaning of the triple product, the denominators $\mathbf{a}\cdot\mathbf{b} \times \mathbf{c}$ which appear in these equations are the volume of the unit (or primitive, if $\mathbf{a}$, $\mathbf{b}$, $\mathbf{c}$ are so chosen) cell of the crystal pattern.

4 Reciprocal vectors and crystal planes

We have seen in (3.1), for example, that $\mathbf{a}^{\#}$ is normal to the plane spanned by $\mathbf{b}$ and $\mathbf{c}$. In a crystal pattern, *stacks of planes* (that is, sets of equidistant parallel planes) are very important in discussing the properties of the crystal, and we shall see that there is a very definite relation between a stack of planes and a specific reciprocal vector. Of course, before we establish such a relation, we shall have to learn how to describe planes and directions in a crystal pattern. We illustrate in Fig. 1 two *consecutive* planes of a stack so that their distance D is the interplanar distance of the given stack. We can always arrange, as we have done in the figure, to take the coordinate origin at one of the planes of the stack, which we shall call the *zeroth plane* of the stack. Thus the top plane of the figure is the *first plane* of the stack. Quite clearly, the orientation of the stack is fully determined by the direction of the unit vector $\mathbf{u}$, normal to any plane of the stack. (Notice that the word '*unit*' here means a vector of unit length and it has not the sense used in discussing a crystal pattern!) We notice that the position vectors $\mathbf{r}_1$ and $\mathbf{r}_2$ of two points of the plane shown form precisely the same scalar products with $\mathbf{u}$:

$$\mathbf{r}_1 \cdot \mathbf{u} = \mathbf{r}_2 \cdot \mathbf{u} = D, \tag{1}$$

so that the equation which characterizes any vector $\mathbf{r}$ as a position vector of a point in the plane is

$$\mathbf{r} \cdot \mathbf{u} = D. \tag{2}$$

It is important to be able to express the scalar product on (L2) by means of components which, as we know, cannot be done without introducing awkward cross terms unless we use reciprocal vectors. As we have discussed in § 2, it is natural to express $\mathbf{r}$ in direct components, so that our 'mixing rule' requires that $\mathbf{u}$ be expressed in reciprocal components:

$$\mathbf{r} = x\mathbf{a} + y\mathbf{b} + z\mathbf{c} =_{\text{def}} [x\,y\,z]. \tag{3}$$

$$\mathbf{u} = u_x\mathbf{a}^{\#} + u_y\mathbf{b}^{\#} + u_z\mathbf{c}^{\#} =_{\text{def}} [u_x\,u_y\,u_z]^{\#}. \tag{4}$$

(Symbols like $[l\,m\,n]$ and $[l\,m\,n]^{\#}$ as used here, must always be understood to designate vectors with components l, m, n, along $\mathbf{a}$, $\mathbf{b}$, $\mathbf{c}$ and $\mathbf{a}^{\#}$, $\mathbf{b}^{\#}$, $\mathbf{c}^{\#}$, respectively.) Therefore,

2.6, 2.7, 3, 4, |2
$$\mathbf{r} \cdot \mathbf{u} = 2\pi(u_x x + u_y y + u_z z) = D. \tag{5}$$

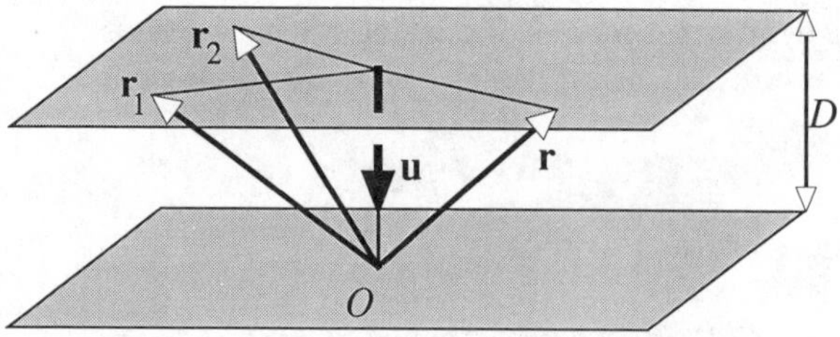

Fig. 4-4.1. Stacks of planes and the equation of the plane.

It is convenient to rewrite (5), first as

$$\frac{2\pi}{D}(u_x x + u_y y + u_z z) = 1. \tag{6}$$

On calling h, k, l the coefficients of x, y, and z in this expression, we obtain our final form:

$$hx + ky + lz = 1. \tag{7}$$

Let us now go back to the original eqn (2). It must be appreciated, from the construction that led to it, that this is the equation of the *first* plane of the stack, $\mathbf{u}$ giving the orientation of this plane (and thus of the whole stack) and D being the distance of this plane to the origin (zeroth plane), and thus the interplanar distance. It is clear that (2) can yield the equation of the n-th plane of the stack, by multiplying its right-hand side by n. In crystallography, however, this amount of detail is rarely necessary and, instead, a symbol is required which merely gives the orientation of the stack of planes. This is given straightaway by the vector $\mathbf{u}$ in (2) but, conventionally, the form (7) of eqn (2) is preferred. In this equation the coefficients h, k, l give precisely the constants $\mathbf{u}$ and D of eqn (2) and we know that, just as in that equation, multiplication of one side of it by an integer merely denotes a different individual plane of the same stack. We can thus agree to take the coefficients h, k, l, with the added convention that they can always be multiplied by any common integer, to denote the orientation of a given stack of planes. These coefficients are written in crystallography in round brackets, $(h\,k\,l)$ and they are called the *Miller indices*.

It is easy to understand the geometrical meaning of the Miller indices. Take the point $\mathbf{r}$ of the first plane of the stack at which this plane cuts the x axis, so that its coordinates y and z along the other two axes must both vanish whereas x must be an integer, since the coordinate of a crystal plane along the $\mathbf{a}$ axis must be an integral multiple of $\mathbf{a}$:

3; 7 $$x = \text{integer} \neq 0,\ y = z = 0 \quad \Rightarrow \quad hx = 1 \quad \Rightarrow \quad h = x^{-1}. \tag{8}$$

This means that h is the inverse of the intersection of the plane with the x axis (along $\mathbf{a}$), and thus a fraction, and similarly for k and l, as illustrated in Fig. 2. Notice that in this figure the zeroth plane of the stack, through the origin O, is not shown, and that the plane depicted is the first plane of the stack (as in Fig. 1), so that eqns (2) and (5) to (7) are all valid for this plane without the need of any integral multiplier. (On the other hand, if h, k, and l are multiplied by their lowest common denominator, integral Miller indices are obtained, as is standard practice in crystallography.) It should be clear that h, k, and l are fully determined in terms of the unit vectors of the direct lattice, so that the planes that we are describing are referred to the crystal translational lattice.

Figure 2 illustrates a most important result, namely that the vector $[h\,k\,l]^{\#}$ in reciprocal components is perpendicular to the stack of planes

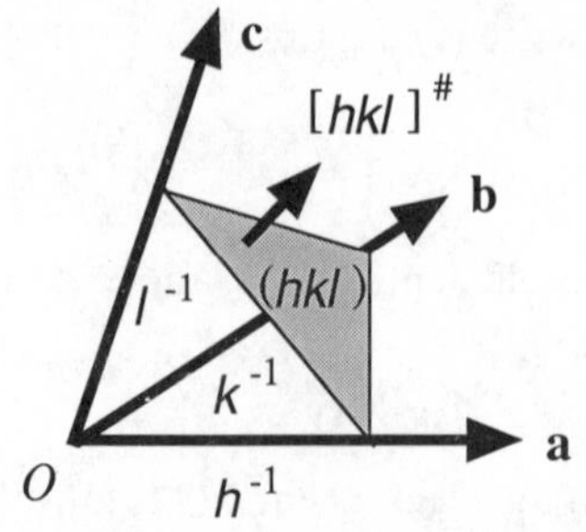

Fig. 4-4.2. The Miller indices as inverse intersections of the given plane with the basic vectors **a**, **b**, **c**, and the relation between (hkl) and $[h\,k\,l]^{\#}$.

denoted with the Miller indices $(h\,k\,l)$, which we shall now prove. The vector **u** is, of course, the vector normal to the stack, and the first thing we have to do is to identify its reciprocal components u_x, u_y, u_z, as defined in (4). On comparison of (6) with (7), we have,

$$u_x = Dh/2\pi, \qquad u_y = Dk/2\pi, \qquad u_z = Dl/2\pi, \tag{9}$$

whence

4; 9|4 $$\mathbf{u} = [u_x\,u_y\,u_z]^{\#} = \frac{D}{2\pi}[h\,k\,l]^{\#}. \tag{10}$$

Since **u** is normal to the stack $(h\,k\,l)$, this means that

$$[h\,k\,l]^{\#} \perp (h\,k\,l), \tag{11}$$

as stated above and illustrated in Fig. 2. Equation (11) provides a geometrical meaning for the *direction* of a reciprocal vector: the vector $[h\,k\,l]^{\#}$ is normal to the stack of planes $(h\,k\,l)$. We shall now show that its *modulus* is 2π times the inverse of the interplanar spacing of this stack. From (10), since $|\mathbf{u}|$ is unity,

$$\frac{D}{2\pi}|[h\,k\,l]^{\#}| = 1 \qquad \Rightarrow \qquad |[h\,k\,l]^{\#}| = \frac{2\pi}{D}. \tag{12}$$

The relations (11) and (12) between a reciprocal vector and a stack of planes are most important. We know that stacks of planes are crucial from the point of view of the propagation of plane waves in the crystal, since plane waves must have constant values over the planes of a stack. Such stacks are responsible, for instance, for the main properties of X-ray diffraction.

We have also learnt a very important idea in this section. One of the strange features of reciprocal vectors is that we determine some vectors in a crystal in terms of components on the crystal unit vectors **a**, **b**, **c**, whereas other vectors are decomposed in terms of the reciprocal vectors $\mathbf{a}^{\#}$, $\mathbf{b}^{\#}$, $\mathbf{c}^{\#}$. How do we decide which choice to take? We have already seen that *position* *vectors*, either of or in the crystal lattice must always be expressed in the direct components. (See eqns 2.8 and 2.9.) We now realize that the sensible thing is to use reciprocal components for all vectors which denote *directions* in the crystal pattern, because these directions can now be most easily identified. If we say, for example, that we have an X-ray beam propagating along the $[111]^{\#}$ direction in the crystal pattern, we know at once that this beam propagates normally to the (111) planes.

A final remark about the Miller indices $(h\,k\,l)$. It should be clear from (8) that in order to denote *planes of the lattice*, that is planes that cut the crystal axes at lattice points, h, k, l must be inverse integers. Multiplication of these indices by a constant does not change the orientation of the plane. If h, k, l are integers larger than unity, the plane cuts the unit vectors *inside* the unit cell. Even in this case, however, the planes in question are parallel to a stack of planes of the lattice. The reader should not find it difficult to verify these statements.

5 Reciprocal lattice and reciprocal space. An example

We show in Fig. 1*a* a hexagonal close-packed lattice in two dimensions, spanned by the vectors **a** and **b**, along the directions x and y respectively. It will be found convenient, so as to simplify the picture, to take $|\mathbf{a}|^2$ and $|\mathbf{b}|^2$ to be both equal to 2π. In order to draw $\mathbf{a}^{\#}$, we take a direction normal to **b** and then intersect it with the normal to **a** which cuts the x axis precisely at **a**. It follows from the figure that $\mathbf{a}^{\#}\cdot\mathbf{a}$ equals $|\mathbf{a}|^2$ and thus 2π, in agreement with (2.6). We proceed likewise for $\mathbf{b}^{\#}$. Once this is done, we can fix a vector $[210]^{\#}$ by taking a component $2\mathbf{a}^{\#}$ along the $\mathbf{a}^{\#}$ axis and a component $\mathbf{b}^{\#}$ along the $\mathbf{b}^{\#}$ axis. We want to compare this vector with the plane (210), which is easy to draw, since, from (4.8), its intersections with **a**, **b**, and **c** must be 1/2, 1/1, and 1/0 (that is ∞) respectively. (We assume **c** here to be normal to **a** and

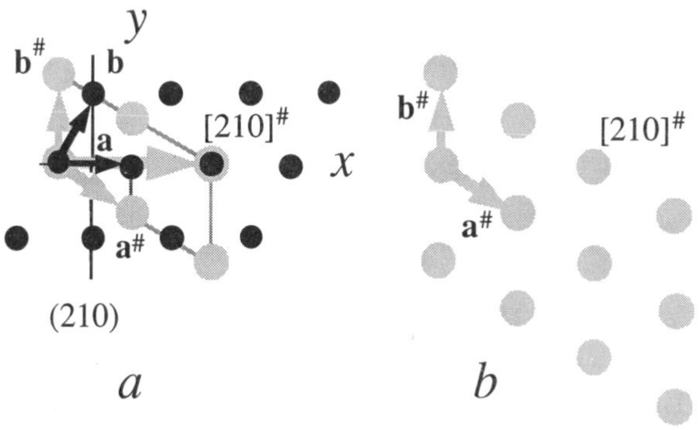

Fig. 4-5.1. The relation between the direct and reciprocal lattices.

b.) We verify immediately, in fact, that $[210]^{\#}$ is normal to (210), as required by (2.11).

An important point must be cleared up by means of this example. When we talk of direct or reciprocal vectors we are talking about precisely the same points in precisely the same space, that of the crystal pattern, which is the one and only space we have. As an example, it can be seen from the figure that $[210]^{\#}$ and [200] are identically the same vector, merely expressed in two different ways. We prefer the former symbol when we want to relate it quickly to the planes (210) to which this vector is normal. (As we have already said, this makes reciprocal vectors, that is, ordinary vectors expressed in reciprocal components, particularly adequate to denote directions in a crystal pattern.)

One serious error must be avoided when dealing with reciprocal vectors. If we want to obtain the modulus of $[210]^{\#}$, say, we must always remember that it is not the square root of the sum of its components squared, a result which is only valid for orthonormal components. What we must do is, for example, to form $[210]^{\#} \cdot [100]$ which equals 4π and notice from the figure that it is equal to $|[210]^{\#}|$ times $|\mathbf{a}|$, which we know is $(2\pi)^{1/2}$. This gives a value of $2(2\pi)^{1/2}$ for $|[210]^{\#}|$ and therefore (4.12) gives D equal to $\frac{1}{2}(2\pi)^{1/2}$, which is precisely $\frac{1}{2}|\mathbf{a}|$, as indeed given by the figure. (In doing this type of work the reader must appreciate that the solid-state convention, which peppers such formulae with factors of 2π, is not calculated to fill everybody's heart with joy.)

Periodic repetition of the reciprocal unit vectors $\mathbf{a}^{\#}$ and $\mathbf{b}^{\#}$ gives a lattice called the *reciprocal lattice*, shown in Fig. 1*a* by means of the grey circles. For clarity, this part of the picture is extended in Fig. 1*b*, where the reciprocal lattice alone is shown. This type of picture, where only reciprocal vectors are shown, is often referred to as depicting the so-called *reciprocal space*. It must be emphatically stated that such a picture represents nothing else than the crystal space of Fig. 1*a*, and that for it to have any meaning, it must be transported back and superimposed, in the correct orientation, on the crystal space of Fig. 1*a*. It is a fact of life, however, that one often works with a reciprocal lattice picture severed from its corresponding direct lattice, which invites a number of interesting mistakes. One common mistake we are already in a position to avoid. Imagine Fig. 1*b* given quite independently from Fig. 1*a* and subsequently *rotated*: if we then take $[210]^{\#}$ in that picture to be a direction in our crystal, sitting in an X-ray goniometer, say, this is wrong, because we have not oriented the reciprocal space in agreement with its direct space. We do not have to bother to do this, however (and this is the reason why people often work entirely in reciprocal space), because we know that the direction $[210]^{\#}$ which we have taken in our figure must be *normal* to the planes (210) in the crystal.

It is important to notice that, because of the manner of its construction, the reciprocal lattice is an entirely translational lattice in the sense that all its points must be translationally equivalent under the reciprocal vectors $\mathbf{a}^{\#}$, $\mathbf{b}^{\#}$,

$\mathbf{c}^{\#}$. In other words, the reciprocal lattice cannot have a basis and must therefore have the form of one of the fourteen Bravais lattices possible in crystallography. In Fig. 1*b* we can see, in fact, that the reciprocal lattice shown is a hexagonal close-packed lattice (like its direct lattice, although in a different orientation; but this coincidence is not general by any means).

If a crystal pattern with primitive or unit vectors **a**, **b**, **c** has a basis, it should be appreciated that the reciprocal lattice is entirely blind to its existence. This is so because the reciprocal lattice is fully determined by **a**, **b**, **c** and there is nothing in its construction that can respond to the existence of a basis. Therefore, if we want to construct the reciprocal lattice of the lattice shown in Fig. **3**-1.1, for example, we must do so from the primitive direct vectors **u**, **v** and not from the unit vectors **a**, **b**. It is a good general rule to choose always direct primitive vectors, rather than unit ones, in order to form the corresponding reciprocal lattice.

6 Summary of vectors

We reproduce in Fig. 1 the same direct and reciprocal lattices of Fig. 5.1, in order to illustrate the notation which we shall always use for vectors *in* and *of* the lattice (see § 2), lattice here meaning either the direct or the reciprocal one. Vectors *in* and *of* the direct lattice are, respectively,

$$\mathbf{r} = x\mathbf{a} + y\mathbf{b} + z\mathbf{c} = [x\,y\,z]; \qquad x,\ y,\ z: \text{real,} \tag{1}$$

$$\mathbf{t} = m\mathbf{a} + n\mathbf{b} + p\mathbf{c} = [m\,n\,p]; \qquad m,\ n,\ p: \text{integers.} \tag{2}$$

In the reciprocal lattice, vectors *in* and *of* the lattice are, respectively,

$$\mathbf{k} = k_x\mathbf{a}^{\#} + k_y\mathbf{b}^{\#} + k_z\mathbf{c}^{\#} = [k_x\,k_y\,k_z]^{\#}; \qquad k_x, k_y, k_z: \text{real,} \tag{3}$$

$$\mathbf{g} = h\mathbf{a}^{\#} + k\mathbf{b}^{\#} + l\mathbf{c}^{\#} = [h\,k\,l]^{\#}; \qquad h,\ k,\ l: \text{integers.} \tag{4}$$

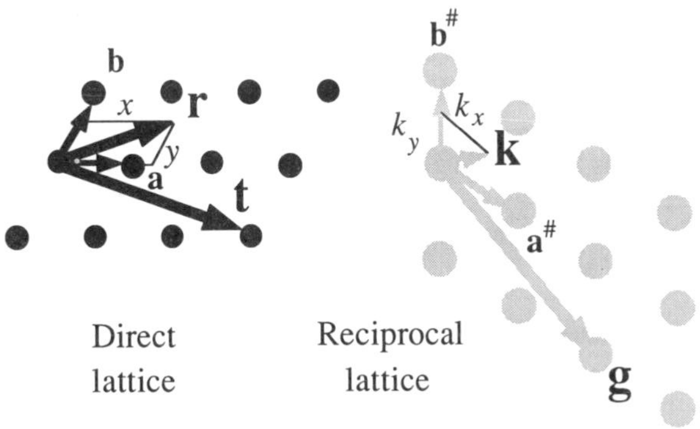

Fig. 4-6.1. Vectors *in* and *of* the direct and reciprocal lattices.

Notice that the components x, y, z of a vector $\mathbf{r}$ of the direct lattice are given in units of $\mathbf{a}$, $\mathbf{b}$, $\mathbf{c}$. Thus in the first unit cell (or primitive cell, as the case might be) x, y, z, range from 0 to 1, and similarly for the components of $\mathbf{k}$ in the reciprocal lattice. Notice also that a vector *of* a lattice (as distinct from a vector *in* the lattice) always means a vector which joins two lattice points.

7 A three dimensional example of a reciprocal lattice

We shall obtain here the reciprocal lattice of the face-centred cubic lattice, with unit cell of lattice constant a, which is depicted in Fig. 1*a*, and which has four lattice points per unit cell. This cell is spanned by the vectors $\mathbf{a}$, $\mathbf{b}$, $\mathbf{c}$, all of modulus a. These vectors, however, are not primitive and must be replaced by the primitive vectors $\mathbf{u}$, $\mathbf{v}$, $\mathbf{w}$, which span the whole of the lattice, all the points of which are indeed translationally equivalent. The components of the vectors $\mathbf{u}$, $\mathbf{v}$, $\mathbf{w}$ in terms of $\mathbf{a}$, $\mathbf{b}$, $\mathbf{c}$ are very easy to obtain from the figure. It is important to remember that the basic vectors are not to be taken of unit length, whence components must be given in terms of the modulus of the basic vectors:

$$\mathbf{u} = [\tfrac{1}{2}a, 0, \tfrac{1}{2}a] = a[\tfrac{1}{2}\ 0\ \tfrac{1}{2}], \qquad \mathbf{v} = a[\tfrac{1}{2}\ \tfrac{1}{2}\ 0], \qquad \mathbf{w} = a[0\ \tfrac{1}{2}\ \tfrac{1}{2}]. \quad (1)$$

Notice also that the order in which the vectors are given in (1) has been carefully chosen so as to ensure that the triple $\mathbf{u}$, $\mathbf{v}$, $\mathbf{w}$ be right-handed, as can be verified, because the triple product

$$\mathbf{u}\cdot\mathbf{v} \times \mathbf{w} = \tfrac{1}{4}a^3 \qquad (2)$$

is positive and not negative. This triple product must of course be the volume of the primitive cell, that is one quarter of the volume of the unit cell (which is a^3), since this contains four atoms, which is in agreement with (R2). The reciprocal vectors are easily obtained from (3.3), (3.5), (3.6), with appropriate changes of notation. On using (2), we have:

$$\mathbf{u}^{\#} = 8\pi a^{-3}(\mathbf{v} \times \mathbf{w}), \qquad \mathbf{v}^{\#} = 8\pi a^{-3}(\mathbf{w} \times \mathbf{u}), \qquad \mathbf{w}^{\#} = 8\pi a^{-3}(\mathbf{u} \times \mathbf{v}). \quad (3)$$

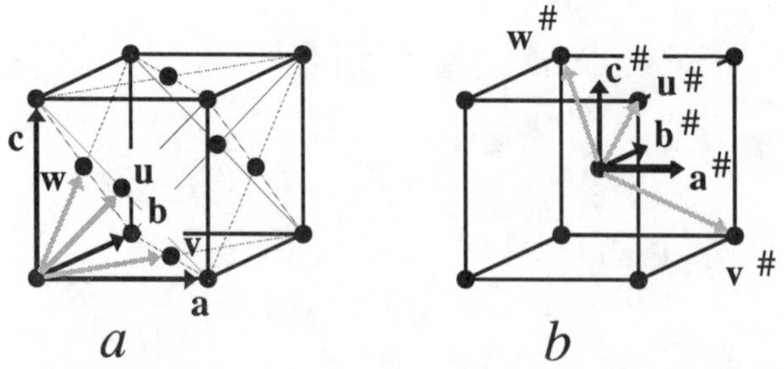

Fig. 4-7.1. The face-centred cubic lattice and its reciprocal lattice.

(Notice in these equations that, ignoring hashes, **u**, **v**, **w** are correctly cyclic.) The result is

$$\mathbf{u}^{\#} = 2\pi a^{-1}[1\,\bar{1}\,1], \quad \mathbf{v}^{\#} = 2\pi a^{-1}[1\,1\,\bar{1}], \quad \mathbf{w}^{\#} = 2\pi a^{-1}[\bar{1}\,1\,1], \tag{4}$$

These vectors, like those in (1), are written in terms of **a**, **b**, and **c**, which, in order to absorb the 2π factors in (4), we replace by corresponding vectors $\mathbf{a}^{\#}$, $\mathbf{b}^{\#}$, and $\mathbf{c}^{\#}$, of modulus $2\pi a^{-1}$, and so on, respectively. These vectors are given at the centre of the cube in Fig. 1*b*, and then the vectors in (4) are easily plotted, as shown. A little thought will persuade the reader that these (grey) vectors span a *body-centred cubic lattice.* Translation of the origin by the sum of $\mathbf{u}^{\#}$ and $\mathbf{w}^{\#}$, for example, takes this point to the centre of the next cubic cell above the one depicted. Thus the reciprocal lattice of a face-centred cubic lattice is a body-centred cubic one. (The inverse result is also true; see Problem 11.4.)

8 Plane waves in crystal structures

We discussed plane waves in § **1**-8, and we shall now write them with their space part only as follows,

1-8.1
$$\varphi_{\mathbf{k}}(\mathbf{r}) = \exp(\mathrm{i}\mathbf{k}\cdot\mathbf{r}). \tag{1}$$

At that stage, however, we did not have a crystal structure at all, so that we must understand what difference the presence of such a structure makes to these functions. On comparison with (4.2) and Fig. 4.1, it is clear that $\mathbf{k}\cdot\mathbf{r}$, and thus $\varphi_{\mathbf{k}}(\mathbf{r})$, are constant over a plane normal to **k**. Just as before, this vector is the *propagation vector* of the wave. In order to get the wave length of the wave, consider two planes of the stack normal to **k** which are separated by a vector **v**. (Parallel, of course, to **k**.) The function (1) must have constant, but in principle different values, over each of these planes. In order to obtain the wave length we assume that the two planes mentioned are the nearest planes of the stack normal to **k** for which $\varphi(\mathbf{r})$ has the same value (i.e. that $|\mathbf{v}|$ equals the wave length λ):

$$\exp(\mathrm{i}\mathbf{k}\cdot\mathbf{r}) = \exp\{\mathrm{i}\mathbf{k}\cdot(\mathbf{r}+\mathbf{v})\} = \exp(\mathrm{i}\mathbf{k}\cdot\mathbf{r})\exp(\mathrm{i}\mathbf{k}\cdot\mathbf{v}). \tag{2}$$

This requires that $\mathbf{k}\cdot\mathbf{v}$ be equal to 2π and, because these two vectors are parallel, and $|\mathbf{v}|$ equals λ,

$$|\mathbf{k}| = 2\pi/\lambda. \tag{3}$$

Thus the propagation vector **k** has two properties. First, it is normal to the planes over which the wave is constant. Secondly, its modulus is the circular wave number (see eqn **1**-8.6).

So far, the planes over which the function (1) is constant are quite arbitrary. We shall now consider the special case when the wave vector coincides with a vector **g** of the reciprocal lattice as defined in (6.4):

$$\mathbf{k} = \mathbf{g} = [hkl]^{\#} \qquad \Rightarrow \qquad \varphi_{\mathbf{k}}(\mathbf{r}) = \varphi_{\mathbf{g}}(\mathbf{r}) = \exp(\mathrm{i}\mathbf{g}\cdot\mathbf{r}). \qquad (4)$$

We know from (4.12) that the modulus of $[hkl]^{\#}$ is 2π divided by the interplanar spacing D of the stack, whence, from (3), λ must coincide with D. We also know from (4.11) that $[hkl]^{\#}$ is normal to the stack (hkl). Thus, when the propagation vector is a vector $[hkl]^{\#}$ of the reciprocal lattice, then the plane wave is constant over the plane (hkl) and its wave length is the interplanar spacing D of the stack. Notice, from the remark at the end of § 4, that, because h, k, l are integers, the plane (hkl), over which the plane wave (4) is constant, cuts the unit axes inside the unit cell and that this plane is parallel to some *plane of the lattice*.

The important result which we have obtained is that the function $\varphi_{\mathbf{g}}(\mathbf{r})$ is a plane wave over the crystal structure, with respect to a single stack of planes in it, namely the one normal to **g** with interplanar spacing D. We also know from (2.12) to (2.15) that this function is periodic with respect to translation by any vector **t** of the direct lattice. Suppose now that we want to find an approximate expression for some arbitrary periodic function, $f(\mathbf{r})$ say, periodic over the same stack and with the same period. By multiplying the exponential in (4) by an appropriate constant, we can arrange that $\varphi_{\mathbf{g}}(\mathbf{r})$ be given the same values as $f(\mathbf{r})$ over the planes of the stack, but we then have no control at all as to the values of the approximation in between them. Consider now the plane wave corresponding to 2**g**. This will be periodic over the same stack but with the interplanar spacing *halved* (see eqn 4.12). It is pretty clear that, if we take as our approximation a linear combination of the two plane waves for **g** and 2**g**, we can now fix the value of the approximation not only at the planes separated by D but also over those separated by $\frac{1}{2}D$. We can clearly go on, by taking **g**, 2**g**, 3**g**, etc. The more we increase our wave vector going out in reciprocal space, the more 'fine structure' can we introduce in our approximation. This leads us to the idea of the Fourier series, to be treated in the next section.

9 The Fourier series

We have seen in § 2 that the plane waves $\exp(\mathrm{i}\mathbf{g}\cdot\mathbf{r})$, further discussed in § 8, are periodic functions in the crystal lattice whenever **g** is a vector of the reciprocal lattice. We shall prove that these functions satisfy a very important property, which will be defined under the name of *orthonormality*, and which will allow us to expand any arbitrary periodic function in terms of them. Such an expression is called the *Fourier series*. In order to do this we must first develop our notation a little.

Notation and definitions

A periodic function is fully known over the crystal if it is determined over the volume of the unit cell (an expression which shall be equally used, when appropriate, for the primitive cell). It is thus convenient to use the following notation:

$$\Omega = \text{volume of the unit cell of the crystal.} \tag{1}$$

Given an arbitrary function $\varphi(\mathbf{r})$, we shall use for it an apparently redundant notation, $|\varphi(\mathbf{r})\rangle$, called a *ket*, which will provide us, nevertheless, with two useful simplifications. First, it will allow us to write complex conjugates without having to use the star superscript. Instead, the *bra* $\langle\varphi(\mathbf{r})|$ will be defined as this conjugate:

$$\langle\varphi(\mathbf{r})| =_{\text{def}} \{\varphi(\mathbf{r})\}^*. \tag{2}$$

Secondly, a bra followed by a ket (*bra-ket*) will entail integration over the unit cell as follows:

$$\langle\varphi(\mathbf{r})|\psi(\mathbf{r})\rangle =_{\text{def}} \int_{\Omega} \varphi(\mathbf{r})^* \psi(\mathbf{r})\, \mathrm{d}\omega, \tag{3}$$

where $\mathrm{d}\omega$ is the volume element.

Two functions $\varphi(\mathbf{r})$ and $\psi(\mathbf{r})$ for which

$$\langle\varphi(\mathbf{r})|\psi(\mathbf{r})\rangle = 0, \tag{4}$$

are said to be *orthogonal* over the unit cell. A single function $\varphi(\mathbf{r})$ for which

$$\langle\varphi(\mathbf{r})|\varphi(\mathbf{r})\rangle = 1, \tag{5}$$

is said to be *normalized* over the unit cell. Two functions $\varphi(\mathbf{r})$ and $\psi(\mathbf{r})$ for which both (4) and (5) are valid are said to be *orthonormal*.

A further and very notable advantage of the bra and ket notation is that we can use a very condensed shorthand notation for the all-important plane waves $\exp(\mathrm{i}\mathbf{g}\cdot\mathbf{r})$, which are periodic over the lattice. In doing this, however, it shall be useful to modify these plane waves slightly. We define:

$$|\mathbf{g}\rangle =_{\text{def}} \Omega^{-1/2} \exp(\mathrm{i}\mathbf{g}\cdot\mathbf{r}). \tag{6}$$

(The usefulness of the factor $\Omega^{-1/2}$ here will be apparent in eqn 11.) It follows from (2) that

$$\langle\mathbf{g}| = \Omega^{-1/2} \exp(-\,\mathrm{i}\mathbf{g}\cdot\mathbf{r}), \tag{7}$$

and, from (3), that

$$\langle\mathbf{g}|\mathbf{g}'\rangle = \Omega^{-1} \int_{\Omega} \exp(-\mathrm{i}\mathbf{g}\cdot\mathbf{r}) \exp(\mathrm{i}\mathbf{g}'\cdot\mathbf{r})\, \mathrm{d}\omega. \tag{8}$$

Orthonormality

We shall assert that for any two vectors **g** and **g**′ of the reciprocal lattice, the functions $|\mathbf{g}\rangle$ and $|\mathbf{g}'\rangle$ are orthonormal over the unit cell:

$$\langle \mathbf{g}|\mathbf{g}'\rangle = \delta(\mathbf{g}, \mathbf{g}'), \tag{9}$$

where the Kronecker delta is

$$\delta(\mathbf{g}, \mathbf{g}') = 0, \qquad \mathbf{g} \neq \mathbf{g}'; \qquad\qquad \delta(\mathbf{g}, \mathbf{g}') = 1, \qquad \mathbf{g} = \mathbf{g}'. \tag{10}$$

It can thus be seen that (9) entails (4) and (5), as required by the definition of orthonormality. We shall prove that (9) is valid whenever **g** and **g**′ are any vectors of the reciprocal lattice, as defined in (6.4). The normalization condition, for **g**′ and **g** equal, is trivial:

8 $$\langle \mathbf{g}|\mathbf{g}\rangle = \Omega^{-1} \int_{\Omega} \mathrm{d}\omega = \Omega^{-1}\Omega = 1. \tag{11}$$

The proof of orthogonality, for **g**′ and **g** not equal, is as follows:

8 $$\langle \mathbf{g}|\mathbf{g}'\rangle = \Omega^{-1} \int_{\Omega} \exp\{\mathrm{i}(\mathbf{g}' - \mathbf{g})\cdot\mathbf{r}\}\, \mathrm{d}\omega, \tag{12}$$

$$= \Omega^{-1} \int_{\Omega} \exp(\mathrm{i}\mathbf{g}''\cdot\mathbf{r})\, \mathrm{d}\omega, \tag{13}$$

because the difference of two vectors of the reciprocal lattice must always be a vector of the reciprocal lattice, which we write as **g**″. If we take its components to be $h^{\#}$, $k^{\#}$, $l^{\#}$, (integers), then **g**″ can be identified with **g** in (2.10) and $\mathbf{g}''\cdot\mathbf{r}$ will then be given by (R2.11). Therefore, in (13),

$$\langle \mathbf{g}|\mathbf{g}'\rangle = \Omega^{-1} \int_0^1\int_0^1\int_0^1 \exp\{2\pi\mathrm{i}(hx + ky + lz)\}\, \mathrm{d}\omega. \tag{14}$$

The volume element $\mathrm{d}\omega$ has to be written here, in principle, in terms of x, y, z, and the limits of integration follow from the fact explained in §6 that, within a unit cell, x, y, z, range from 0 to 1. At least one of the components h, k, l in (14), say h, must be different from zero, so that we can integrate over the corresponding variable, which will give the following term in the integral:

$$\Omega^{-1} \frac{1}{2\pi\mathrm{i}h} \exp(2\pi\mathrm{i}h)\Big|_0^1 = 0, \tag{15}$$

because h is an integer. Thus, (R14) must be zero for $\mathbf{g}$ and $\mathbf{g}'$ not equal, which completes the proof of (9).

Expansion of a periodic function

Consider an arbitrary function $V(\mathbf{r})$ periodic in the crystal pattern:

$$V(\mathbf{r}) = V(\mathbf{r} + \mathbf{t}), \tag{16}$$

for any vector $\mathbf{t}$ of the direct lattice. (See eqn 6.2.) Such a function could be, for example, the crystal potential. We shall show that any function such as (16) can be expanded in terms of the periodic functions $|\mathbf{g}\rangle$, with some coefficients $V_\mathbf{g}$ which depend on $\mathbf{g}$:

$$V(\mathbf{r}) = \sum_{\mathbf{g}} V_\mathbf{g} |\mathbf{g}\rangle. \tag{17}$$

It should be clear that, because all the $|\mathbf{g}\rangle$ are periodic in $\mathbf{t}$, (R17) is also correctly periodic in $\mathbf{t}$, as required by (16). The coefficients $V_\mathbf{g}$ can be very simply obtained by multiplying both sides of (17) with $\langle \mathbf{g}'|$. This will automatically imply integration on the right-hand side of the equation, through the term $\langle \mathbf{g}'|\mathbf{g}\rangle$, whence we must also integrate on its left. This is most simply achieved by replacing $V(\mathbf{r})$ on (L17) by its identical form $|V(\mathbf{r})\rangle$ (Remember that a function can always be written as a ket.) We thus obtain

17,9 $$\langle \mathbf{g}'|V(\mathbf{r})\rangle = \sum_{\mathbf{g}} V_\mathbf{g}\langle \mathbf{g}'|\mathbf{g}\rangle = \sum_{\mathbf{g}} V_\mathbf{g}\delta(\mathbf{g}', \mathbf{g}) = V_{\mathbf{g}'}, \tag{18}$$

since, from (10), the only non vanishing term in the last summation over $\mathbf{g}$ is the one for $\mathbf{g}$ equal to $\mathbf{g}'$, for which the δ is unity. Therefore, the coefficients $V_\mathbf{g}$ in (17) are given as follows:

18,7,3 $$V_\mathbf{g} = \langle \mathbf{g}|V(\mathbf{r})\rangle = \Omega^{-1/2}\int_{\Omega} \exp(-\,i\mathbf{g}\cdot\mathbf{r})\,V(\mathbf{r})\,d\omega. \tag{19}$$

The expansion (17) is the *Fourier series* in the crystal pattern, and the coefficients $V_\mathbf{g}$ are called the *Fourier coefficients*. Notice that for any given function $V(\mathbf{r})$ in (17) they are *uniquely defined* for each $\mathbf{g}$, as follows from (19). If the functions used in (17) were not orthogonal, there would be no guarantee that this would be the case, so that the coefficients could not have any physical meaning.

Notice two important features of the Fourier series, as given in (17). First, the Fourier coefficients (see eqn 19) are entirely defined over the unit cell. Secondly, as it follows from (8.4) the plane waves $|\mathbf{g}\rangle$ which appear in the expansion are constant over planes which all cut the crystal axes inside the unit cell. It is in this way that the periodic function is fully determined from its values inside the first unit cell.

Fourier series for a lattice with basis

Consider a lattice with a basis, the position vectors of which will be denoted with vectors $\boldsymbol{\rho}$ with respect to the origin of the first unit cell. This vector $\boldsymbol{\rho}$ will be understood to run over the values $\boldsymbol{\rho}_1, \boldsymbol{\rho}_2, \ldots, \boldsymbol{\rho}_n$ for the n pattern points of the basis. We assume that a function, such as a potential function of the crystal, is given which at the point $\mathbf{r}$ depends only on the vector $\mathbf{r} - \boldsymbol{\rho}$ from the basis point at $\boldsymbol{\rho}$ to the point $\mathbf{r}$ in question. We shall thus write this function as $V(\mathbf{r} - \boldsymbol{\rho})$ and shall assume that it is periodic in the pattern:

$$V(\mathbf{r} - \boldsymbol{\rho}) = V(\mathbf{r} - \boldsymbol{\rho} + \mathbf{t}), \tag{20}$$

as indeed would be the case for the potential. On adding up such functions over all $\boldsymbol{\rho}$ over the first unit cell, we shall have a function that is periodic over the crystal, which would reproduce, for example, its potential over the first unit cell:

$$V(\mathbf{r}) = \sum_{\boldsymbol{\rho}} V(\mathbf{r} - \boldsymbol{\rho}). \tag{21}$$

Since $V(\mathbf{r} - \boldsymbol{\rho})$ is periodic, it can be expanded in the Fourier series (17), but the functions $|\mathbf{g}\rangle$ in it, which are really of the form $|\mathbf{g}(\mathbf{r})\rangle$ defined in (6), must be replaced by functions $|\mathbf{g}(\mathbf{r} - \boldsymbol{\rho})\rangle$. Likewise, the Fourier coefficient $V_{\mathbf{g}}$ in (19) will depend on the value of $\boldsymbol{\rho}$ chosen and must therefore be rewritten as $V_{\mathbf{g}\boldsymbol{\rho}}$:

17 $$V(\mathbf{r} - \boldsymbol{\rho}) = \sum_{\mathbf{g}} V_{\mathbf{g}\boldsymbol{\rho}} |\mathbf{g}(\mathbf{r} - \boldsymbol{\rho})\rangle \tag{22}$$

6|22 $$= \sum_{\mathbf{g}} V_{\mathbf{g}\boldsymbol{\rho}} \exp(-i\mathbf{g}\cdot\boldsymbol{\rho}) |\mathbf{g}(\mathbf{r})\rangle. \tag{23}$$

23|21 $$V(\mathbf{r}) = \sum_{\mathbf{g}} \left\{ \sum_{\boldsymbol{\rho}} V_{\mathbf{g}\boldsymbol{\rho}} \exp(-i\mathbf{g}\cdot\boldsymbol{\rho}) \right\} |\mathbf{g}(\mathbf{r})\rangle. \tag{24}$$

24 $$=_{\text{def}} \sum_{\mathbf{g}} \mathcal{V}_{\mathbf{g}} |\mathbf{g}(\mathbf{r})\rangle. \tag{25}$$

We notice in (25) that we recover the standard form (17) of the Fourier series, merely by changing the Fourier coefficient $V_{\mathbf{g}}$ by the new coefficient $\mathcal{V}_{\mathbf{g}}$, which is called the *structure factor*:

25, 24 $$\mathcal{V}_{\mathbf{g}} = \sum_{\boldsymbol{\rho}} V_{\mathbf{g}\boldsymbol{\rho}} \exp(-i\mathbf{g}\cdot\boldsymbol{\rho}), \tag{26}$$

where the sum is over all the position vectors $\boldsymbol{\rho}$ of the basis.

An example of the structure factor: hexagonal close-packed structure

A hexagonal close-packed structure is shown in Fig. 1, which for greater generality can also be considered to depict the corresponding crystal pattern. The primitive cell in this pattern is the parallelepiped shown in Fig. 1*b*, spanned by **a**, **b** and **c**, the base of which is shown in Fig. 1*a*. The eight points at the corners of the primitive cell are shared by eight primitive cells and thus contribute only one point to it. With the internal pattern point in the second layer, the primitive cell contains thus two pattern points and we therefore have a lattice with basis. With reference to the axes shown, the basis consists of the point at the origin *O* plus the internal point, the components of which along **a**, **b** and **c** can be read from the two parts of the figure. It follows that the two $\boldsymbol{\rho}$ vectors for the two points of the basis are:

$$\boldsymbol{\rho}_1 = [000], \qquad \boldsymbol{\rho}_2 = [\tfrac{2}{3}\tfrac{2}{3}\tfrac{1}{2}]. \tag{27}$$

We shall calculate the structure factor $\mathscr{V}_{\mathbf{g}}$ for the following vector of the reciprocal lattice,

$$\mathbf{g} = [001]^{\#} \perp (001), \tag{28}$$

which corresponds to a plane wave constant over the planes of the stack (001), to which the top face of the primitive cell belongs. Since the two points of the basis are equivalent (not translationally, of course, but under appropri-

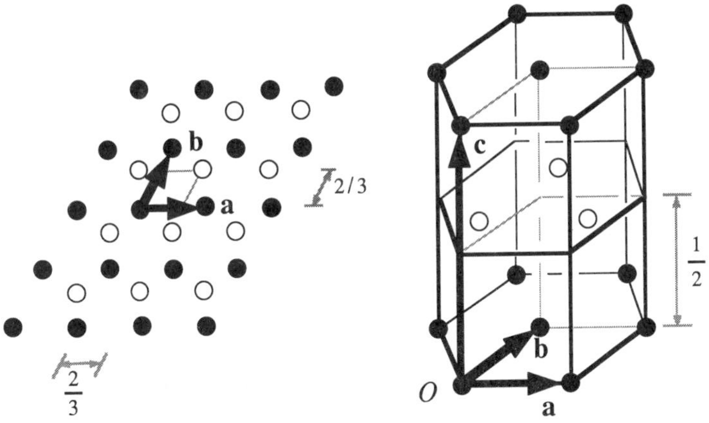

a *b*

Fig. 4-9.1. The hexagonal close-packed structure. The black circles correspond to the first and third layers of the structure, and the open circles to the second layer. Notice that all three layers are hexagonal close-packed, the pattern points being disposed in centred hexagons.

ate covering operations), the function $V_{\mathbf{g}\boldsymbol{\rho}}$ must have the same value irrespective of $\boldsymbol{\rho}$, and can be written as $V_{\mathbf{g}}$. Thus, from (26),

$$\mathscr{V}_{[001]^*} = V_{[001]^*}\{\exp(0) + \exp(-2\pi \mathrm{i}\,\tfrac{1}{2})\} = 0. \tag{29}$$

This result will later on be seen to be physically significant, indeed, to be most important.

10 Plane-wave notation and properties

It will be useful to modify the expressions (8.1) and (8.4) which we have given for plane waves, since it is convenient to adopt for them the normalization and notation used in (9.6). In analogy with the terminology for vectors, we shall say that

8.4, 9.6
$$|\mathbf{g}\rangle = \Omega^{-1/2}\exp(\mathrm{i}\mathbf{g}\cdot\mathbf{r}), \qquad \mathbf{g} = [hkl]^{\#}, \tag{1}$$

with h, k, l integers, is a plane wave *of* the lattice, constant over the planes of the stack (hkl), which is parallel to planes of the lattice, that is to planes that cut the crystal axes at lattice points. (See eqn 8.4 and its discussion). For an arbitrary vector $\mathbf{k}$ in the reciprocal lattice, we shall have instead

8.1
$$|\mathbf{k}\rangle = \Omega^{-1/2}\exp(\mathrm{i}\mathbf{k}\cdot\mathbf{r}), \qquad \mathbf{k} = [k_x\, k_y\, k_z]^{\#}, \tag{2}$$

with k_x, k_y, k_z real numbers, which we shall call a plane wave *in* the lattice, with propagation vector $\mathbf{k}$ and constant over planes (normal to $\mathbf{k}$) which are not, in general, parallel to planes of the lattice.

It is easy to identify the momentum associated with the plane wave (2), by acting on it with the momentum operator $\mathbb{p}$:

$$\mathbb{p}|\mathbf{k}\rangle = \frac{\hbar}{\mathrm{i}}\frac{\partial}{\partial\mathbf{r}}\{\Omega^{-1/2}\exp(\mathrm{i}\mathbf{k}\cdot\mathbf{r})\} = \hbar\mathbf{k}|\mathbf{k}\rangle. \tag{3}$$

The eigenvalue p of $\mathbb{p}$ on (R 3) is $\hbar\mathbf{k}$, for which reason the $\mathbf{k}$ vector (or wave vector, or propagation vector) is also called the *momentum vector*.

11 Problems

1. Verify that the vectors $\mathbf{a}^{\#}, \mathbf{b}^{\#}, \mathbf{c}^{\#}$, defined in (3.3), (3.5), (3.6), satisfy the conditions (2.6) and (2.7).
2. Prove that, if Ω and $\Omega^{\#}$ are the volumes spanned by the vectors $\mathbf{a}$, $\mathbf{b}$, $\mathbf{c}$, and $\mathbf{a}^{\#}, \mathbf{b}^{\#}, \mathbf{c}^{\#}$, respectively, then $\Omega^{\#}$ equals $(2\pi)^3\Omega^{-1}$.
3. Prove that, if a is the lattice constant of a face-centred cubic lattice, then the lattice constant of its reciprocal, body-centred cubic lattice, is $4\pi a^{-1}$.
4. Prove that the reciprocal lattice of a body-centred cubic lattice of lattice constant a is a face-centred cubic lattice of lattice constant $4\pi a^{-1}$.

5. A *trigonal lattice* is defined by three primitive vectors **a**, **b**, **c**, all of equal length a and such that the angle θ between any two of these vectors is constant. Verify that the following vectors, listed on an orthonormal basis, satisfy these conditions:

$$\mathbf{a} = [m\,n\,p], \qquad \mathbf{b} = [p\,m\,n], \qquad \mathbf{c} = [n\,p\,m]. \tag{1}$$

Prove hence that the reciprocal lattice of the trigonal lattice defined above is another trigonal lattice.

Further reading

Good discussions on the reciprocal lattice can be found in Ashcroft and Mermin (1976) and Kelly and Groves (1973). Further discussion on the Fourier series can be found in many books on mathematics, such as Arfken (1985), Boas (1983), and Sneddon (1961).

5

Bloch functions and Brillouin zones

We have proved in § **1**-3 that, *in free space*, the momentum eigenfunctions are eigenfunctions of the energy (see eqn **1**-3.6), and these momentum eigenfunctions have just been written as the plane waves $|\mathbf{k}\rangle$ in (**4**-10.2), labelled by a propagation vector **k** in the reciprocal lattice. We want to consider in this chapter the electrons moving in the crystal not just as free particles but rather under the influence of the crystal potential field. In these circumstances, we cannot expect the energy eigenfunctions to continue being the plane waves $|\mathbf{k}\rangle$, although, when we obtain their general form, we shall see that they are deeply related to them. (It is in fact for this reason that we have paid so much attention to these plane waves.)

The argument which we shall follow in this and in the next chapter in order to study the energy eigenfunctions in the crystal is based on the general principle discussed in Chapter **2**, whereby the energy eigenfunctions must be bases for the irreducible representations of the group: each irreducible basis must correspond to one energy eigenvalue. So, our task is to find the irreducible bases of the representations of the space groups, these being the symmetry groups of crystals. This is a tall order and we must necessarily proceed in easy stages. We know that every space group G admits of a subgroup T, the translation group, which is an invariant subgroup of G. We shall mainly be concerned in this chapter with finding the irreducible representations of T as well as their bases, which are called the *Bloch functions*, in the hope that they might play a part in the determination of the irreducible bases of G. We shall find in Chapter **6** (see §§ **6**-2, **6**-3), in fact, that these irreducible bases can fully be determined from the Bloch functions. With this plan in mind, we shall first of all discuss the properties of the translation group T.

1 The translation group *T*

Because a symmetry operation is always defined as a covering transformation of a copy of the system, we were forced to assume in § **3**-1, when defining translations, that the crystal and its copy are infinite: a translation of the copy would otherwise leave a row of points hanging over empty space. This means that we also have an infinite number of translations in T, which is most inconvenient. This difficulty is easily remedied, however, by using the periodic, Born–von Karman, boundary conditions already discussed in § **1**-2. In that section, however, the lump of crystal under study was supposed to have no internal structure at all, whereas we must now consider the effect of the crystal structure. For simplicity, we illustrate in Fig. 1 the way in which the periodic boundary conditions affect the translation group of a one-dimensional lattice, along the x axis, with lattice constant a, and N_x primitive cells. (We have taken N_x to be six in the figure, for convenience of the example.) We represent in the first row of the figure the actual lattice (shaded) extended periodically to infinity. We must assume that a copy of this system is identically superimposed on it. In the second row, which must be imagined to be superimposed on the first (as is also the case for all the successive rows), we depict the *copy of the system* translated by a, or, in the Seitz notation, by $\{E\,|\,\mathbf{a}\}$. In the third row, the copy has been translated by $2a$, which is the same as

3-8.2 $$\{E\,|\,2\mathbf{a}\} = \{E\,|\,\mathbf{a}\}\{E\,|\,\mathbf{a}\} \equiv \{E\,|\,\mathbf{a}\}^2. \qquad (1)$$

In the same manner, the next translation (not illustrated) would be $\{E\,|\,\mathbf{a}\}^3$, and so on.

Let us now see what happens to the transformation of the lattice point coordinates effected by the covering operations described. We take the origin of coordinates at the point 1 in the first row, which, by periodicity, means any of the (infinite in number) points labelled 1 in that row. We immediately see, from the position of the copy in the second row of the figure, that $\{E\,|\,\mathbf{a}\}$ changes the coordinate of the point labelled 1 from 0 to a. Likewise, $\{E\,|\,\mathbf{a}\}^2$ changes the same coordinate into $2a$, and so on. When we finally perform $\{E\,|\,N_x\mathbf{a}\}$ (that is, a translation by $6a$ in the figure), the point 1 returns to the origin and all the other points recover their original coordinates: nothing

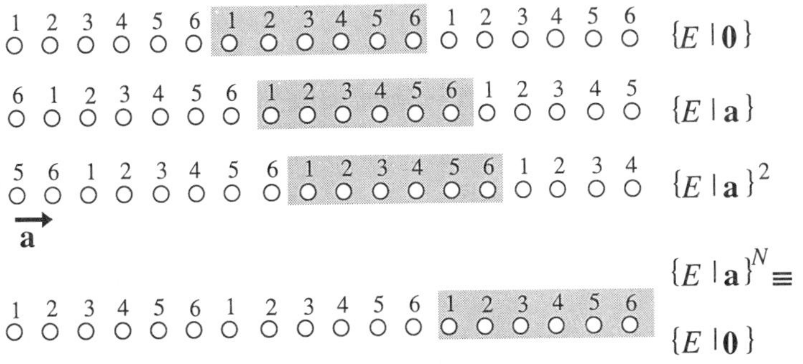

Fig. 5-1.1. Translations with periodic boundary conditions.

has changed from the original configuration and this operation is thus the identity:

1 $$\{E\,|\,N_x\mathbf{a}\} = \{E\,|\,\mathbf{a}\}^{N_x} = \{E\,|\,\mathbf{0}\}. \tag{2}$$

It should be clear that, after the translation in (2) has been performed, the next translation, by $(N_x + 1)\mathbf{a}$, coincides identically with $\{E\,|\,\mathbf{a}\}$, which is already accounted for. The result of this exercise is thus most satisfying: by using the periodic boundary conditions we have cut down the number of translations in our group from infinity to N_x. If we now consider the full group T in three dimensions, it is clear (but you must remember that translations commute, so that any pair of operations appears as a single product and thus only once in the group) that the total number of operations in T, that is its order $|\,T\,|$ will be

$$|\,T\,| = N_x N_y N_z = N. \tag{3}$$

Here, N_x, N_y, N_z, are the number of primitive cells along the axes **a**, **b**, **c**, respectively, so that N is the total number of cells in the lump of crystal under study.

We shall now discuss the structure of T. The elements of T are the translation operators.

$$\{E\,|\,\mathbf{t}\} = \{E\,|\,m\mathbf{a} + n\mathbf{b} + p\mathbf{c}\}, \qquad m, n, p\text{: integers}, \tag{4}$$

for all the vectors **t** of the lattice (see eqn **4**-6.2). It is clear that all the translations along one axis alone form a subgroup of T:

$$\{E\,|\,m\mathbf{a}\} \in_{\text{def}} T_x, \qquad m \quad \text{integral}, \qquad 1 \le m \le N_x, \tag{5}$$

and similarly for the other two axes. Because the operations of T_x, T_y, T_z, all commute (all translations commute!), T is the direct product of these three groups (see eqn **2**-1.10),

$$T = T_x \otimes T_y \otimes T_z, \tag{6}$$

an expression which will simplify our work considerably. It is also clear from (4) that the general operation $\{E\,|\,\mathbf{t}\}$ of T can be written so as to reflect this structure:

4, **3**-6.6 $$\{E\,|\,\mathbf{t}\} = \{E\,|\,m\mathbf{a}\}\,\{E\,|\,n\mathbf{b}\}\,\{E\,|\,p\mathbf{c}\}, \tag{7}$$

the factors on (R7) belonging respectively to the subgroups on (R6).

Finally, we must count the number of classes of T because, as discussed at the beginning of §**2**-5, we can then discover the number and dimensions of the irreducible representations of the group. There are two properties of T which we must remember for this purpose. One is that its order is N and the other that, from eqns (**3**-8.3) and (**3**-8.4), it is commutative (that is abelian). This latter property has an important consequence, namely that each operation of T is its own class, whence the number of classes in the group equals its order

N. In order to prove this result we recall (see §2-1) that the class of one element, $\{E\,|\,\mathbf{t}\}$, say, is obtained by conjugating this element under all the elements $\{E\,|\,\mathbf{t}'\}$ of the group:

$$\{E\,|\,\mathbf{t}'\}\,\{E\,|\,\mathbf{t}\}\,\{E\,|\,\mathbf{t}'\}^{-1} = \{E\,|\,\mathbf{t}'\}\,\{E\,|\,\mathbf{t}'\}^{-1}\,\{E\,|\,\mathbf{t}\} = \{E\,|\,\mathbf{t}\}. \tag{8}$$

It is thus clear that the conjugate of $\{E\,|\,\mathbf{t}\}$ is always $\{E\,|\,\mathbf{t}\}$ itself, so that no other element comes up in its class, which proves our assertion. It should be clear that, in the same manner, the number of classes in T_x is N_x and likewise for the other subgroups of T.

2 The representations of T

The results obtained at the end of last section together with the rules stated in §2-5 give us at once the number and dimensions of the irreducible representations of T. Their number must equal the number of classes in the group, which is its order N. The sum of the squares of the dimensions of these N representations must equal the order of the group N, whence *all the irreducible representations, N in number, must be one-dimensional.*

We shall first be concerned with finding the irreducible representations of T_x, N_x in number. The elements of this subgroup are of the form $\{E\,|\,m\mathbf{a}\}$, so that the irreducible representations should be written in principle as $\hat{T}\{E\,|\,m\mathbf{a}\}$. We must add to this symbol, however, a label, which we shall write as k_x, in order to distinguish each of the different N_x irreducible representations. (It will soon be clear why we choose for this label the same symbol which we have so far used for a component of a vector in the reciprocal lattice.) Thus, the matrix (number, in our case, because the representation is one-dimensional) corresponding to $\{E\,|\,m\mathbf{a}\}$ in the k_x representation of T_x will be written as ${}_{k_x}\hat{T}\{E\,|\,m\mathbf{a}\}$, where we dispense with the now redundant suffix on the right of the T. We are now ready to obtain these matrices, remembering that they must multiply precisely like the corresponding operations:

3-8.2 $$\{E\,|\,m\mathbf{a}\}\,\{E\,|\,m'\mathbf{a}\} = \{E\,|\,(m+m')\mathbf{a}\}. \tag{1}$$

2-4.6 $${}_{k_x}\hat{T}\{E\,|\,m\mathbf{a}\}\;{}_{k_x}\hat{T}\{E\,|\,m'\mathbf{a}\} = {}_{k_x}\hat{T}\{E\,|\,(m+m')\mathbf{a}\}. \tag{2}$$

Equation (2) suggests that the one-dimensional matrices (i.e. numbers) in that equation must be exponentials in m multiplied by some arbitrary constant, which we choose as follows:

$${}_{k_x}\hat{T}\{E\,|\,m\mathbf{a}\} = \exp(-2\pi i k_x m). \tag{3}$$

Clearly, the constant chosen, as it is the case in (3), must be related to the label k_x of the representation, in such a way that (R3) changes with this label, as it should. The other details of the constant must be chosen after trial and error, so as to make the work which follows as simple as possible (§ 4), as well as to ensure that the labels of the representations have the correct physical meaning. (See § 6.) In the same manner, we obtain representations for the other two subgroups of T:

$$_{k_y}\hat{T}\{E\,|\,n\mathbf{b}\} = \exp(-2\pi \mathrm{i}k_y n), \qquad _{k_z}\hat{T}\{E\,|\,p\mathbf{c}\} = \exp(-2\pi \mathrm{i}k_z p). \quad (4)$$

We now exploit the direct product form of T from (1.6). We know that, as a consequence of this form, the general operation of T, $\{E\,|\,\mathbf{t}\}$, is written, as in (1.7), as a product of the three operations which appear in (3) and (4). The direct product result for the representations, in eqn (**2**-5.9), tells us that the representatives of $\{E\,|\,\mathbf{t}\}$ are themselves products of the representatives of the operations which, from (1.7), appear in $\{E\,|\,\mathbf{t}\}$. Thus:

$$_{k_xk_yk_z}\hat{T}\{E\,|\,\mathbf{t}\} = {}_{k_x}\hat{T}\{E\,|\,m\mathbf{a}\}\,{}_{k_y}\hat{T}\{E\,|\,n\mathbf{b}\}\,{}_{k_z}\hat{T}\{E\,|\,p\mathbf{c}\}, \quad (5)$$

where we have constructed the label for $\hat{T}$ from those which appear on (R5). Therefore,

3,4 | 5 $$_{k_xk_yk_z}\hat{T}\{E\,|\,\mathbf{t}\} = \exp\{-2\pi \mathrm{i}(k_x m + k_y n + k_z p)\}. \quad (6)$$

The expression in the round bracket in (6) looks like a scalar product, so that we now exploit our old reciprocal lattice trick. Because $\mathbf{t}$ is a vector *of* the translation lattice, it has the form

4-6.2 $$\mathbf{t} = m\mathbf{a} + n\mathbf{b} + p\mathbf{c}, \qquad m, n, p: \text{integers}. \quad (7)$$

It will therefore be convenient to build the labels k_x, k_y, k_z, into a vector *in* the reciprocal lattice:

4-6.3 $$\mathbf{k} = k_x\mathbf{a}^{\#} + k_y\mathbf{b}^{\#} + k_z\mathbf{c}^{\#}, \quad k_x, k_y, k_z: \text{reals}, \quad (8)$$

which allows us to write the exponent in (6) as $-\mathrm{i}\mathbf{k}\cdot\mathbf{t}$, as it follows at once from the by now usual work with eqns (**4**-2.6) and (**4**-2.7). Once this is done, it is natural to consolidate the labels on (L6) into the vector $\mathbf{k}$, which now becomes the label of the irreducible representations of T. (This is a sensible but momentous step. Although we appear to complicate things by introducing a rather artificial vector instead of its three simple components, this will allow for a graphical depiction of the irreducible representations of T which pervades the whole of solid-state theory through the concept of the Brillouin zone.) We thus write

6 $$_{\mathbf{k}}\hat{T}\{E\,|\,\mathbf{t}\} = \exp(-\mathrm{i}\mathbf{k}\cdot\mathbf{t}). \quad (9)$$

As implied in (8), the label $\mathbf{k}$ of the irreducible representations can be any vector *in* the reciprocal lattice. This gives us an infinite number of such labels, instead of the finite N we know we must have. We shall show in the next two sections that account must be taken of various restrictions on $\mathbf{k}$ which ultimately will provide us with the known correct number of irreducible representations.

3 Equivalent k vectors and the Brillouin zone

Let us define two **k** vectors as *equivalent* whenever they differ by a vector **g** *of* the reciprocal lattice:

4-6.4 $$\mathbf{g} = h\mathbf{a}^{\#} + k\mathbf{b}^{\#} + l\mathbf{c}^{\#} = [hkl]^{\#}; \qquad h, k, l\text{: integers.} \qquad (1)$$

We shall prove that two such vectors, **k** and **k** + **g**, label identical representations of T:

2.9 $${}_{\mathbf{k}+\mathbf{g}}\hat{T}\{E|\mathbf{t}\} = \exp\{-\mathrm{i}(\mathbf{k}+\mathbf{g})\cdot\mathbf{t}\} = \exp(-\mathrm{i}\mathbf{k}\cdot\mathbf{t})\exp(-\mathrm{i}\mathbf{g}\cdot\mathbf{t}). \qquad (2)$$

From (1) and (2.7), in the now usual way,

$$\mathbf{g}\cdot\mathbf{t} = 2\pi(hm + kn + lp) = 2\pi v, \qquad v\text{: integer,} \qquad (3)$$

so that $\exp(-\mathrm{i}\mathbf{g}\cdot\mathbf{t})$ in (2) equals unity. Therefore,

3,2.9|2 $${}_{\mathbf{k}+\mathbf{g}}\hat{T}\{E|\mathbf{t}\} = {}_{\mathbf{k}}\hat{T}\{E|\mathbf{t}\}, \qquad (4)$$

for all $\{E|\mathbf{t}\}$ in T, as we had stated.

Condition (4) is a periodicity condition in the reciprocal lattice, which means that all the non-equivalent **k** vectors required in order to label *all* the irreducible representations of T can be obtained from the primitive cell in the reciprocal lattice, as illustrated in Fig. 1. It can be seen in Fig. 1*b* that, given an arbitrary **k** vector outside the primitive cell, **k**′, we can always find an equivalent **k** vector (equal to **k**′ + **g**), inside the primitive cell, for a conveniently chosen **g**. (All this is nothing more than a repetition, for the reciprocal

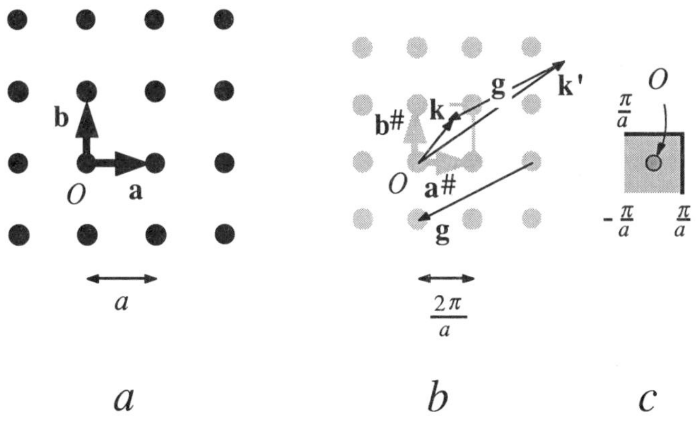

Fig. 5-3.1. *a*: A simple square lattice. *b*: Its reciprocal lattice. *c*: The first Brillouin zone in *b*. All three figures must be understood as superimposed after a parallel displacement that brings the three points labelled *O* into coincidence. Notice that $\mathbf{a}^{\#}$ is perpendicular to **b**, as required by (**4**-2.7) and that its modulus follows from (**4**-2.6), since **a** and $\mathbf{a}^{\#}$ are parallel.

lattice, of the fundamental property of primitive cells discussed in § **3**-1, of being maximal sets of translationally inequivalent points.)

Although the primitive cell of the reciprocal lattice is entirely sufficient for labelling all the irreducible representations of T, it is not very convenient, for reasons which we shall soon discuss, and it is in practice replaced by the *centred primitive cell* (see § **3**-2) shown in Fig. 1*c*. This centred primitive cell in the reciprocal lattice is called the *Brillouin zone* or, sometimes, the *first Brillouin zone*, because Brillouin zones of higher order will be defined later on.

The reason why the centred primitive cell is preferable to the ordinary primitive cell is the same one that was illustrated in Fig. **3**-2.2 in the direct lattice, namely that the centred primitive cell reflects much better the symmetry of the lattice. This is most important in the present case for the following reason. We have seen in § **4**-8 that the **k** vectors (vectors in the reciprocal lattice), are the propagation vectors of plane waves. Although in the present chapter **k** vectors and the reciprocal space have turned up in an entirely different way, as a geometrical device for labelling irreducible representations of T, we must expect that these two manifestations of reciprocal space be deeply related, as we shall soon find. (This is the reason why we have used the same notation in both cases.) If we thus accept that reciprocal space will be, even in the present case, basically the space of some propagation vectors, it would be silly for any depiction of it to contain only positive **k** vectors and no negative vectors at all, as is the case with the standard primitive cell in Fig. 1*b*. In the centred primitive cell of Fig. 1*c*, (that is, in the Brillouin zone), we see that for every positive **k** vector its negative also appears in the cell, which is much more satisfactory. For free electrons, for example, we know that the energy as a function of **k**, $E(\mathbf{k})$, is such that $E(\mathbf{k})$ equals $E(-\mathbf{k})$, a property which could not be displayed in the cell of Fig. 1*b* but which could easily be put in evidence in the cell of Fig. 1*c*. Although, of course, our work is far from completed, we must keep a careful eye on these requirements. After all, if we use reciprocal space to label the irreducible representations of T, it is because we expect these labels to be useful for labelling the irreducible representations of the space group, and thus the energy eigenvalues in the crystal. In other words, we must expect that some functions of the energy as a function of our present labels **k** will emerge. Even though the new functions $E(\mathbf{k})$ cannot be expected to coincide with their free-electron counterparts, we shall have to be able to plot them in a sensible way. We shall see in § **6**-5, in fact, that the symmetry in **k** and $-\mathbf{k}$ of the energy is valid not just for free electrons but that it is entirely general. We must therefore use a centred cell in **k** space, in order to be able to display such an important symmetry of the functions which we are seeking.

There is another question about which we must be careful. Remember that what we are doing is basically counting up irreducible representations (as a move towards counting states). It is thus utterly important to make sure that our book-keeping is right. This requires two things. First, that we have included in our count all the possible labels of all the irreducible representations. (We already know that we got this right by taking just the centred unit cell or Brillouin zone.) Secondly, we must ensure that no representation is counted twice, and it is this point that requires attention. If, for simplicity of the description, we go back to the standard cell in Fig. 1*b*, and we take any point along the side of it which is parallel to the $\mathbf{b}^{\#}$ axis, this point, under translation with $-\mathbf{a}^{\#}$ has always another point equivalent to it in the same cell. This means that the side in question must not be counted as part of the cell, and the same is the case for the side parallel to $\mathbf{a}^{\#}$. This is a situation which we have already met in §**3**-1, when unit and primitive cells were introduced and we discovered that they have to be *open* surfaces (in two dimensions) or open polyhedra (in three). It is thus clear that the Brillouin zone of Fig. 1*c* must, like all such cells, be open. The usual convention is to take the sides shown in full black lines, and only those sides, as part of the cell, because they correspond to positive **k**.

Finally, one important point about the symmetry of the Brillouin zone, a subject which easily leads into confusion. The problem is this. The Brillouin zone, like the reciprocal lattice, is entirely determined by the Bravais lattice and not by the crystal pattern or, what is the same, by its space group. We know that, although the point group operations are not necessarily symmetry operations of the space group, they are covering (that is, symmetry) operations of the Bravais lattice. (See §**3**-5.) Therefore, all the operations of the point group must be covering operations of the Brillouin zone. This is very mystifying because, as a result, the Brillouin zone will often appear to have higher symmetry than the crystal. One must then be very careful because some of the covering operations of the Brillouin zone are not in the space group and are not symmetry operations of the Hamiltonian. This is important, for example, because such operations cannot create degeneracies in the energy. (See §**6**-3.) The fact that the point group is a covering group of the Brillouin zone, however, has significant consequences. (§**6**-2.) In case of doubt, however, the only really safe rule is to assume that the Brillouin zone has no symmetry of its own (however visible its geometrical symmetry might be) and appeal in each case to the symmetry operations of the space group (see the end of §**6**-4 for more details).

4 Quantization of k

Let us now go back to the subgroup T_x, for which we got the following representations:

2.3, 2.8 $\quad {}_{k_x}\hat{T}\{E\,|\,m\mathbf{a}\} = \exp(-2\pi i k_x m), \qquad k_x \text{ real}, \qquad 0 < k_x \leq 1.$ (1)

Notice that we also include here the restriction that k_x must be taken within the primitive cell, that is in the open region from $\mathbf{0}$ to $\mathbf{a}^{\#}$, or, in the centred primitive cell (Brillouin zone), in the open region from $-\frac{1}{2}\mathbf{a}^{\#}$ to $\frac{1}{2}\mathbf{a}^{\#}$. (See, for example, Fig. 3.1*c*.) Let us apply this result for m equal to N_x in which case we know from (1.2) that $\{E \mid N_x\mathbf{a}\}$ is the identity, for which the matrix representative must be unity. (Otherwise, the corresponding operation would be changing the basis functions, which the identity cannot do.) Therefore:

1 $$_{k_x}\hat{T}\{E \mid N_x\mathbf{a}\} = \exp(-2\pi i k_x N_x) = 1, \tag{2}$$

which requires

$$k_x N_x = \kappa_x, \qquad \kappa_x = 1, 2, \ldots, N_x. \tag{3}$$

In principle, of course, all that follows from (2) is that κ_x must be an integer, but it is also clear from (3) that if we go above or below the range therein given, then k_x would go outside the range in (1). The result of (3) is that we have only N_x irreducible representations in T_x and, in the same way, N_y and N_z representations in T_y and T_z respectively. That is, the total number of irreducible representations of T is the product $N_x N_y N_z$, which, from (1.3), is equal to N, as it should be. (See the top of § 2). The wisdom of having included the factor $2\pi i$ in the exponent of the representatives (2.3) should now be apparent.

5 The number of states in the Brillouin zone

What we have so far done is to construct a graphical device in order to label, that is identify, the irreducible representations of the translation group T of the crystal. First, we have seen that the labels of these representations must be **k** vectors in the reciprocal lattice. Secondly, we have seen that these **k** vectors must be constrained to the first primitive cell of the reciprocal lattice. Thirdly, we have replaced this cell by an entirely equivalent one, centred around the origin, which is the first Brillouin zone. Fourthly, we have seen that out of the infinite number of **k** vectors in the Brillouin zone, a *uniformly spaced* grid (see eqn 4.3) must be constructed which contains precisely N points, N being the number of primitive cells in the direct lattice: it is the points of this grid that label the irreducible representations of T. We show this grid in Fig. 1, for the Brillouin zone given in Fig. 3.1*c*. It should be appreciated that, when as usually N is very large the discrete grid in the Brillouin zone is exceedingly fine and we can consider that the distribution of **k** vectors inside it is *quasi-continuous*.

All this may be very good but, what is the use of it? And this is an important question to ask: if we had done the above job for *all* the irreducible representations of G, we know that we would have classified completely all the energy levels in the crystal, but we have merely dealt with a subgroup of G. It is therefore a remarkable fact that the N permitted **k** vectors in the Brillouin zone do indeed label *all* the permitted states of the energy in the crystal, a point which we shall now discuss.

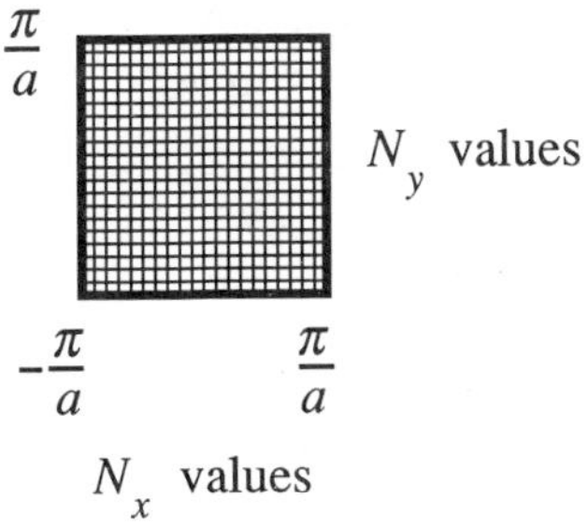

Fig. 5-5.1. The Brillouin zone of the simple square lattice (Fig. 3.1) displaying the discrete grid of permitted **k** vector values in its interior.

Let us denote, for simplicity, the translation $\{E|\mathbf{t}\}$ as t. We must recall three facts. First, that all the irreducible representations of T are one-dimensional and are labelled by the **k** vectors, so that, if we call $\psi_{\mathbf{k}}(\mathbf{r})$ their basis functions, we must have a relation of the following form,

$$t\,\psi_{\mathbf{k}}(\mathbf{r}) = \psi_{\mathbf{k}}(\mathbf{r})\,{}_{\mathbf{k}}\hat{T}(t) =_{\text{def}} A_{\mathbf{k}}\,\psi_{\mathbf{k}}(\mathbf{r}). \tag{1}$$

This is nothing else than the standard definition of a representation (**2**-4.5) with some slight changes of notation. As already said, we are now dropping the circlets. Also, in the last term, we write the matrix, in a new notation, on the left (which we can do: it is merely a number) to stress that the functions $\psi_{\mathbf{k}}(\mathbf{r})$ are now eigenfunctions of the translations, $A_{\mathbf{k}}$ being the eigenvalue.

The label **k** of the eigenvalues and eigenfunctions can take only N values, over the Brillouin zone. We shall assume that the functions $\psi_{\mathbf{k}}(\mathbf{r})$ vary continuously over the Brillouin zone, that is that small changes in **k** entail small changes in $\psi_{\mathbf{k}}(\mathbf{r})$. Therefore, if we start with $\psi_{\mathbf{0}}(\mathbf{r})$ for **k** equal to the origin of the Brillouin zone and then let **k** vary over the whole of it, we determine a set of N continuously varying functions $\psi_{\mathbf{k}}(\mathbf{r})$ which form what we shall call a *band* (see §6). The 'matrix' ${}_{\mathbf{k}}\hat{T}(t)$, that is $A_{\mathbf{k}}$, is given in (2.9), from which it follows that all the N values of $A_{\mathbf{k}}$ in (1) are distinct. There can be, therefore, only one eigenfunction $\psi_{\mathbf{k}}(\mathbf{r})$ for each eigenvalue. This, as long as we keep to the same band, provides us with the second fact:

the eigenvalues $A_{\mathbf{k}}$ *are not degenerate.* (2)

The translations, of course, belong to the Schrödinger group, which gives us our third fact:

2-3.2 $$\mathbb{H}t = t\mathbb{H}. \tag{3}$$

We can now obtain the consequences of all these facts.

1 $$\mathbb{H}t\psi_{\mathbf{k}}(\mathbf{r}) = A_{\mathbf{k}}\mathbb{H}\psi_{\mathbf{k}}(\mathbf{r}). \tag{4}$$

3 $$t\mathbb{H}\psi_{\mathbf{k}}(\mathbf{r}) = A_{\mathbf{k}}\mathbb{H}\psi_{\mathbf{k}}(\mathbf{r}). \tag{5}$$

From (1) and (5), $\mathbb{H}\psi_{\mathbf{k}}(\mathbf{r})$ is an eigenfunction of t with $A_{\mathbf{k}}$ as eigenvalue, but from (2) it *must* be linearly dependent on $\psi_{\mathbf{k}}(\mathbf{r})$:

5, **2**-3.10 $$\mathbb{H}\psi_{\mathbf{k}}(\mathbf{r}) = c\psi_{\mathbf{k}}(\mathbf{r}) \qquad \Rightarrow \qquad c \equiv E. \qquad (6)$$

Several things should be clear from this result. One is that the eigenfunctions of the translations, labelled by **k**, provide eigenfunctions of the energy. The other is that the energy eigenvalue E in (6) should also be labelled by the **k** vector. That is, each value of **k** labels one eigenstate $E_{\mathbf{k}}$ of the energy. Therefore, we do have N eigenstates of the energy labelled by the N values of **k** in the Brillouin zone. It should be absolutely clear that we are not saying that these N values of $E_{\mathbf{k}}$ are all different. When the symmetry operations of G not in T are introduced, degeneracies may appear whereby several $\psi_{\mathbf{k}}(\mathbf{r})$ with different **k** may belong to the same energy, so that the corresponding $E_{\mathbf{k}}$ become equal. Although this is important because, as we have repeatedly said, the degeneracies determine many properties of the electron system in a crystal, their existence does not affect the counting of states. We shall still have precisely N states, although some of them may be degenerate from the point of view of the energy.

Though the above argument is a fair one, we must admit that it is not entirely complete for the following reason. We know that the final classification of the energy levels in the crystal must be done by the irreducible representations of its space group G. When the operations of G not in T are introduced, two problems might arise. One is that the transforms of the eigenfunctions of T under such operations might cease to be eigenfunctions of T, in which case the eigenfunctions of T would not be any good in order to form bases for G. (We shall see in §**6**-1 that this difficulty does not arise.) The second problem is this. It could be possible that, when all the irreducible representations of G are formed, some at least will entail bases which have no relation whatsoever to the eigenfunctions of the translations $\psi_{\mathbf{k}}(\mathbf{r})$, in which case new states would arise over and above those so far counted, N in number. What we are saying, of course, is that there is no guarantee that the functions $\psi_{\mathbf{k}}(\mathbf{r})$ will give us *all* the irreducible representations of G. It is a difficult theorem of group theory to prove that, in fact, all the irreducible representations of G can be constructed from the eigenfunctions of T, and we shall not prove it, although an example in Chapter **7** will help to persuade the reader that this result is true. The important thing is that this theorem validates the argument which we have given above.

We may now go back to more practical things. The Brillouin zone of Fig. 1, as all Brillouin zones, contains N states, N being the number of primitive cells in the crystal. Therefore, *the Brillouin zone contains one state per primitive cell of the crystal or, if spin is included, two states.* In order to understand the physical meaning of this statement, suppose that the metal calcium were to have the simple cubic structure for which the Brillouin zone of Fig. 1 is a cross-section. (See Fig. 3.1.) In this structure there is only one pattern point per primitive cell, so that, if Ca atoms were sited there, we would have two valence electrons per primitive cell. The result of our work so far is that these two electrons would generate precisely the number of states in the crystal required to fill the whole of the Brillouin zone. If, instead, we were to have sodium atoms, we would have only one valence electron per primitive cell, and the states generated would fill in precisely one half the volume of the Brillouin zone. Remember that the density of the grid over the Brillouin zone is entirely uniform, so that counting states merely amounts to determining volumes in the Brillouin zone. Of course, we cannot know at this stage how the Brillouin zone would fill for sodium, because we do not know how the energy goes as a function of **k**. If, for the sake of an example, **k** were to behave exactly like a propagation vector for a free wave, the corresponding energy would be proportional to $|\mathbf{k}|^2$, which means that the Brillouin zone would start filling at the centre, this being the point of lowest energy, and then grow outwards, in spherical shells (because these are the surfaces of constant energy) until a sphere centred at the origin with a volume equal to half that of the Brillouin zone is filled.

6 The bases for the representations of T: the Bloch functions

We know from (5.1) that the bases of the representations of T are eigenfunctions of the translations $\{E\,|\,\mathbf{t}\}$:

5.1 $$\{E\,|\,\mathbf{t}\}\psi_{\mathbf{k}}(\mathbf{r}) = \psi_{\mathbf{k}}(\mathbf{r})\,{}_{\mathbf{k}}\hat{T}(t). \qquad (1)$$

We go back here to the full notation for the translation operator, but, as in the previous section, we expect the reader to recognize that the operator on the left is meant to be a function space operator and not a configuration space operator. It is important to remember that the label **k** of the eigenfunctions in (1) ranges over the Brillouin zone and only over it: this label denotes the translation representation ${}_{\mathbf{k}}\hat{T}(t)$ to which the eigenfunctions belongs, and we have seen in (3.4) that adding a vector **g** of the reciprocal lattice (that is a vector which would take **k** outside the Brillouin zone) does not change this representation.

The value of ${}_{\mathbf{k}}\hat{T}(t)$ in (1) is well known:

2.9|1 $$\{E\,|\,\mathbf{t}\}\psi_{\mathbf{k}}(\mathbf{r}) = \psi_{\mathbf{k}}(\mathbf{r})\exp(-\mathrm{i}\mathbf{k}\cdot\mathbf{t}). \qquad (2)$$

The eigenfunctions $\psi_{\mathbf{k}}(\mathbf{r})$ of the translations, which satisfy (2), are called *Bloch functions* and they can be determined by trial and error from the functional relation entailed by (2), very much as a function is determined by its differential equation. We assert the following result:

$$\psi_{\mathbf{k}}(\mathbf{r}) = \exp(\mathrm{i}\mathbf{k}\cdot\mathbf{r})\,u_{\mathbf{k}}(\mathbf{r}), \qquad (3)$$

where $u_{\mathbf{k}}(\mathbf{r})$ is a periodic function in the lattice:

$$u_{\mathbf{k}}(\mathbf{r}) = u_{\mathbf{k}}(\mathbf{r} + \mathbf{t}), \qquad \forall\, \mathbf{t}. \tag{4}$$

It follows from (4) that the functions $u_{\mathbf{k}}(\mathbf{r})$ are fully defined over the crystal if they are known over its primitive cell, and for this reason they are called *cell functions*. We shall now prove that (3) satisfies (2):

2-2.7 $$\{E\,|\,\mathbf{t}\}\psi_{\mathbf{k}}(\mathbf{r}) = \psi_{\mathbf{k}}(\{E\,|\mathbf{t}\}^{-1}\,\mathbf{r}) = \psi_{\mathbf{k}}(\{E\,|-\mathbf{t}\}\,\mathbf{r}). \tag{5}$$

(Notice that only the first operator here is in function space: the others are in configuration space.) From (**3**-6.2), $\{E\,|-\mathbf{t}\}$ acting on $\mathbf{r}$ changes it into $\mathbf{r} - \mathbf{t}$, whence,

5,3 $$\{E\,|\,\mathbf{t}\}\psi_{\mathbf{k}}(\mathbf{r}) = \psi_{\mathbf{k}}(\mathbf{r} - \mathbf{t}) = \exp\{\mathrm{i}\mathbf{k}\cdot(\mathbf{r} - \mathbf{t})\}\, u_{\mathbf{k}}(\mathbf{r} - \mathbf{t}) \tag{6}$$

4|6 $$= \exp(\mathrm{i}\mathbf{k}\cdot\mathbf{r})\exp(-\,\mathrm{i}\mathbf{k}\cdot\mathbf{t})\, u_{\mathbf{k}}(\mathbf{r}) \tag{7}$$

3|7 $$= \exp(-\,\mathrm{i}\mathbf{k}\cdot\mathbf{t})\, \psi_{\mathbf{k}}(\mathbf{r}), \tag{8}$$

which agrees with (2), thus completing the proof.

It is useful at this stage to consider the form of the representation involved in (2), which comes from (2.9). It follows from eqn (2.3), however, that at that stage it would have been perfectly possible to invert the sign of the exponent. Suppose we had taken it as positive: if we go over the above proof, it is easy to see (Problem 7.2) that in order to get such a sign in (8), the sign of the exponent in the Bloch function (3) would have to be *negative*. This would be mathematically correct but a physical nonsense. Looking at the Bloch functions (3) we can already see that they must be deeply related to plane waves and we shall confirm in the next sub-section that this is so. Therefore, the label $\mathbf{k}$ of the representation will have a meaning of a propagation vector and, if the exponent in (3) were to be given a change of sign, then the Bloch function $\psi_{\mathbf{k}}(\mathbf{r})$ would be propagating in the direction opposite to $\mathbf{k}$! This completes our discussion of the details of eqn (2.3). Readers should beware that there are papers and even tables in which these details are not treated in the manner given here.

Physical considerations. Bands

On comparing the Bloch functions (3) to plane waves in the lattice, as given by eqn (**4**-10.2), they can be written in the following form:

$$\psi_{\mathbf{k}}(\mathbf{r}) = |\,\mathbf{k}\rangle\, u_{\mathbf{k}}(\mathbf{r}), \tag{9}$$

where we ignore the normalization factor in the plane wave which, in any case, can easily be adjusted. The Bloch functions in the form (9) can be recognized as plane waves in the lattice, modulated by the cell function $u_{\mathbf{k}}(\mathbf{r})$, the latter reflecting the perturbation of the free-electron waves introduced by the periodic lattice. This picture will allow us to intepret the physical meaning of the label **k** of the irreducible representations of T, as well as that of the cell functions $u_{\mathbf{k}}(\mathbf{r})$, for which purpose we shall have to consider two limiting cases.

(i) $V(\mathbf{r}) \to 0$.

In this case, because the potential vanishes, the eigenfunctions (9) must go into the free-electron eigenfunctions $|\mathbf{k}\rangle$, which requires that the cell functions $u_{\mathbf{k}}(\mathbf{r})$ become constants. This shows that our so far purely abstract label **k** of the representations must have an asymptotic meaning (that is a meaning in a limiting case) of a momentum for vanishing fields. This explains why we have given that label the same symbol as that of the momentum. For non-vanishing potentials, however, the **k** vector of a Bloch function is not identical to the momentum, and it is sometimes referred to as a *pseudomomentum.*

(ii) $\mathbf{a} = \mathbf{b} = \mathbf{c} \to \infty$.

This means that the atoms of the crystal are at infinite separation, in which case the whole of the reciprocal space collapses into its origin $\mathbf{k} = \mathbf{0}$, as we shall now show:

$$\mathbf{a} \cdot \mathbf{a}^{\#} =_{\text{def}} |\mathbf{a}|\,|\mathbf{a}^{\#}| \cos\alpha = 2\pi \qquad \Rightarrow \qquad \lim_{|\mathbf{a}| \to \infty} |\mathbf{a}^{\#}| = 0. \tag{10}$$

Similar results for $\mathbf{b}^{\#}$ and $\mathbf{c}^{\#}$ confirm our assertion. On the other hand, because of the term $\exp(\mathrm{i}\mathbf{k}\cdot\mathbf{r})$, the free-electron eigenfunctions $|\mathbf{k}\rangle$ become constant for **k** equal to zero, and the whole of the Bloch function (9) becomes the cell function $u_{\mathbf{0}}(\mathbf{r})$. We immediately realize that something must be missing here. For, in this limiting case of isolated atoms, the wave function must be allowed to become one of the manifold atomic orbitals ϕ^j (where the index j denotes one of the states 1s, 2s, etc.) rather than the single function $u_{\mathbf{0}}(\mathbf{r})$. This deficiency is easily corrected: we must write the cell functions as $u^j_{\mathbf{k}}(\mathbf{r})$, so that in the limiting case of isolated atoms we get $u^j_{\mathbf{0}}(\mathbf{r})$, which must coincide with one atomic orbital ϕ^j. This new label j must, of course, also be introduced into the Bloch functions $\psi_{\mathbf{k}}(\mathbf{r})$ in (9):

$$\psi^j_{\mathbf{k}}(\mathbf{r}) = |\mathbf{k}\rangle\, u^j_{\mathbf{k}}(\mathbf{r}). \tag{11}$$

This is a most important result. Remember that **k** can take N values (over the Brillouin zone). Thus, for each value of j we have N states (or $2N$ states including spin), which will be said to form the j-th *band*. What characterizes such a band is that, in the limiting case of separated atoms, all the states of the band collapse into the single atomic state labelled with j. If we had a crystal made up of N hydrogen atoms with only one atom in each primitive cell, then each of the atomic states 1s, 2s, 2p, etc. will spread out into a band of N states (without spin) in the solid. The successive bands in the solid are called the 1s, 2s, 2p, bands and so on, and they are illustrated in a very schematic way in Fig. 1.

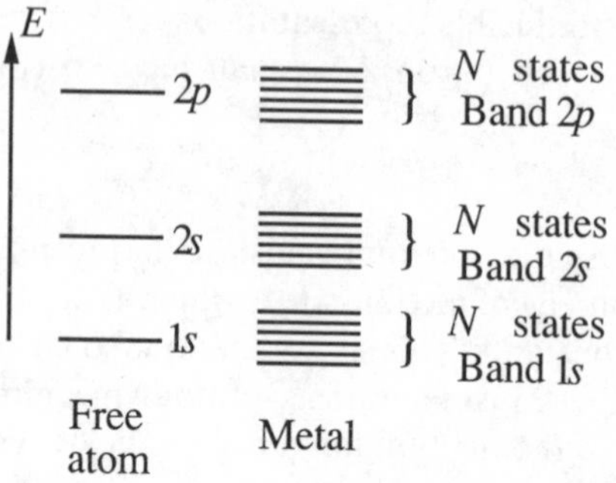

Fig. 5-6.1. Schematic depiction of the relation between the atomic levels in a free atom and their corresponding bands in the solid.

Various examples of bands will be discussed in some detail in Chapter 7. Before we do this, however, we must consider the representations of the space group, which we shall do in the next chapter.

7 Problems

1. Verify that the translation group representations (2.9) satisfy the group multiplication rules.
2. On changing the sign of the exponent in (2.9) find, by the method of § 6, the corresponding form of the Bloch functions.
3. Prove that the plane waves $|\mathbf{k}\rangle$ in (6.9) are eigenfunctions of the translations with the same eigenvalue as the Bloch functions.
4. Prove that the plane waves $|\mathbf{k}\rangle$ in (6.9), with $\mathbf{k}$ ranging over the N values in the Brillouin zone, are orthogonal over the crystal volume for different values of $\mathbf{k}$. Compare this result with the orthogonality of the plane waves $|\mathbf{g}\rangle$ in **4**-9.9.
5. Given a one-dimensional chain with period a and N atoms, show that its Brillouin zone, and the N states in it, are given as follows for N even or odd.

N even

$$-\frac{1}{2}\mathbf{a}^{\#} < \mathbf{k} \leq \frac{1}{2}\mathbf{a}^{\#}, \qquad |\mathbf{a}^{\#}| = \frac{2\pi}{a}, \qquad \mathbf{k} = \frac{\kappa}{N}\mathbf{a}^{\#}, \tag{1}$$

$$\kappa = -\frac{N}{2} + 1, \ -\frac{N}{2} + 2, \ldots, \frac{N}{2}. \tag{2}$$

N odd

$$-\frac{1}{2}\mathbf{a}^{\#} < \mathbf{k} < \frac{1}{2}\mathbf{a}^{\#}, \qquad |\mathbf{a}^{\#}| = \frac{2\pi}{a}, \qquad \mathbf{k} = \frac{\kappa}{N}\mathbf{a}^{\#}, \tag{3}$$

$$\kappa = -\frac{N-1}{2}, \ -\frac{N-1}{2} + 1, \ldots, \frac{N-1}{2}. \tag{4}$$

Notice that for N odd, as given in (3), neither of the two edges of the Brillouin zone is a permitted state. (See eqn 4.) For N large, however, the difference between the two cases considered here is trivial. Also, for N large, the occupied states in the Brillouin zone can be taken to cover the whole of it.

Further reading

The pioneering classic by Mott and Jones (1936) is still good reading on Bloch functions and Brillouin zones, as is Brillouin (1946). Jones (1960) provides a very comprehensive discussion.

6

Space group representations

We have obtained the Bloch functions $\psi_{\mathbf{k}}(\mathbf{r})$ in Chapter **5** as eigenfunctions of the translations (eqn **5**-5.1) but their real importance stems from three properties of the N Bloch functions which correspond to the N internal points in the Brillouin zone (see §**5**-5), where N is the number of primitive cells in the crystal. (i) They form bases for representations of the space group G. (ii) These bases are irreducible. (iii) These bases span *all* the irreducible bases of G. We shall go a long way in this chapter in establishing (i) and (ii) fairly decently, but (iii) is beyond our scope. In Chapter **7**, however, we shall work out a simple example in which we shall be able to satisfy ourselves that these properties are all true. The treatment that we shall give in this and the following chapters will mostly be concerned with symmorphic space groups, but its extension for non-symmorphic ones is possible without major changes in the basic concepts which we shall discuss.

The first thing we must do in order to pass from the translation group T, as fully discussed in Chapter **5**, towards the space group G, is to find out how the space group operations act upon the Bloch functions.

1 Effect of the space group operators on the Bloch functions

We shall now prove that the general space group operator $\{p|\mathbf{w}\}$, on acting upon the Bloch function that corresponds to the vector $\mathbf{k}$, transforms it into the Bloch function that corresponds to the vector $p\mathbf{k}$:

$$\{p|\mathbf{w}\}\psi_{\mathbf{k}} = \psi_{p\mathbf{k}}. \quad (1)$$

In order to prove this result we shall need the relation that defines the Bloch functions as translation eigenfunctions:

5-6.2 $$\{E|\mathbf{t}\}\psi_{\mathbf{k}} = \exp(-\mathrm{i}\mathbf{k}\cdot\mathbf{t})\psi_{\mathbf{k}}, \quad (2)$$

as well as the following operator product,

$$\{E|\mathbf{t}\}\{p|\mathbf{w}\} = \{p|\mathbf{w}\}\{E|p^{-1}\mathbf{t}\}, \quad (3)$$

which can easily be verified by applying on both sides the multiplication rule (**3**-6.6) of the Seitz operators.

In order to find the $\mathbf{k}$ index of the function $\{p|\mathbf{w}\}\psi_{\mathbf{k}}$ on (L1), we must obtain, in accordance with (2), the exponential factor which appears when $\{E|\mathbf{t}\}$ acts on that function which, for emphasis, we enclose in square brackets:

3 $$\{E|\mathbf{t}\}[\{p|\mathbf{w}\}\psi_{\mathbf{k}}] = \{p|\mathbf{w}\}\{E|p^{-1}\mathbf{t}\}\psi_{\mathbf{k}} \tag{4}$$

2|4 $$= \{p|\mathbf{w}\}\exp(-\mathrm{i}\mathbf{k}\cdot p^{-1}\mathbf{t})\psi_{\mathbf{k}}. \tag{5}$$

We must massage a little the scalar product on (R5). Point group operators p must not change a scalar product such as $\mathbf{r}\cdot\mathbf{r}'$ between two vectors of the configuration space since, otherwise, in forming the new product $p\mathbf{r}\cdot p\mathbf{r}'$, lengths of vectors and the angles between them would change, which symmetry operators must not do. Therefore,

$$\mathbf{k}\cdot p^{-1}\mathbf{t} = p\mathbf{k}\cdot pp^{-1}\mathbf{t} = p\mathbf{k}\cdot\mathbf{t}. \tag{6}$$

6|5 $$\{E|\mathbf{t}\}[\{p|\mathbf{w}\}\psi_{\mathbf{k}}] = \{p|\mathbf{w}\}\exp(-\mathrm{i}p\mathbf{k}\cdot\mathbf{t})\psi_{\mathbf{k}} \tag{7}$$

7 $$= \exp(-\mathrm{i}p\mathbf{k}\cdot\mathbf{t})[\{p|\mathbf{w}\}\psi_{\mathbf{k}}]. \tag{8}$$

(Remember that the Seitz operators act on functions of $\mathbf{r}$, which the exponential on R7 is not.) Comparison of (8) with (2) shows that the function in the square bracket in (8) pertains to the value $p\mathbf{k}$ of the $\mathbf{k}$ vector, thus verifying (1).

We must remember in all this work that the functions $\psi_{\mathbf{k}}$ are functions of $\mathbf{r}$, $\psi_{\mathbf{k}}(\mathbf{r})$, and that the functions $\{p|\mathbf{w}\}\psi_{\mathbf{k}}$ in (8) may therefore be multiplied by any factor independent of $\mathbf{r}$ without affecting the validity of that equation. A difference in approach to the left-hand sides of (1) and (2) should thus be understood. Whereas in (1) we are not interested in possible $\mathbf{r}$-independent factors, such factors are essential in (2) because they provide the eigenvalues.

A special case of (1), which is important, arises for the inversion operator $\{\mathbb{i}|\mathbf{0}\}$, which is the inversion at the coordinates origin. The inversion, clearly, must change the sign of all vectors in direct as well as in reciprocal space. (Remember that, after all, both spaces are identically the same real space of the crystal.) Thus $\mathbb{i}\mathbf{k}$ equals $-\mathbf{k}$. Therefore,

1 $$\{\mathbb{i}|\mathbf{0}\}\psi_{\mathbf{k}} = \psi_{\mathbb{i}\mathbf{k}} = \psi_{-\mathbf{k}}, \tag{9}$$

a result which will be much used later on.

2 The bases of the space group

In trying to understand the significance of (1.1), you must remember that what characterizes the functions of a basis of a representation is that they are transformed one into another (within, perhaps, a linear combination) by the symmetry operations of the group. Equation (1.1) means therefore that $\psi_{\mathbf{k}}$ and $\psi_{p\mathbf{k}}$, where p is any operation of the point group P, belong to the same basis for a representation of the space group G. We have seen at the end of §5-3 that P is a covering group of the Brillouin zone. Therefore, since we must assume that **k** belongs to the Brillouin zone (since any vectors outside it are always equivalent to a vector internal to the zone), $p\mathbf{k}$ must also be a vector of the first Brillouin zone. Given a Bloch function $\psi_{\mathbf{k}}$ we thus generate one basis of G as follows:

$$\langle\psi_{\mathbf{k}}| =_{\text{def}} \langle\psi_{p\mathbf{k}}|, \qquad \forall\, p \in P, \qquad \forall\, p\mathbf{k} \in \text{Brillouin zone}. \tag{1}$$

Notice that identically the same basis is generated by anyone of its functions. Take for example $\psi_{p'\mathbf{k}}$, where p' is in P: because of the closure of P, the set $\{\psi_{pp'\mathbf{k}}\}$ for all p in P coincides with the set $\{\psi_{p\mathbf{k}}\}$ which appears in (1). It follows that, although we can start with any vector **k** in the Brillouin zone in order to generate a basis of G, not all such vectors will generate a different basis.

We shall now prove a second and most important result, namely that the same vector $p\mathbf{k}$ cannot appear in two different bases, such as $\langle\psi_{\mathbf{k}}|$ and $\langle\psi_{\mathbf{k}'}|$. If p and p' are in P, we must have:

$$\psi_{p\mathbf{k}} \in \langle\psi_{\mathbf{k}}|, \qquad \psi_{p'\mathbf{k}'} \in \langle\psi_{\mathbf{k}'}|. \tag{2}$$

Suppose now that $p\mathbf{k}$ and $p'\mathbf{k}'$ coincide:

$$p\mathbf{k} = p'\mathbf{k}' \qquad \Rightarrow \qquad \mathbf{k} = p^{-1}p'\mathbf{k}' \qquad \Rightarrow \qquad \mathbf{k} =_{\text{def}} p''\mathbf{k}', \tag{3}$$

since the product of two operations of P must be another operation of P. Equation (3) means that **k** belongs to the basis $\langle\psi_{\mathbf{k}'}|$ and therefore that $p\mathbf{k}$, for all p in P, belongs to $\langle\psi_{\mathbf{k}'}|$. Thus the *whole* of $\langle\psi_{\mathbf{k}}|$ (the functions of which are of the form $\psi_{p\mathbf{k}}$) belongs to $\langle\psi_{\mathbf{k}'}|$. We can prove in the same manner, conversely, that the whole of $\langle\psi_{\mathbf{k}'}|$ must belong to $\langle\psi_{\mathbf{k}}|$, whence these two bases must coincide. The net result is that two bases $\langle\psi_{\mathbf{k}}|$ and $\langle\psi_{\mathbf{k}'}|$ either are identical or have no **k** vector in common.

What these two results mean is both simple and very important. We know that the irreducible representations of T are labelled by the N distinct **k** vectors in the Brillouin zone. When we now consider the bases of the space group, these N distinct vectors separate in *disjoint bases* of G, that is bases which have no **k** vector in common. The nature of these bases will be studied in the next section but, before we do this, let us return to (1.1). It is this remarkable equation that tells us that we can construct bases for the whole space group G starting from the bases of its subgroup T, that is from the Bloch functions. Why have we been so lucky in our choice of subgroup? We cannot get here into the finer details of this question, but a short answer is that T does so well for us because it is a privileged subgroup, being invariant (see eqn 3-8.6). We shall illustrate this property by considering a non invariant subgroup of the space group of Fig. 3-5.2c, the subgroup $\mathbf{C}_s$ made up of E and σ_y. Imagine now that we had started by obtaining eigenfunctions of this subgroup. Because this subgroup has only two irreducible representa-

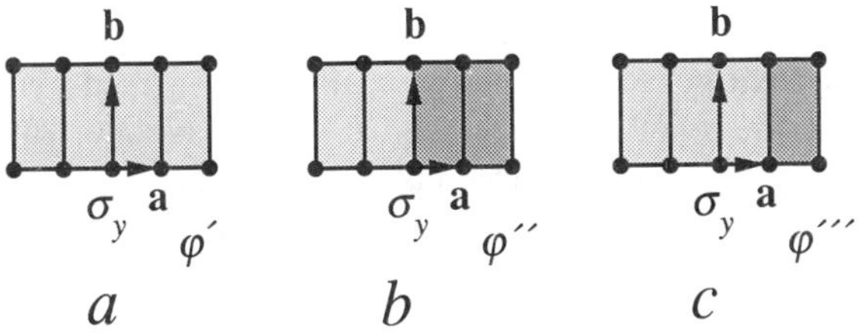

Fig. 6-2.1. This picture represents schematically functions defined over the whole of a crystal pattern, of which only four primitive cells are shown. Thus, the tints must be imagined to be extended over the whole pattern. The functions are assumed to have a constant value over the whole region covered by the light tint, and its negative over the region covered with the dark tint. The functions φ' and φ'' depicted in *a* and *b* are symmetric and anti-symmetric, respectively, with respect to σ_y. The function φ''' depicted in *c* has been obtained by translation of φ'' by **a** and it is neither symmetric nor antisymmetric with respect to σ_y.

tions (see any table of such representations), it is well known that its eigenfunctions are of two types, either symmetrical (φ') or antisymmetrical (φ'') with respect to the symmetry plane. They are illustrated in Fig. 1, *a* and *b* respectively. It is clear from Fig. 1*c* that when, for example, we translate φ'' by **a**, a new function is obtained which no longer is either symmetrical or antisymmetrical with respect to σ_y. Whereas the eigenfunctions of the translations, in accordance to (1.1) remain eigenfunctions of the translations under *all* the operations of G, this is not so for the eigenfunctions of $\mathbf{C}_s$ displayed. This is why the consequences of our use of the subgroup T in order to build up bases of G from its eigenfunctions are not just the result of accidental good luck but of a very careful initial choice.

3 The star, the group of the k vector, and the small representation

We shall now refine considerably the two major results from the last section. First, we had proved that, given a **k** vector and its Bloch function $\psi_{\mathbf{k}}$, then the functions $\psi_{p\mathbf{k}}$, for all p in the point group P, form a basis for the representation of the space group G. This, however, does not mean that this basis is irreducible and we shall show that, on the contrary, it often contains redundancies in the sense that the same Bloch function may appear more than once in it. The consideration of such redundancies will lead in this section to the concept of the *star*.

Secondly, we have seen that the same **k** vector cannot belong to two distinct bases. It is possible, however, for the same **k** vector to be repeated more than once in the same basis, as it will follow from the concepts of the *group of the* **k** *vector* and the *small representation.*

The Brillouin zone provides a very simple graphical device in order to visualize the obtention, for a given $\psi_{\mathbf{k}}$, of the basis formed by the functions $\psi_{p\mathbf{k}}$, for all p in the point group, as defined in (2.1). We illustrate this in Fig. 1 with reference to a two-dimensional square lattice depicted in Fig. 1*a*. (Notice that in this case the crystal pattern coincides with its lattice.) In order to show the point group symmetry more clearly, we have constructed here the centred unit cell in the direct lattice, and the point group is

$$\mathbf{C}_{4v}: E, C_2, C_4^+, C_4^-, \sigma_{v1}, \sigma_{v2}, \sigma_{d1}, \sigma_{d2}. \tag{1}$$

The rotations C_2, C_4^+, C_4^- are all around the four-fold axis of rotation shown at O and the σ are the reflection planes shown. (Notice that *strict two-dimensionality* is assumed, which means that no operation that entails the third dimension, like a rotation by π around $\mathbf{a}$, is permitted.) We show in Fig. 1*b* the Brillouin zone which, as in Fig. **5**-3.1, must be read as superimposed on the crystal lattice of Fig. 1*a*.

It is usual in this work to label certain characteristic $\mathbf{k}$ vectors with capital (Latin or Greek) letters and in Fig. 1*b* we consider a $\mathbf{k}$ vector G_1 in the interior of the Brillouin zone, on which we act with the eight operations listed in (1) which, in the precise order therein used, produce the following $\mathbf{k}$ vectors:

$$G_1, G_5, G_3, G_7, G_2, G_6, G_4, G_8. \tag{2}$$

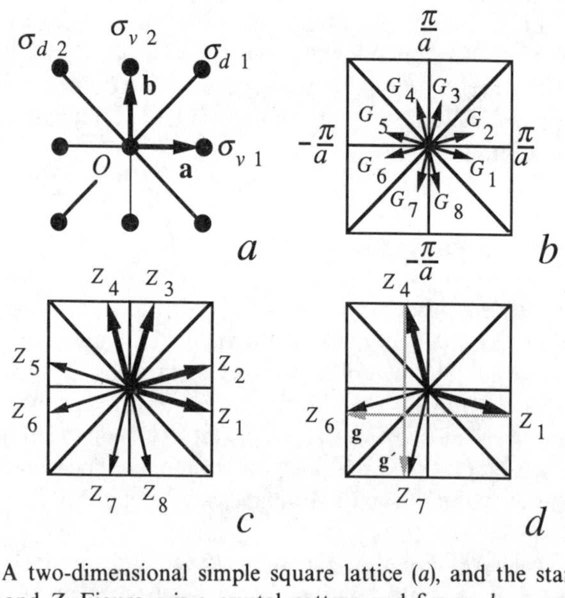

Fig. 6-3.1. A two-dimensional simple square lattice (*a*), and the stars of the two $\mathbf{k}$ vectors G and Z. Figure *a* is a crystal pattern and figures *b*, *c*, and *d* are all in reciprocal space. For convenience of the picture $|\mathbf{a}|$ has been arbitrarily chosen to be $\sqrt{\pi}$, so that $|\mathbf{a}^{\#}|$ equals $2|\mathbf{a}|$, as shown.

Notice that any one of these **k** vectors transforms into all the others under the operations of the point group. As an example, G_3 transforms into G_4 and G_2 under the reflections σ_{v2} and σ_{d1}, respectively. From (2.1), as discussed at the beginning of the present section, the set of eight Bloch functions which correspond to the eight vectors in (2) form a basis for the representation of the space group. It is plausible to accept that this basis is irreducible, since none of the **k** vectors in it are identical or equivalent, so that the corresponding Bloch functions are all linearly independent. Such a set of **k** vectors as in (2) is called a *star*, but it must not be believed that the star is always formed by the transform $p\mathbf{k}$ of a given **k** under all p in P, as we shall now discuss.

We consider in Fig. 1*c* a **k** vector Z_1, derived from G_1 by lengthening it until it hits the Brillouin zone edge. Eight **k** vectors are generated precisely as in (2), but we shall see that they are not all distinct. The vector Z_6, for example, (which is obtained by acting with σ_{v2} on Z_1) is equivalent to Z_1, since it differs from it by a vector **g** of the reciprocal lattice, as shown in Fig. 1*d*. It follows from §**5**-3 that the Bloch functions labelled by Z_1 and Z_6 are identical. Likewise, Z_7 is equivalent to Z_4, Z_5 to Z_2, and Z_8 to Z_3. This means that in order to form an irreducible basis for the space group, corresponding to the **k** vector Z_1, the only **k** vectors which are required are those shown in thick lines in Fig. 1*c*:

$$Z_1, Z_2, Z_3, Z_4. \tag{3}$$

Such a set of **k** vectors is called a *star*. The full definition of this object, therefore, is as follows. *The star of the* **k** *vector* **k**′, *say, is the set of all mutually inequivalent* **k** *vectors of the form p***k**′, *for p ranging over all the operations of the point group P*. It is clear that a star will lead to an irreducible basis of the space group G, but in order to demonstrate this a little more work is required.

The group of the **k** *vector*

Consider the **k** vector Z_1 in Fig. 2*b*. It is left identically invariant under the operation E, of course, but the operation σ_{v2} also leaves it invariant, in the sense that it transforms it into an equivalent **k** vector. Clearly, the product of two operations which leave a **k** vector invariant in this manner must also be another operation with the same property, so that the set of all operations which leave a **k** vector invariant forms a group, called the *group of the* **k** *vector*. In the case of Z_1 this group is very simple, since it contains only the operations E and σ_{v2}, and it is the point group $\mathbf{C}_s$. It will be instructive to discuss the more involved case of the vector M_1, shown in Fig. 2*c*. If we consider the point group operations in (1) we see that not only E but also σ_{d1} leave M_1 *identically* invariant. The operation σ_{v2}, on the other hand, transforms it into M_2, which is equivalent to M_1, from which it differs by the vector **g** of the reciprocal lattice shown in Fig. 1*b*. The product $\sigma_{v2}\sigma_{d1}$ is C_4^+, which also leaves M_1 invariant, since it transforms it into its equivalent M_4.

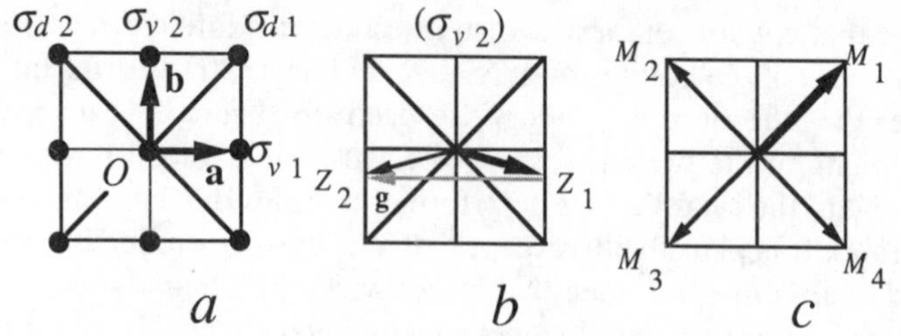

Fig. 6-3.2. The group of the **k** vector. This figure illustrates the two-dimensional simple square lattice of Fig. 1 with the same choice of units. Figure *a* is the crystal pattern (direct lattice) whereas *b* and *c* are in the reciprocal lattice and depict the Brillouin zone. The symmetry operations mentioned in the text are always symmetry operations of the direct lattice and must be obtained when operating in *b* and *c* by imagining that these figures are given parallel translations until their centre coincides with *O*. It is in this way that the vertical axis in *b* coincides with the σ_{v2} plane of the crystal pattern, for which reason this element is shown in brackets in *b*. (See the end of §4 for a discussion of the symmetry operations in the reciprocal lattice.)

In the same manner, it can be seen that all the operations of $\mathbf{C}_{4v}$ listed in (1) leave M_1 invariant, so that this is the group of this **k** vector. It should be noticed, therefore, that the star of M_1 contains the single vector M_1.

The small representation

We shall illustrate the way in which the group of the **k** vector helps us to obtain irreducible bases of *G* by considering the basis (2.1) for **k** equal to the vector Z_1 of Fig. 1*c*, and displayed again in Fig. 3. For convenience, the Bloch

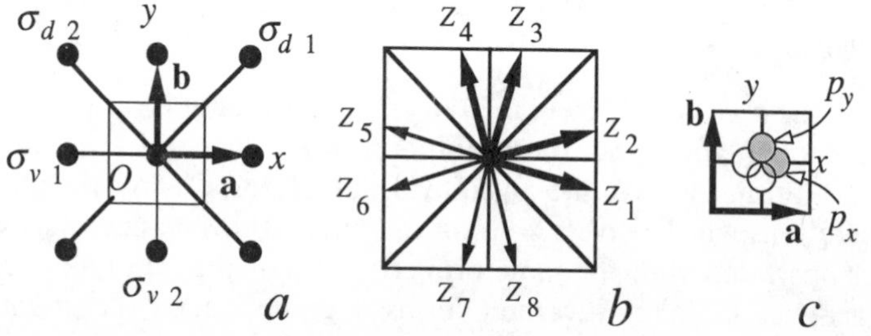

Fig. 6-3.3. The star of *Z*. The operations of the group of *Z* in the reciprocal space picture in *b* must be read from those in the crystal pattern (*a*) as explained in the caption to Fig. 2. The unit cell in *c* (direct lattice) is the centred unit cell shown by the square in *a*. Its centre, therefore, contains the atom at *O* for which the two orbitals p_x and p_y are illustrated. The grey lobes are positive and the white lobes have the corresponding opposite (negative) values. The symmetry plane σ_{v2} in *a* is the one used in the text in order to classify functions as symmetric (′) and antisymmetric (″) with respect to it.

function corresponding to each of the vectors Z_i therein shown will be written as ψ_i. The basis (2.1) is, therefore,

$$\langle \psi_1 \psi_2 \psi_3 \psi_4 \psi_5 \psi_6 \psi_7 \psi_8 | . \tag{4}$$

Of course, from our discussion about the group of the **k** vector, we know that we can throw away the last four of these functions. We shall now be a bit more clever than this: armed with the knowledge that the group of this **k** vector is $\mathbf{C}_s$, as shown above, we shall reduce the basis in (4) from first principles. The irreducible representations of $\mathbf{C}_s$ are shown in Table 1 and we see that there are only two one-dimensional representations, A' and A'', symmetrical and antisymmetrical, respectively, with respect to the reflection plane. In order to reduce (4) we have to realize that, for example, neither ψ_1 nor ψ_6 are symmetrical or antisymmetrical with respect to σ_{v2}, but that the combinations $\psi_1 + \psi_6$, $\psi_1 - \psi_6$ belong to A' and A'' respectively. Likewise, we can symmetrize and antisymmetrize with respect to σ_{v1}. It is thus clear that our job will be done by replacing the eight functions in (4) by eight linear combinations of them chosen in the manner described:

$$\langle \psi_1 + \psi_6, \psi_2 + \psi_5, \psi_3 + \psi_8, \psi_4 + \psi_7, \\ \psi_1 - \psi_6, \psi_2 - \psi_5, \psi_3 - \psi_8, \psi_4 - \psi_7 | . \tag{5}$$

Notice that we have been careful to ensure that in each of the eight linear combinations in (5) the two functions involved have exactly the same **k** vector, Z_8, say, being the same as Z_3. We can thus call $\psi_3 + \psi_8$ and $\psi_3 - \psi_8$, respectively, ψ_3' and ψ_3'', because they both belong to the same **k** value Z_3 and to the representations A', A'', respectively. With this change of notation, we rewrite (5),

$$\langle \psi_1', \psi_2', \psi_3', \psi_4', \psi_1'', \psi_2'', \psi_3'', \psi_4'' | . \tag{6}$$

It is pretty clear that no symmetry operation can transform a symmetric function (primed) into an antisymmetric one (doubly primed). Such functions, therefore, should not belong in the same basis; that is, the basis (6) reduces into two new bases:

$$\langle \psi_1', \psi_2', \psi_3', \psi_4' |, (A'); \qquad \langle \psi_1'', \psi_2'', \psi_3'', \psi_4'' |, (A''). \tag{7}$$

Table 6-3.1. The irreducible representations of $\mathbf{C}_s$.

The x and y directions are perpendicular and parallel, respectively, to the plane σ.

	E	σ	bases
A'	1	1	p_y
A''	1	-1	p_x

It is easy to accept now that these bases are irreducible and, also, that they are the only bases which can be constructed from the **k** vector Z_1. Notice that the Bloch functions which appear in these two bases are precisely those that correspond to the star of Z_1. (See eqn 3.) This shows that a star determines an irreducible basis, provided that the functions of the star are symmetrized with respect to the irreducible representations of the group of the **k** vector, which are called the *small representations.* With this, we have finished in principle our quest for the irreducible representations of the space group. The ideas discussed, however, need a bit of tidying up, which we shall now do. A little later (§4) we shall discuss in more detail the physical significance of what we have done.

The first thing we must notice is that the group of the **k** vector is not identically the same for all the Bloch functions of the same star. For Z_1, as we have seen, it is $\mathbf{C}_s$ as given by E and σ_{v2}, whereas for Z_4 it is $\mathbf{C}_s$ as given by E and σ_{v1}. (See Fig. 3.) This, however, gives no trouble, as we shall now show. If we consider the two bases in (7), we must ensure that they are orthogonal, as functions that belong to different irreducible representations are. Of course, two functions with different suffix are automatically orthogonal by virtue of belonging to different irreducible representations of T. If we consider ψ'_4, all that we need to prove, therefore, is that it is orthogonal to ψ''_4. But these two functions have been symmetrized with respect to the group of Z_4 (E and σ_{v1}) and they are therefore orthogonal.

The next thing we must do is to understand what is the difference between ψ'_1, $(\psi_1 + \psi_6)$, and ψ''_2, $(\psi_1 - \psi_6)$, say, in the bases in (7). In these functions, because the factor $|\mathbf{k}\rangle$ in the Bloch functions is the same (see eqn **5**-6.11), it is clear that we are forming different linear combinations of the cell functions, which can be called u'_1 and u''_2 respectively for ψ'_1 and ψ''_2. That is, for the **k** value Z_1, we have two bands one symmetrical and the other antisymmetrical with respect to the reflection plane in the group of the **k** vector of Z_1 and the difference between these two bands, as always, is determined by the difference in the cell functions. As an example, the primed and doubly-primed cell functions will behave respectively like the p_y and p_x functions shown in Fig. 3. (This does not mean, of course, that the cell functions actually coincide with these functions.) Notice in (7) that each value of **k** appears only once for each basis, a result which we shall see is very significant physically, although it is not valid in general. In order to discuss this we must realize that, in the example above, all the small representations are one-dimensional, and it is important to see what happens otherwise. (Remember in this discussion that an n-dimensional representation means that the corresponding basis contains n functions.)

Consider the **k** vector M_1 in Fig. 2*c*, the **k** group of which is $\mathbf{C}_{4v}$ listed in (1). The irreducible representations of this group are given in Table 2. This means that there are four one-dimensional bands at M_1, corresponding respectively to the four one-dimensional representations and that there is a two-dimensional basis corresponding to the representation E. If we denote the Bloch

Table 6-3.2. Characters of the irreducible representations of $\mathbf{C}_{4v}$.

	E	$2C_4$	C_2	$2\sigma_v$	$2\sigma_d$
A_1	1	1	1	1	1
A_2	1	1	1	-1	-1
B_1	1	-1	1	1	-1
B_2	1	-1	1	-1	1
E	2	0	-2	0	0

function corresponding to M_1 with the symbol χ_1, the two-dimensional basis must comprise two such functions. Obviously, they could not be identically repeated, so that they must belong to *different* bands, and we shall write this basis as

$$\langle \chi_1', \chi_1'' | . \tag{8}$$

Notice that we use the same band index as that employed for the two bands at Z_1, spanned by p_y-type and p_x-type cell functions respectively. This is not without reason: it is easy to prove that two such functions span a representation which agrees precisely with the E representation in Table 2, which, when necessary, will be denoted as E_M. (See Problem 6.1.)

It is pretty obvious that the null $\mathbf{k}$ vector, origin of the reciprocal space, which is always conventionally denoted with the letter Γ, is left invariant by all the operations of the point group $\mathbf{C}_{4v}$. Thus, its little group and small representations coincide with those of M. In particular, therefore, we must also have here a two-dimensional representation E (denoted when necessary as E_Γ), spanned by two functions, one of p_x and the other of p_y type, both leading to the same energy eigenvalue. The s functions, instead, span the A_1 representation, which is of course non degenerate.

The result of all this is a very important rule. In general, there is a succession of bands in ascending order of energy for each $\mathbf{k}$ vector in the Brillouin zone. As $\mathbf{k}$ varies along the Brillouin zone, the corresponding group of its $\mathbf{k}$ vector also varies and, if this group contains a multiply-dimensional representation, then two or more of these bands may coincide in energy, which means that these otherwise separate bands touch or cross at this particular value of $\mathbf{k}$. Examples of this situation will be discussed in the next section.

4 The irreducible representations of the space group and the energy as a function of k

The very important result of § 3 is that each irreducible representation of the space group is fully determined by two objects: the star of the $\mathbf{k}$ vector and the small representation (irreducible representation of the group of the **k** vector). We shall now discuss in more detail the implications of these results as regards the behaviour of the energy bands which are defined over the Brillouin zone. Of course, the behaviour of the bands depends on their type and in order to have a sufficiently complex example we shall deal in this section with *p* bands. The *s* band will be discussed in §**8**-7.

We first recall that both **k** and **k** + **g**, where **k** is in the Brillouin zone and **g** is a vector of the reciprocal lattice, give identically the same irreducible representations, because they label identical Bloch functions. (See eqn **5**-3.4 and the discussion after eqn **5**-6.1.) Each irreducible representation of the space group, labelled by **k** (or, identically, by **k** + **g**) denotes an energy eigenvalue *E* which can thus be written as *E*(**k**), or, identically, as *E*(**k** + **g**). As **k** varies quasi-continuously (which means continuously in practice) over the Brillouin zone, the function *E*(**k**) will move over different surfaces corresponding to the different bands and must be labelled as $E^j(\mathbf{k})$. Since, as we have seen, **k** and **k** + **g** label the same irreducible representations and therefore the same eigenvalues, we have

$$E^j(\mathbf{k}) = E^j(\mathbf{k} + \mathbf{g}), \tag{1}$$

so that *the energy is periodic in the reciprocal space.* In order to express $E^j(\mathbf{k})$ it is therefore sufficient to let **k** range over its *N* quantized values in the Brillouin zone (*N* being the number of primitive cells in the crystal lattice). The superscript on the letter *E* denotes the *j* band, that is the set of *N* values of the energy obtained when **k** varies *continuously* over the Brillouin zone. (Although, as we know, the permitted **k** values over the Brillouin zone form a discrete grid, with *N* points, *N* is so large and thus the grid is so dense, that the concept of continuity over **k** space is still significant.) Our purpose now is to study the behaviour of $E^j(\mathbf{k})$ when **k** varies continuously over the Brillouin zone. With this object in view, we depict in Fig. 1*a* the Brillouin zone of Figs 3.1 and 3.2. The **k** vectors *G*, *Z*, and *M* correspond to the vectors so named in those figures and *Γ* and *X* correspond respectively to the centre of the Brillouin zone (**k** = **0**) and to that of one of its faces. We want to see how the energy varies when we move continuously from *Γ* to *G* and then to *Z*, and from there to *M*, as depicted in Figs 1*b* and 1*c*.

Let us first consider the general point *G*. We know that we have here an eight-dimensional basis corresponding to the star of *G* in (3.2). The meaning of this statement as far as $E^j(\mathbf{k})$ is concerned, is that precisely the same value of the energy marked in the figure for, say, the lower of the two curves at *G* would obtain for any of the eight values of *G* listed in (3.2) and depicted in Fig. 3.1*b*. At *G*, because there is only one small representation (the identical one), we may choose only one cell function, $u_\mathbf{k}^j(\mathbf{r})$ the continuous variation of which over the Brillouin zone will define the *j* band under study. Of course there are infinite many functions which we could choose, but we shall proceed as follows. Move along the horizontal axis of Fig. 1*b*, from *G* to *Γ*, following

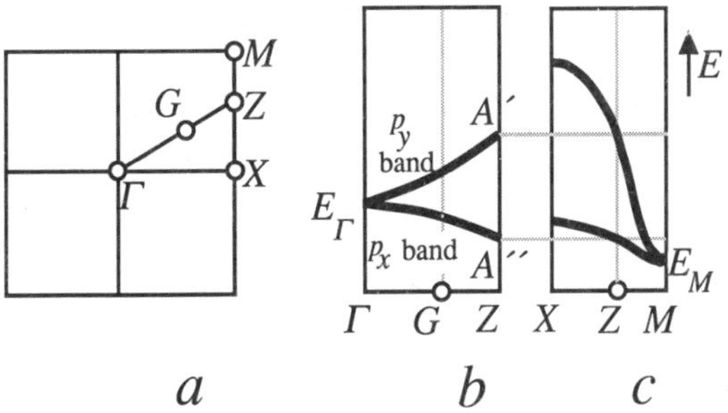

Fig. 6-4.1. The energy as a function of $\mathbf{k}$ for p bands. The Brillouin zone given in a corresponds to the simple square lattice of Fig. 3.1. In b and in c the energy is given as a function of the $\mathbf{k}$ vector in the directions ΓZ and XM respectively. The horizontal scales in b and c are identical with the linear scale in a. See Problem **8**-8.10 for some details of the bands drawn here.

the lower curve shown in the figure, until the eigenvalue E_Γ is reached. (Remember that E here is not the energy but that it denotes a doubly-degenerate irreducible representation. See § 3.) What happens is this: as we reduce continuously the modulus of G, the cell function varies continuously, and with it the energy levels, following the curve shown, until we reach Γ, when the cell function must coincide with one of the two functions of the basis, which are either p_x or p_y type and which, in the limit of separate atoms, would indeed become atomic p_x or p_y orbitals respectively. For reasons to be discussed in the next subsection, we have assumed that the cell function, in the limiting process described, becomes p_x type. This means that the whole of the band of states which we are considering may be described as a p_x band, this band label, as we know, having always a meaning as a limit. In the same manner, the upper curve at G may be labelled a p_y band. Quite correctly, when they reach Γ, these two bands join together in the level E_Γ, for which the small representation is doubly degenerate. Let us see what happens to these two bands when we reach Z. There are here two small representations, both one-dimensional, A' and A'', symmetrical and antisymmetrical, respectively, with respect to σ_{v2} in Fig. 3.2. We must, as explained in § 3, have here two cell functions, one p_x type (A'') and the other p_y type (A'). They must lead to two different energy levels, because they belong to different irreducible representations of the space group. They are the levels shown in Fig. 1b where, given our previous assumption that the p_x band is the lower of the two at G, we have to assume that the A'', p_x-type level, is also the lower one in energy. It is clear that it is only in this way that we can ensure that the p_x and p_y bands vary continuously from G to Z. The behaviour of these two bands along the XM direction is very easy to guess, (Fig. 1c) since at Z along this direction we must have precisely the values obtained in Fig. 1b and since we know that these two bands must touch at M, because there they span the same, two dimensional small representation.

It should be firmly stressed that the bands described here are all, in principle, above the s bands which originate from the one-dimensional A_1 (totally symmetrical) small representation of Γ (see Table 3.2). Naturally, the degeneracy shown in Fig. 1 at Γ and M disappears for these bands since the small representations are all one-dimensional.

Band trends: guesswork

Once particular forms of the cell functions are assumed, it is often both possible and useful to produce a very rough picture of the Bloch functions for particular **k** vectors which will allow us to guess the way in which the bands vary over the Brillouin zone. We shall do this for cell functions of p_x and p_y types and for the following **k** vectors:

$$\Gamma = [00]^{\#}, \qquad X = [\tfrac{1}{2}0]^{\#}. \tag{2}$$

(The coordinates of X are immediate from Fig. 1*a* when it is remembered that the sides of the Brillouin zone are $\mathbf{a}^{\#}$ and $\mathbf{b}^{\#}$.) Of course, in order to interpret Fig. 1 we would need Z, but X will do for getting the general trend of the curves, and it is easier to work with X than with Z. It is clear from eqn (**5**-6.3) that for **k** null (point Γ), the whole of the Bloch function coincides with the cell function, which is totally periodic. The two cell functions p_x and p_y are illustrated, in a highly idealized way, in Fig. 2*a*, thus depicting the Bloch functions for Γ over the whole of the crystal (of which only four primitive cells are shown). In order to depict the Bloch function for X we require two things. First, from (**5**-6.3), we know that at the origin (**r** equal to **0**) of the first cell, it coincides with the cell function. We thus fill in the first cell in Fig. 2*b*. (Remember that all this is exceedingly rough: we know that the cell functions, of course, must be deformed in changing the **k** vector, but we are ignoring this

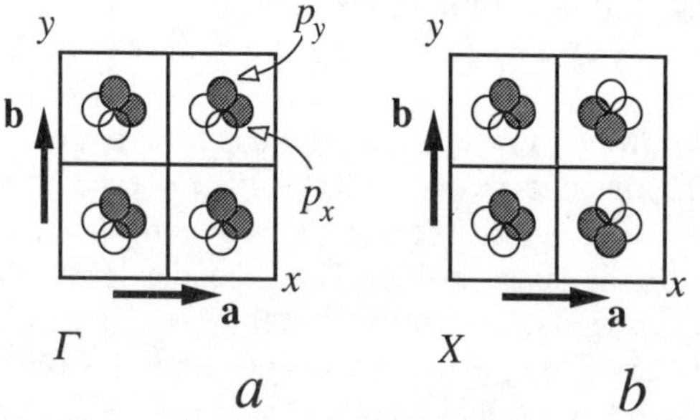

Fig. 6-4.2. Disposition of the cell functions for **k** vectors at the centre (Γ) and the edge (X) of the Brillouin zone of Fig. 3.3. This figure is in the direct lattice (crystal pattern), in which centred primitive cells are drawn.

effect, as well as the $\mathbf{r}$ dependence over the first cell.) To fill the other cells, we use eqn (1.2), in accordance to which translation by $\{E|\mathbf{t}\}$ multiplies the Bloch function $\psi_{\mathbf{k}}$ by the factor $\exp(-\,\mathrm{i}\mathbf{k}\cdot\mathbf{t})$. It is very easy to see that for $\mathbf{t}$ either [10] or [01], (translations by $\mathbf{a}$ or $\mathbf{b}$ respectively), $\mathbf{k}\cdot\mathbf{t}$ is π or 0 and thus that the exponential factor is -1 or $+1$ respectively, which allows us to fill in the remaining three cells in Fig. 2*b*.

We learn three things from Fig. 2. First, the rotation C_4^+ is a covering operation of Fig. 2*a* which takes, in the first cell, p_x into p_y. Also, C_4^+ is in the group of Γ: thus these two functions are degenerate, in agreement with Fig. 1*b*. (As shown in Problem 6.2, C_4^+ is not in the group of the vector X, which explains why this degeneracy does not exist at this point.) Secondly, if we look at the Bloch function corresponding to p_x, as artistically depicted in Fig. 2*b*, we can see that it shows no nodes, except at the atomic cores, whereas the p_y function is full of them. Since it is a useful rule of thumb that the higher the number of nodes in a wave function the higher its energy, we may conclude that the p_x band is lower in energy than the p_y band at X. Of course, from Fig. 1, it is clear that the same must be true at Z. Let us consider, thirdly, the way in which the energy changes in these bands when going from Γ to X. On comparing Figs 2*a* and 2*b*, it is pretty clear that p_x losses nodes in this change, and should therefore go down in energy, whereas the opposite is true for p_y. We have now explained all that happens in Fig. 1*b*. The explanation of Fig. 1*c* is left to the reader. (Problem 6.3.) Although our arguments here are very crude, and thus have to be taken with a salubrious pinch of salt, it is remarkable that they indeed work in a not unreasonable way when dealing with real materials. (See §**9-1** for an application to silicon.)

Properties of the $E(\mathbf{k})$ surfaces over the Brillouin zone

Two important ideas have to be learnt from the previous example. One is that the degeneracy introduced by the star is of comparatively little interest in describing the energy bands. (In other problems, however, such as phase changes, it is vital!) As an example: in Fig. 3.1*c*, on considering the star of Z, we deduced that there are four degenerate states for the $\mathbf{k}$ values Z_1, Z_2, Z_3, Z_4 so that any band $E^j(\mathbf{k})$ must have identical values at all these four points. It is simpler in practice, however, to think of $E^j(\mathbf{k})$ as having the full point group symmetry for all eight values Z_1 to Z_8 (as it is indeed the case for G in Fig. 3.1*b*), and to keep in mind, whenever necessary, that Z_5, Z_6, Z_7, Z_8 do not contribute any new states to the band but, rather, that they are identical to the states originated from Z_2, Z_1, Z_4, and Z_3, respectively. Because of this point group symmetry thus imposed on $E^j(\mathbf{k})$, it is clear that, in the example of Fig. 3.1, it is not necessary to give $E^j(\mathbf{k})$ over the whole of the Brillouin zone but rather, to allow $\mathbf{k}$ to range over the *basic domain* shown in Fig. 3.

The second and most important idea which has to be grasped is that, for a given $\mathbf{k}$, we shall never have two bands with the same energy unless there is

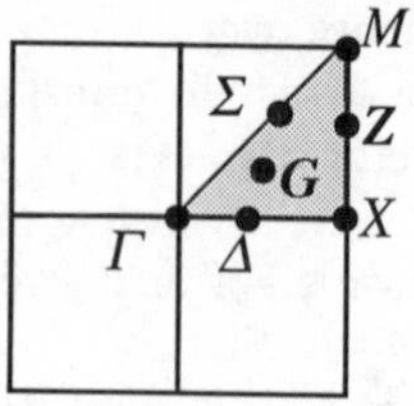

Fig. 6-4.3. Basic domain for the Brillouin zone of Fig. 3.1. Conventional labels are given to the **k** vectors that characterize the different types of stars.

a multi-dimensional small representation in the group of **k**. (This is so because it is only in this case that we can have two or more Bloch functions with the same **k** in the same irreducible representation.) Bands can never cross or touch except for **k** values for which multi-dimensional small representations exist. Naturally, for a *general* **k** vector (vector of no symmetry) such representations cannot exist and thus touching or crossing of bands at general points cannot occur.

There are two questions that remain as regards the symmetry of the $E^j(\mathbf{k})$ surfaces. We have seen that they have the symmetry given by the space group, via its point group. How does this compare with the geometrical symmetry of the Brillouin zones? The second question is whether this is all the symmetry possessed by the $E^j(\mathbf{k})$ surfaces. We shall see in the next section that, indeed, a very important symmetry of non-geometrical origin exists. But before we discuss this, it is important to answer our first question.

The symmetry of the Brillouin zone and of the reciprocal lattice: a warning

This is a subject so fraught with danger that, in case of worry, it is useful to remember one drastic statement, already advanced in §**5**-3: *the reciprocal lattice, and thus the Brillouin zone, have no geometrical symmetry of their own.* It is extremely easy to look at the square Brillouin zone of Fig. 4*b* and to 'see' that it has a binary axis C_2, two fourfold rotations, C_4^+, C_4^-, and the four reflection planes σ 'shown', that is all the operations of the point group $\mathbf{C}_{4v}$ of (3.1), which is the symmetry group of a strictly two-dimensional square. To use your eyes this way, however, is entirely wrong and extremely dangerous. It must be remembered that the reciprocal lattice has no existence of its own but, rather, that it is another manifestation of direct space. This is what we meant when we said, with reference to Fig. 3.2*b*, that it must be read as superimposed to its crystal lattice of Fig. 3.2*a*. *All the symmetry operations we ever talk about are the symmetry operations of the crystal and of nothing else.* This is most important because the Brillouin zone is constructed purely and entirely from the reciprocal vectors $\mathbf{a}^{\#}$, $\mathbf{b}^{\#}$, $\mathbf{c}^{\#}$, that is purely and entirely from **a**, **b**, **c**. This means that the contents of the primitive cell, and thus the full crystal symmetry, is ignored by the Brillouin zone, which is only concerned with translational symmetry and it is thus blind to anything else. To see what really happens, consider the crystal lattice depicted in Fig. 4*c*, which has precisely the same primitive vectors **a** and **b** as the square lattice in Fig. 4*a*. Whereas the point group of the latter, as we know, is $\mathbf{C}_{4v}$, that of the former is $\mathbf{C}_4$: because of the basis introduced in the primitive cell all reflection symmetry has been destroyed. Yet, because the patterns in Figs 4*a* and 4*c* have identical translational (Bravais) lattices, they both generate identically the same reciprocal space and thus the same Brillouin zone in Fig. 4*b*. Thus, when this is used as the Brillouin zone of the space group of Fig. 4*c*, the point group is only $\mathbf{C}_4$ and, for example, the star of G contains only G_1, G_3, G_5, and G_7. It is thus clear that this Brillouin zone, despite all appearances, *does not have any symmetry planes.*

The question we have just discussed of the symmetry of the Brillouin zone is very important for the following reason. The purpose of the Brillouin zone, for us, is that of depicting graphically the irreducible representations of the space group and thus the energy as a function over the Brillouin zone. If the energy is calculated for a crystal with the pattern given in Fig. 4*a*, σ_{v2} is a symmetry plane both for the crystal and for its Brillouin zone, whence the energy for G_3 and G_4 must be equal. This is not the case, however, if we are plotting the energy, over the same Brillouin zone, but computed for a crystal with the pattern depicted in Fig. 4*c*, for which there are no symmetry planes. Obviously, because energy eigenvalue calculations are lengthy, one wants to restrict the calculated values to some basic domain of the Brillouin zone, such as the one depicted in Fig. 3, and it should now be clear that such a domain cannot be chosen from consideration of the Brillouin zone alone, but that a careful study must be made of the space group of the corresponding crystal.

We can now safely pass on to the study of a most important non-geometrical symmetry. In a way, this will restore to reciprocal space the distinction of having some symmetry entirely its own.

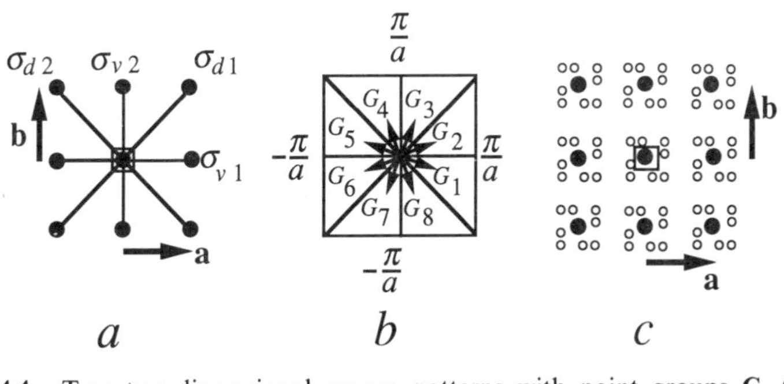

Fig. 6-4.4. Two two-dimensional square patterns with point groups $\mathbf{C}_{4v}(a)$ and $\mathbf{C}_4(c)$, respectively, with exactly the same Brillouin zone (*b*). The two small squares in *a* and *c* indicate a four-fold rotation axis.

5 Symmetry of $E(\mathbf{k})$ and $E(-\mathbf{k})$: the conjugator operator

If the space group G contains the inversion $\mathbb{i}$ then, in order to form the star of a $\mathbf{k}$ vector, $\mathbb{i}\mathbf{k}$, which equals $-\mathbf{k}$, must be constructed (see eqn 1.9). Two things may happen: either $-\mathbf{k}$ is inequivalent to $\mathbf{k}$, in which case $E(\mathbf{k})$ equals $E(-\mathbf{k})$ and $\mathbf{k}$, $-\mathbf{k}$ are independent states, or $-\mathbf{k}$ is equivalent to $\mathbf{k}$, in which case $-\mathbf{k}$, $\mathbf{k}$ correspond to identically the same state (that is, the same Bloch function). In this latter case, however, we can still think, if we do not worry too much about the redundance, that $E(\mathbf{k})$ equals $E(-\mathbf{k})$. That is, if the inversion exists, we can always take $E(-\mathbf{k})$ as equal to $E(\mathbf{k})$ and then sort out, if necessary, whether or not a new state is entailed by the corresponding Bloch function $\psi_{-\mathbf{k}}$. This means that we can always reckon $\psi_{-\mathbf{k}}$ to be either degenerate with or identical to $\psi_{\mathbf{k}}$, as long as the inversion exists. The remarkable result will now be proved, however, that this relation is always valid, whether inversion exists or not in G, as follows from the use of the conjugator operator $\mathbb{j}$ (see eqn **2**-3.11). We shall in fact prove that

$$\mathbb{j}\psi_{\mathbf{k}}(\mathbf{r}) = \psi_{-\mathbf{k}}(\mathbf{r}). \tag{1}$$

The proof is based on the fact that $\mathbf{k}$ is defined as the label of the eigenvalue by which $\psi_{\mathbf{k}}(\mathbf{r})$ is multiplied when acted upon by $\{E|\mathbf{t}\}$:

5-6.2
$$\{E|\mathbf{t}\}\psi_{\mathbf{k}}(\mathbf{r}) = \exp(-\mathrm{i}\mathbf{k}\cdot\mathbf{t})\psi_{\mathbf{k}}(\mathbf{r}). \tag{2}$$

We now operate with $\mathbb{j}$ on both sides of (2) and introduce on the left the commutation of $\mathbb{j}$ with all symmetry operations (see eqn **2**-3.17):

$$\{E|\mathbf{t}\}\,\mathbb{j}\psi_{\mathbf{k}}(\mathbf{r}) = \mathbb{j}\{\exp(-\mathrm{i}\mathbf{k}\cdot\mathbf{t})\psi_{\mathbf{k}}(\mathbf{r})\} \tag{3}$$

2-3.11|3
$$= \exp(\mathrm{i}\mathbf{k}\cdot\mathbf{t})\mathbb{j}\psi_{\mathbf{k}}(\mathbf{r}). \tag{4}$$

Comparison of (4) with (2) shows that $\mathbb{j}\psi_{\mathbf{k}}(\mathbf{r})$ belongs to $-\mathbf{k}$ rather than $\mathbf{k}$, thus verifying (1). If we now consider the Schrödinger equation,

$$\mathbb{H}\psi_{\mathbf{k}}(\mathbf{r}) = E(\mathbf{k})\psi_{\mathbf{k}}(\mathbf{r}), \tag{5}$$

it follows from §**2**-3 (see discussion of eqn **2**-3.13) that, as long as $\mathbb{H}$ is real, $\psi_{\mathbf{k}}(\mathbf{r})$ and $\mathbb{j}\psi_{\mathbf{k}}(\mathbf{r})$ must be degenerate, and therefore they must both belong to the same eigenvalue $E(\mathbf{k})$. Since $\mathbb{j}\psi_{\mathbf{k}}(\mathbf{r})$, from (1), is $\psi_{-\mathbf{k}}(\mathbf{r})$ and must therefore belong to $E(-\mathbf{k})$, it follows that

$$E(\mathbf{k}) = E(-\mathbf{k}). \tag{6}$$

The important result is that this relation will always be valid whether or not the inversion belongs to the space group, as long as the Hamiltonian be real. (Remember, however, that if $\mathbf{k}$ and $-\mathbf{k}$ happen to be equivalent, relation (6) does not entail a degeneracy since, in this case, $\psi_{\mathbf{k}}(\mathbf{r})$ and $\psi_{-\mathbf{k}}(\mathbf{r})$ are identical.) We find, therefore, that although the $\mathbf{k}$ space has no *geometrical* symmetry of its own, it has an intrinsic *inversion* symmetry for real Hamiltonians, as shown by eqn (6), which it derives from the conjugator operator. The remarkable result is that the origin of the reciprocal space is always a centre of inversion, even when this operation is not in the space group. This property is well known in the theory of X-ray diffraction.

We have now done all the important work on the irreducible representations of a space group G. We have not proved, of course, that the method given provides always irreducible representations, or that they are all the possible representations of G, but an example in Chapter **7** will show that this is indeed the case. The example will also help the reader in getting a firmer grasp on the concepts so far introduced. Before going into these niceties, however, many readers might prefer to consider in more detail the properties of the Brillouin zones and of the bands therein defined, so as to acquire a more practical familiarity with band theory. In this case they can proceed straightaway, without lack of continuity, to Chapter **8**.

6 Problems

1. Show that the functions p_x and p_y span a two-dimensional representation of $\mathbf{C}_{4v}$ which behaves like the E representation in Table 3.2.
2. Find the group of the **k** vector X, as defined in Fig. 4.1*a*.
3. Draw, in the loose style of Fig. 4.2, the Bloch functions pertaining to the **k** vector M of Fig. 4.1*a*, and give plausible arguments for the following two statements: (i) the bands p_x and p_y are degenerate at this point; (ii) in going from X to M the energy of the p_x band goes up while that of the p_y band goes down.

Further reading

The treatment given here is a highly simplified version of that of Altmann (1977). The book by Streitwolf (1971) is particularly useful as regards space group representations, and a comprehensive treatment is given by Bradley and Cracknell (1972). The fundamental concepts of the star, group of the **k** vector, and small representation were all introduced in the seminal paper by Bouckaert, Smoluchowski, and Wigner (1936), which is still one of the most readable expositions on the subject. A reprint of this paper appears in Knox and Gold (1964), where a useful discussion can also be found.

7

The representation of space groups: an example

A remarkable property of space groups was discovered in Chapters **5** and **6**. If, given a space group G, we obtain bases for all the irreducible representations of its translation subgroup T, then from these bases (which are the Bloch functions), bases of the representations of G can be constructed which have two properties: they are irreducible and they are complete (that is, they span *all* the irreducible representations of G). This result is of major importance since the bases for the irreducible representations of G in which we are interested are the energy eigenfunctions, but it is a difficult result to prove. We shall provide in this chapter an example which will allow the reader to verify that the method given in Chapters **5** and **6** actually works, and which will help to consolidate some important concepts such as those of stars, small representations, degeneracies and bands. Except insofar as it might be required in order to build up experience in these topics, this chapter will not be necessary for the rest of the book, so that the reader who feels already sufficiently confident about the methods so far developed can proceed straightaway to Chapter **8** and return to the present chapter at any later time.

1 The space group model

We want to consider as an example of a space group one which is sufficiently simple for the final answer to be known beforehand. We shall thus be able to compare this answer with that obtained by forming bases constructed from those of the translation group. The model which we shall consider is the linear chain of six atoms with periodic boundary conditions illustrated in Fig. **5**-1.1, a representative part of which is shown in Fig. 1. As it follows from § **5**-1, the translation subgroup T consists of the following six operations:

$$T\colon \{E|m\mathbf{a}\}, \qquad m = 0, 1, \ldots, 5; \qquad \{E|6\mathbf{a}\} = \{E|\mathbf{0}\}. \tag{1}$$

We shall assume strict one-dimensionality, so that the point group is

$$\mathbf{C}_i\colon \; E, \mathbb{i}, \tag{2}$$

where $\mathbb{i}$ is the inversion at the point labelled 1 in the figure, that is, the

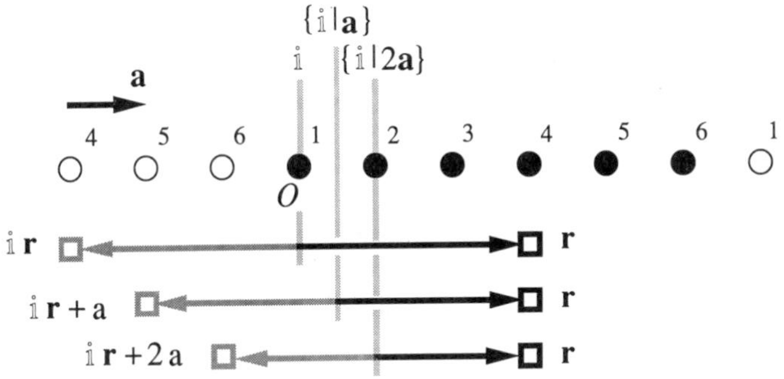

Fig. 7-1.1. The linear chain and the operations $\bar{\imath}$, $\{\bar{\imath}|a\}$, $\{\bar{\imath}|2\mathbf{a}\}$. The three lines below the linear chain should all be read as superimposed on it; they give the effect of the above-mentioned operations on a vector **r** in the linear chain.

operation which leaves the first centred primitive cell invariant. Since the space group operations which contain E already appear in (1), the remaining operations of the space group G are:

$$\{\bar{\imath}|m\mathbf{a}\}, \qquad m = 0, 1, \ldots, 5. \tag{3}$$

We shall identify the first two of these operations:

3-6.2 $$\{\bar{\imath}|\mathbf{a}\}\mathbf{r} = \bar{\imath}\mathbf{r} + \mathbf{a}, \tag{4}$$

3-6.2 $$\{\bar{\imath}|2\mathbf{a}\}\mathbf{r} = \bar{\imath}\mathbf{r} + 2\mathbf{a}, \tag{5}$$

where **r** is an arbitrary point in the chain. The results given in (R4) and (R5) are formed in Fig. 1 and it follows at once that $\{\bar{\imath}|\mathbf{a}\}$ and $(\bar{\imath}|2\mathbf{a}\}$ are the inversions at the mid-point of the 1-2 bond and at the point 2, respectively. It can be proved in the same manner (see Problem 4.1) that $\{\bar{\imath}|3\mathbf{a}\}$ and $\{\bar{\imath}|5\mathbf{a}\}$ are the inversions at the mid-points of the bonds 2-3 and 3-4, respectively, whereas $\{\bar{\imath}|4\mathbf{a}\}$ and $\{\bar{\imath}|6\mathbf{a}\}$ are the inversions at the points 3 and 4, respectively. The reader may wonder what happens to the inversion at the point 5, say, not so far listed. In fact, it takes the vector labelled **r** in the figure into the point labelled with a black circle at 6. This point, however, is identical under the periodicity condition to the point labelled with an open circle 6. Since **r** goes into this point under $\{\bar{\imath}|2\mathbf{a}\}$, the inversion at 5 coincides with this operation and thus need not be listed.

The twelve operations of G listed in (1) and (3) can easily be re-identified when we remember that the cyclic boundary conditions are equivalent to bending the basic period (that is the length of the chain from 1 to 6 shown in full circles in Fig. 1) into a circle, as shown in Fig. **1**-2.1 and done in Fig. 2. The six translations in (1) become the six positive rotations by $2\pi/6, 2\pi/3, \ldots, 2\pi$ in Fig. 2. We mean by this statement that these rotations effect the same permutations of the labels 1 to 6 as the translations do in

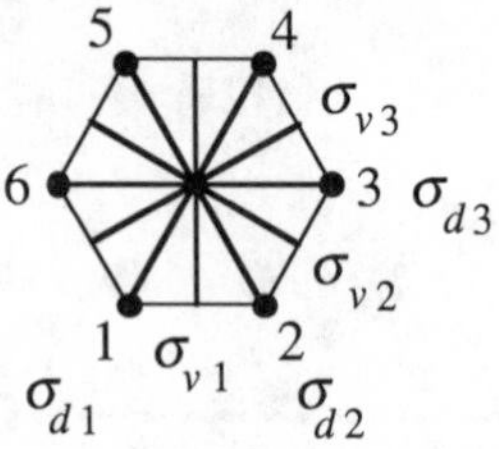

Fig. 7-1.2. The linear chain of Fig. 1 in cyclic form.

Table 7-1.1. The classes of G.

Class of $\mathbf{C}_{6v}$	Class of G
E	$\{E\|\mathbf{0}\}$
$2C_6$	$\{E\|\mathbf{a}\}, \{E\|5\mathbf{a}\}$
$2C_3$	$\{E\|2\mathbf{a}\}, \{E\|4\mathbf{a}\}$
C_2	$\{E\|3\mathbf{a}\}$
$3\sigma_d$	$\{i\|\mathbf{0}\}, \{i\|2\mathbf{a}\}, \{i\|4\mathbf{a}\}$
$3\sigma_v$	$\{i\|\mathbf{a}\}, \{i\|3\mathbf{a}\}, \{i\|5\mathbf{a}\}$

Fig. 1. Likewise, $\{i|\mathbf{0}\}$ in Fig. 1 changes 2 into 6, 3 into 5, and so on, precisely as the reflection σ_{d1} does in Fig. 2. It is thus very easy to establish the correspondence between the operations in (1) and (3) and those in Fig. 2, which we do in the following lists:

$$\{E|\mathbf{0}\}, \quad \{E|\mathbf{a}\}, \quad \{E|2\mathbf{a}\}, \quad \{E|3\mathbf{a}\}, \quad \{E|4\mathbf{a}\}, \quad \{E|5\mathbf{a}\}. \qquad (6a)$$

$$E \quad C_6^+ \quad C_3^+ \quad C_2 \quad C_3^- \quad C_6^- \qquad (6b)$$

$$\{i|\mathbf{0}\}, \quad \{i|\mathbf{a}\}, \quad \{i|2\mathbf{a}\}, \quad \{i|3\mathbf{a}\}, \quad \{i|4\mathbf{a}\}, \quad \{i|5\mathbf{a}\}. \qquad (7a)$$

$$\sigma_{d1} \quad \sigma_{v1} \quad \sigma_{d2} \quad \sigma_{v2} \quad \sigma_{d3} \quad \sigma_{v3} \qquad (7b)$$

(Notice how obvious it now is from the figure that the inversions at the points 5 and 2 are the same operation, as it was argued before.)

The correspondence displayed above means that the group G of the twelve operations shown in (6*a*) and (7*a*) is for all uses and purposes the same as the group of the twelve operations listed in (6*b*) and (7*b*). (In group-theoretical language one says that the two groups are *isomorphic*.) This latter group, as illustrated in Fig. 2, is the same (with a slight change of notation) as the group $\mathbf{C}_{6v}$ in Fig. **2**-6.1. Thus, as long as we use the correspondences established in (6) and (7), the irreducible representations of G are those of $\mathbf{C}_{6v}$ listed in Table **2**-6.1. It is useful to compare the classes of $\mathbf{C}_{6v}$ given in that table with

(6) and (7) in order to obtain the classes of G listed in Table 1. (See also Problem 4.2.)

What we now want to do is to try and construct the irreducible bases of the representations of G in terms of the irreducible bases of the representations of its subgroup T of (1). We shall then be able to compare with the correct results of Table **2**-6.1 to see whether the method works. The first thing we have to do, therefore, is to obtain the irreducible representations of T.

2 The irreducible representations of *T*

We had

5-2.9 $$_{\mathbf{k}}\hat{T}\{E|\mathbf{t}\} = \exp(-\,\mathrm{i}\mathbf{k}\cdot\mathbf{t}). \tag{1}$$

so that, in one dimension, for $\mathbf{k}$ equal to $k\mathbf{a}^{\#}$ and $\mathbf{t}$ equal to $m\mathbf{a}$,

4-2.6|1 $$_{\mathbf{k}}\hat{T}\{E|m\mathbf{a}\} = \exp(-\,2\pi\mathrm{i}km), \qquad m = 0, 1, \ldots, 5. \tag{2}$$

We also know that k is quantized and, with a slight change of notation, it is given by (**5**-4.3) as an integer κ divided by N, the number of primitive cells, 6 in our case. The integer κ goes from 1 to 6, but since six can be added to or subtracted from any of these values, we take it in the range from 0 to 5:

$$k = \kappa/6, \qquad \kappa = 0, 1, \ldots, 5. \tag{3}$$

On labelling the representations with the more convenient index κ rather than k, we can now rewrite (2):

$$_{\kappa}\hat{T}\{E|m\mathbf{a}\} = \exp(-\,2\pi\mathrm{i}\kappa m/6), \qquad m = 0, 1, \ldots, 5. \tag{4}$$

Let us call

$$\omega = \exp(-\,2\pi\mathrm{i}/6), \tag{5}$$

so that

4 $$_{\kappa}\hat{T}\{E|m\mathbf{a}\} = (\omega^{\kappa})^{m}, \qquad \kappa = 0, 1, \ldots, 5; \qquad m = 0, 1, \ldots, 5. \tag{6}$$

The bracket around ω^{κ} here is entirely irrelevant, but it helps us to recognize that we must first of all identify the first six powers of ω, which we do in the Argand diagram of Fig. 1. These powers of ω are plotted at once and, inside the circle shown, they are identified in terms of ω, ω^*, and the unity. The irreducible representations are formed in Table 1 from eqn (6). The second column of the table identifies ω^{κ} from Fig. 1 and each representation is obtained by raising ω^{κ} to the powers from 0 to 5. This is most simply done from Fig. 1, by multiplying the argument of ω^{κ} by $0, 1, \ldots, 5$. The last column of the table merely establishes an obvious notation for the bases of the representations.

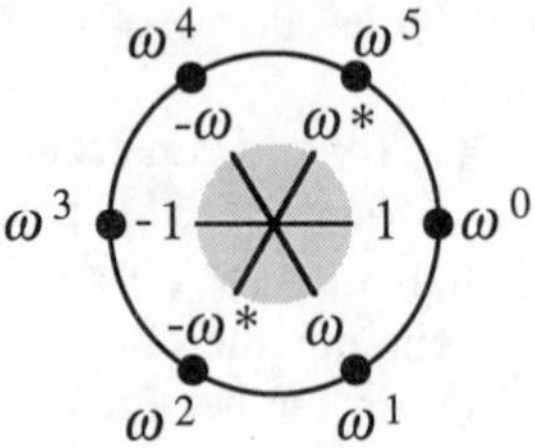

Fig. 7-2.1. The powers ω^κ for κ from 0 to 5 plotted in an Argand diagram.
$\omega = \exp(-2\pi i/6)$, $\omega + \omega^* = 1$.

Table 7-2.1. The irreducible representations ${}_\kappa\hat{T}\{E|m\mathbf{a}\}$.
$\kappa, m = 0, 1, 2, 3, 4, 5$. $\omega = \exp(-2\pi i/6)$.

		m						bases
κ	ω^κ	0	1	2	3	4	5	ψ^κ
0	1	1	1	1	1	1	1	ψ_0
1	ω	1	ω	$-\omega^*$	-1	$-\omega$	ω^*	ψ_1
2	$-\omega^*$	1	$-\omega^*$	$-\omega$	1	$-\omega^*$	$-\omega$	ψ_2
3	-1	1	-1	1	-1	1	-1	ψ_3
4	$-\omega$	1	$-\omega$	$-\omega^*$	1	$-\omega$	$-\omega^*$	ψ_4
5	ω^*	1	ω^*	$-\omega$	-1	$-\omega^*$	ω	ψ_5

3 The irreducible representations of *G*

The functions ψ_0 to ψ_5 of Table 2.1 are the Bloch functions, that is the eigenfunctions of the translations:

5-6.2
$$\{E|\mathbf{t}\}\,\psi_{\mathbf{k}}(\mathbf{r}) = \exp(-\,i\mathbf{k}\cdot\mathbf{t})\,\psi_{\mathbf{k}}(\mathbf{r}). \tag{1}$$

In fact, when an operation $\{E|m\mathbf{a}\}$ acts on ψ_κ, it multiplies it by the factor listed in the column corresponding to m in Table 2.1.

In going now from T to G, we know that, because $\{\mathbf{i}|\mathbf{0}\}$ is a symmetry operation of G, ψ_κ and $\{\mathbf{i}|\mathbf{0}\}\,\psi_\kappa$ must be degenerate and must therefore belong to the same irreducible basis of G. We can easily obtain $\{\mathbf{i}|\mathbf{0}\}\,\psi_\kappa$:

6-1.9
$$\{\mathbf{i}|\mathbf{0}\}\,\psi_\kappa = \psi_{-\kappa} = \psi_{-\kappa+6}, \tag{2}$$

(see the remark just above eqn 2.3). The result of (2) for all the bases ψ_κ of T is given in the third column of Table 1. The second column gives the trivial action of E on the same functions, and from these two columns the bases for

Table 7-3.1. The bases of the representations of G.
ψ_κ are the bases of T. E and $\mathbb{i}$ are written for $\{E|\mathbf{t}\}$ and $\{\mathbb{i}|\mathbf{0}\}$ respectively.

ψ_κ	$E\psi_\kappa$	$\mathbb{i}\psi_\kappa$	bases	group of **k**	irred. bases
ψ_0	ψ_0	ψ_0	$\langle\psi_0|$	$\mathbf{C}_i$	$\langle\psi_0^g|\langle\psi_0^u|$
ψ_1	ψ_1	ψ_5	$\langle\psi_1\psi_5|$	$\mathbf{C}_1$	$\langle\psi_1\psi_5|$
ψ_2	ψ_2	$\psi_{-2}=\psi_4$	$\langle\psi_2\psi_4|$	$\mathbf{C}_1$	$\langle\psi_2\psi_4|$
ψ_3	ψ_3	$\psi_{-3}=\psi_3$	$\langle\psi_3|$	$\mathbf{C}_i$	$\langle\psi_3^g|\langle\psi_3^u|$
ψ_4	ψ_4	$\psi_{-4}=\psi_2$	$\langle\psi_4\psi_2|$		
ψ_5	ψ_5	$\psi_{-5}=\psi_1$	$\langle\psi_5\psi_1|$		

the representations of G are formed in the fourth column of the table. It is clear that the last two rows of this column merely repeat the bases generated in the third and second rows, respectively, so that the bases which we must expect for the irreducible representations of G must be obtained from those listed in the first four rows of the fourth column of the table. We list in the fifth column of the table the group of the **k** vector, given by the operations in each row which leave ψ_κ invariant.

Before we go any further in analyzing the rest of Table 1, it will be useful to compare what we have done so far with our discussion in Chapter **6** on stars, groups of the **k** vector, and small representations. We must first draw the Brillouin zone for our linear chain, which goes in **k** from $-\frac{1}{2}\mathbf{a}^\#$ to $\frac{1}{2}\mathbf{a}^\#$. We had written

,2.3 $$\mathbf{k} = k\mathbf{a}^\# = \tfrac{1}{6}\kappa\mathbf{a}^\#, \qquad \kappa = 0, 1, \ldots, 5, \tag{3}$$

but, since six can be added to or subtracted from κ we take for its range the six values $-2, -1, 0, 1, 2, 3$, which coincides with the *open* range in **k** for the stated Brillouin zone. (Remember that $-\frac{1}{2}\mathbf{a}^\#$ does not belong to the zone, being equivalent to $\frac{1}{2}\mathbf{a}^\#$.) This Brillouin zone is displayed in Fig. 1 and it can immediately be seen that, by acting on the various **k** vectors with the operations of the point group, E and $\mathbb{i}$, four stars are formed. Since the κ pairs $1, -1$ and $2, -2$ are equivalent (on addition of six to the second member of each pair) to 1, 5 and 2, 4, respectively, we can see that the four stars in Fig. 1 correspond precisely to the four distinct bases obtained in the fourth column of Table 1.

The next thing we notice from the figure is that the **k** vectors corresponding to κ equal to 0 and 3 are invariant under the two operations of $\mathbf{C}_i$, which it is thus the group of the **k** vector, as already derived in Table 1. It follows from the treatment given in § **6**-3 that, in order to generate the final irreducible bases for G, the four independent bases in the fourth column of Table 1 have to be symmetrized with respect to the small representations, that is, the irreducible representations of the corresponding group of the **k** vector. Since

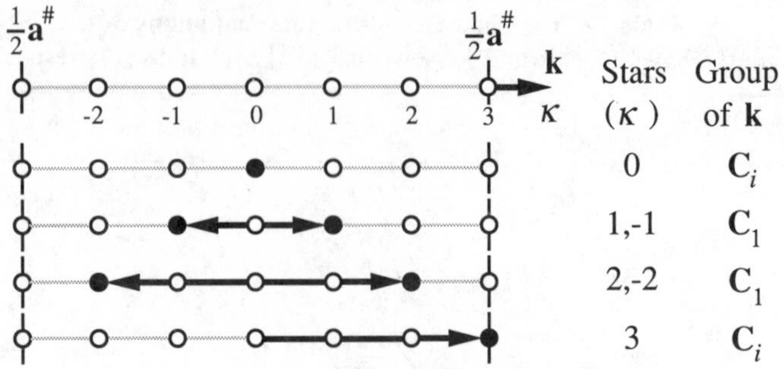

Fig. 7-3.1. The Brillouin zone, shown at the top of the picture, and the stars for the linear chain of Fig. 1.1.

Table 7-3.2. The irreducible representations of $\mathbf{C}_i$.

$\mathbf{C}_i$	E	$\bar{i}$	bases
A_g	1	1	s
A_u	1	-1	p

the latter is the trivial group $\mathbf{C}_1$ for $\langle\psi_1\psi_5|$ and $\langle\psi_2\psi_4|$, these bases require no further symmetrization, whereas $\langle\psi_0|$ and $\langle\psi_3|$ have to be symmetrized with respect to the irreducible representations of $\mathbf{C}_i$. These are displayed in Table 2. Conventionally, A_g and A_u are symmetric (*gerade*) and antisymmetric (*ungerade*) representations, respectively, with respect to the inversion and they are spanned by *s*-type or *p*-type functions. Thus, if we choose *s*-type or *p*-type cell functions we shall have the irreducible bases $\langle\psi_0^g|$ or $\langle\psi_0^u|$, respectively, and similarly for $\langle\psi_3|$, as shown in the last column of Table 1. This means that we shall have six representations, four one-dimensional and two two-dimensional.

It is very easy, with the information so far available, to obtain the predicted irreducible representations, which are listed in Table 3, the construction of which will be discussed with a few examples.

Consider the basis $\langle\psi_3^u|$. From Table 2.1 it is multiplied by -1 under $\{E|\mathbf{a}\}$, as shown in Table 3. In order to consider the action of operations of the form $\{\bar{i}|m\mathbf{a}\}$, these operations must first be written as follows:

3-6.6 $$\{\bar{i}|m\mathbf{a}\} = \{E|m\mathbf{a}\}\{\bar{i}|\mathbf{0}\}. \qquad (4)$$

Consider m equals 1. From Table 2.1 the first term on (R4) gives a factor of -1. Because the basis is ungerade we know that the second term on (R4) will also entail a factor -1. Thus, the corresponding representative (character,

Table 7-3.3. Characters of the irreducible representations of G.

G $\mathbf{C}_{6v}$	$\{E\|\mathbf{0}\}$ E	$2\{E\|\mathbf{a}\}$ $2C_6$	$2\{E\|2\mathbf{a}\}$ $2C_3$	$\{E\|3\mathbf{a}\}$ C_2	$3\{\mathbb{i}\|\mathbf{0}\}$ $3\sigma_d$	$3\{\mathbb{i}\|\mathbf{a}\}$ $3\sigma_v$	bases
A_1	1	1	1	1	1	1	$\langle\psi_0^g\|$
A_2	1	1	1	1	-1	-1	$\langle\psi_0^u\|$
B_1	1	-1	1	-1	1	-1	$\langle\psi_3^g\|$
B_2	1	-1	1	-1	-1	1	$\langle\psi_3^u\|$
E_1	2	1	-1	-2	0	0	$\langle\psi_1\psi_5\|$
E_2	2	-1	-1	2	0	0	$\langle\psi_2\psi_4\|$

because the representation is one-dimensional) is $+1$, as shown in the table. In the same manner all the other characters for this basis are obtained and the same applies to the other one-dimensional representations. Consider now the basis $\langle\psi_1\psi_5|$. From Table 2.1,

$$\{E|\mathbf{a}\}\langle\psi_1\psi_5| = \langle\omega\psi_1, \omega^*\psi_5| = \langle\psi_1\psi_5|\begin{bmatrix}\omega & \\ & \omega^*\end{bmatrix}. \tag{5}$$

Because here the sum of the diagonal element of the matrix is unity (see Fig. 2.1) the character is $+1$ as given in the table. For $\{\mathbb{i}|\mathbf{a}\}$, say, we use eqn (4) with the matrix in (5) multiplied by the matrix for $\{\mathbb{i}|\mathbf{0}\}$. From Table 3.1:

$$\{\mathbb{i}|\mathbf{0}\}\langle\psi_1\psi_5| = \langle\psi_5\psi_1| = \langle\psi_1\psi_5|\begin{bmatrix} & 1\\ 1 & \end{bmatrix}, \tag{6}$$

which, after the stated multiplication, gives zero for the character. It is very easy to verify that each of the representations listed in Table 3 is irreducible and that they form a complete set of representations, their number being correctly the number of classes of the group. Thus no other irreducible representations are possible. Of course, the table also coincides with the well-known table for the group $\mathbf{C}_{6v}$ quoted in Table **2**-6.1. All this amply verifies that the methods of Chapters **5** and **6** do deliver the goods.

The bands

Let us now consider the bands which arise from these results, for which purpose we shall guess, in the manner of § **6**-4, the energy relations between the functions $\langle\psi_0^g|$ and $\langle\psi_0^u|$, at the centre of the Brillouin zone, and those at its edge, $\langle\psi_3^g|$ and $\langle\psi_3^u|$. It follows from Table 2, that the cell functions for the g and u bases must be respectively symmetric (s-type) and antisymmetric (p-type) under the inversion. We can now draw a picture like Fig. **6**-4.2 (but in one dimension) and observe that $\langle\psi_0^g|$ and $\langle\psi_0^u|$ entail no node and one node per unit cell respectively (see Problem 4.5). This shows that at the centre of the Brillouin zone there are two bands, the gerade one (*s*-type) being the lowest and the ungerade one (*p*-type) being the highest in energy. It is easy to see that in going from $\langle\psi_0^g|$ to $\langle\psi_3^g|$ the number of nodes increases from 0 to 1 per unit cell, which indicates that the lowest band, which we shall call the *s* band, goes up in energy towards the edge of the Brillouin zone. In the same manner, the upper ungerade or *p* band must go down in energy from the centre towards the edge of the zone (see Problem 4.5). It is sensible to label the first and second bands with the superscripts *s* and *p* respectively: they will contain each four energy levels, two singly and two doubly degenerate:

$$\langle\psi_0^s|,\ \langle\psi_1^s\psi_{-1}^s|,\ \langle\psi_2^s\psi_{-2}^s|,\ \langle\psi_3^s|, \tag{7}$$

$$\langle\psi_0^p|,\ \langle\psi_1^p\psi_{-1}^p|,\ \langle\psi_2^p\psi_{-2}^p|,\ \langle\psi_3^p|. \tag{8}$$

It must be strongly emphasized that whereas $\langle\psi_0^s|$ and $\langle\psi_0^p|$ belong to different irreducible representations of G, $\langle\psi_1^s\psi_{-1}^s|$ and $\langle\psi_1^p\psi_{-1}^p|$ belong identically to the same representation. The inversion does not belong to the group of this **k** vector and symmetrization with respect to it is meaningless. (See Problem 4.4.) Another way to look at this is to recognize that, although $\{\mathrm{i}|\mathbf{0}\}$ belongs to G, this inversion symmetry is broken when a wave with **k** vector corresponding to κ equal to 1 propagates along the chain, and thus loses meaning, whereas this is not so for the **k** vectors corresponding to κ equal to 0 or 3. (See Fig. 3.1.) It must be appreciated thus that the difference between $\langle\psi_1^s\psi_{-1}^s|$ and $\langle\psi_1^p\psi_{-1}^p|$ is given not by the symmetry of their cell functions but rather by the fact that the cell functions of these bases are 'continuously' related to the cell functions of $\langle\psi_0^s|$ and $\langle\psi_0^p|$, respectively. (We write 'continuously' here in quotes because the **k** space of our model is very discrete indeed but in a real metal there are so many values of **k** in the Brillouin zone that continuity makes sense.)

4 Problems

1. Prove from Fig. 1.1 that $\{\mathrm{i}|3\mathbf{a}\}$ and $\{\mathrm{i}|5\mathbf{a}\}$ are the inversions at the mid-points of the bonds 2-3 and 3-4, respectively, whereas $\{\mathrm{i}|4\mathbf{a}\}$ is the inversion at the point 3.
2. Prove by conjugation that the operations $\{E|\mathbf{a}\}$, and $\{E|5\mathbf{a}\}$ defined in § 1 belong to the same class.
3. Use the method of eqns (3.5) and (3.6) in order to obtain the matrix of one operation in each class of G for the representation spanned by $\langle\psi_1\psi_5|$ and compare with the result in Table 3.3.
4. Consider as in Problem 3 two bases $\langle\psi_1^g\psi_5^g|$ and $\langle\psi_1^u\psi_5^u|$ the functions of which are respectively gerade and ungerade with respect to the inversion and show that the corresponding representations have the same characters and are therefore equivalent.

5. On taking s-type and p-type cell functions (in the latter case with axis oriented along the linear chain of Fig. 1.1) for the functions $\langle\psi_0^s|$, $\langle\psi_3^s|$, $\langle\psi_0^p|$, and $\langle\psi_3^p|$, show that their corresponding number of nodes per unit cell are 0, 1, 2, and 1, respectively. Hence draw plausible pictures for the bands (3.7) and (3.8) in the Brillouin zone of Fig. 3.1. Explain why these two bands cannot touch or cross at any point of the Brillouin zone.
6. Compare the first band obtained for the model in this chapter, eqn (3.7) with the energy levels for benzene given in Fig. **2**-6.2. (Note that in the strictly two-dimensional model used in this chapter, the conventional p_z orbitals of the carbon atoms in benzene are gerade with respect to the inversion, although, of course, this is not so in three dimensions.)

Further reading

A detailed treatment of the general linear chain may be found in Altmann (1977).

8

Brillouin zones and energy bands

We shall start this chapter by considering a three-dimensional Brillouin zone, which will serve us as an example in order to study the properties of the Brillouin zone faces. After this, we shall go back to a one-dimensional example so as to be able to discuss in detail various schemes which are used in order to plot energy bands and Fermi surfaces in **k** space.

1 The Brillouin zone for the face-centred cubic lattice

The reciprocal lattice of a face-centred cubic lattice was constructed in Fig. **4**-7.1, where it was shown to be a body-centred cubic lattice. Part of this lattice is shown in Fig. 1 where we give the lattice point at the origin (labelled Γ), its eight first neighbours (labelled 1) at the corners of a cube centred at the origin, and a shell of six second neighbours (labelled 2), of which only three are shown in the figure. The prescription required for the construction of the first centred primitive cell in this lattice, or Brillouin zone, is as follows. (See § **3**-2.) We must consider all *lattice directions* through the origin, that is, all the lines which join the origin to first, second, third neighbours and so on. Then, along each such direction, we must construct the normal bisector planes of the segment which joins the origin with the *first* lattice point along the direction chosen. The smallest polyhedron which is bounded by all such normal bisector planes is the Brillouin zone. It turns out in the case of the reciprocal lattice illustrated in Fig. 1 that the Brillouin zone is bounded by bisectors to first and second neighbours only. Consider the six second neighbours first: clearly, the bisector planes concerned are the faces of the cube shown in the figure. The corresponding lattice directions are $[100]^{\#}$, $[010]^{\#}$, and $[001]^{\#}$, which can be seen in Fig. **4**-7.1 (look at the direct lattice please!), to be four-fold axes. Therefore, the six Brillouin zone faces on these bisector planes must be polygons compatible with this symmetry, that is polygons of $4n$ sides, for any integer n. By the same argument, the faces on the eight bisector planes of the $\Gamma 1$ directions must be polygons of $3m$ sides (m integral), because the $[111]^{\#}$ direction, for example, is a three-fold axis. It can be seen that, in order to form a polyhedron with these two different types of polygonal faces, n and m must be 1 and 2 respectively, leading to the six square and the eight hexagonal faces shown in the figure.

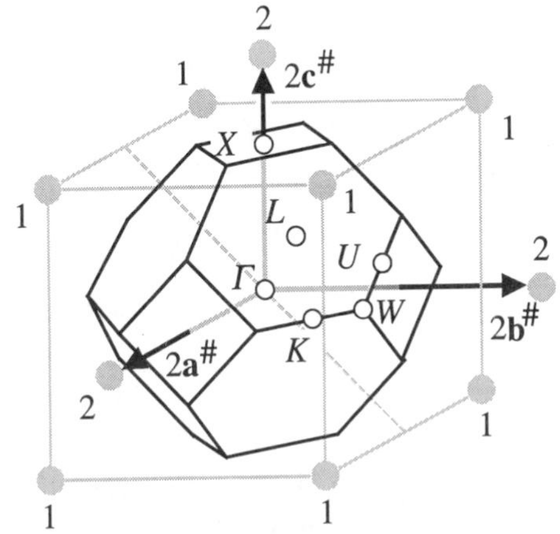

Fig. 8-1.1. The Brillouin zone for the face-centred cubic lattice. The grey circles indicate the reciprocal lattice (a body-centred lattice). The small open circles designate some important **k** vectors in a conventional but standard notation. The broken line joining the mid-points of two opposite edges of the cube is a binary axis which can be seen to be also a binary axis of the direct lattice. The vectors $\mathbf{a}^{\#}$, $\mathbf{b}^{\#}$, $\mathbf{c}^{\#}$ here are the same as in Fig. 4-7.1*b*, except that they are shown in a different orientation.

2 Properties of the Brillouin zone faces

As always, we consider **k** vectors with their tails at the origin of the reciprocal space, that is, at the centre of the Brillouin zone, and we shall be interested in such of those vectors as end on a Brillouin zone face and thus characterize these faces. We have two different conditions which involve these vectors, namely conditions which must be satisfied by the vectors themselves and conditions which obtain for the derivative of $E(\mathbf{k})$ with respect to **k**, when calculated at the tip of the vector, that is at the point at which the vector touches the Brillouin zone face.

Conditions on the k vectors

It is important to realize that Brillouin zone faces must always come in parallel pairs, separated by a vector **g** of the reciprocal lattice, as illustrated in Fig. 1. Consider in fact the direction from the origin Γ to some neighbour n, (this integer indicating the order of the neighbour) separated from Γ by a vector **g** of the reciprocal lattice. Because of translational symmetry, another neighbour n' must exist along this direction, separated from Γ by $-\mathbf{g}$ (see Fig. 1). Thus, the two normal bisectors (to which the Brillouin zone faces must belong) are parallel and separated by **g**, as shown in the figure. (The region between the two planes so constructed forms a slab of translationally

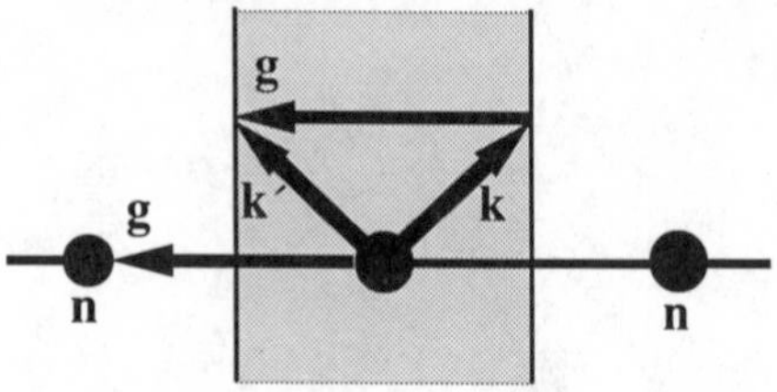

Fig. 8-2.1. Condition for **k** vectors that end on the Brillouin zone surfaces.

inequivalent points, as defined in Fig. **3**-2.1, and the intersection of the various slabs corresponding to the different lattice directions determines the set of translationally inequivalent points which is the primitive cell.)

Because of this property, for every **k** vector which ends on the surface of the Brillouin zone, another vector $\mathbf{k}'$ must exist of the same length, and which differs from **k** by a vector of the reciprocal lattice:

$$|\mathbf{k}'| = |\mathbf{k}|, \qquad \mathbf{k}' = \mathbf{k} + \mathbf{g}, \tag{1}$$

as shown in Fig. 1.

Conditions (1) are the so-called *von Laue conditions* which must be satisfied by the **k** vector of a plane wave which is reflected by a crystal. Consider, in fact, in Fig. 1, a plane wave with vector $\alpha\mathbf{k}$, for some $\alpha < 1$. Nothing happens to this plane wave but, if α is increased until it equals unity, then the **k** vector of the wave must change into $\mathbf{k}'$, as shown in Fig. 1. Why should this be so? We can reason as follows. First, imagine in Fig. 1 that α were increased infinitesimally over 1. Then the **k** vector of the wave would go out of the first primitive cell, and would have to be taken back to it, by addition with **g**, thus giving a vector infinitesimally near $\mathbf{k}'$. The same must be the case when α is precisely unity, because we know that the primitive cell must be an open polyhedron, so that of every pair of faces only one belongs to the cell. As usual, the face along the negative direction is eliminated from the polyhedron, so that **k** must be replaced by $\mathbf{k}'$. (Notice in Fig. 1 that we have taken **g** to be positive and that therefore the positive direction is to the left.) It is pretty clear that a reflection has taken place: consider the momenta **p** of the waves involved, which are $\hbar\mathbf{k}$ and $\hbar\mathbf{k}'$ respectively. It follows at once from the figure that the component of $\hbar\mathbf{k}'$ normal to the Brillouin zone face in Fig. 1 is equal and opposite to the same component of the original wave $\hbar\mathbf{k}$, which entails a reflection. This reflection is usually called a *Bragg reflection* and a more physical description of it will be discussed in § 3 and § **11**-3.

Condition for $\partial E/\partial \mathbf{k}_n$

The energy E is a function $E(\mathbf{k})$ of the **k** vector and its general derivative must thus be written as $\partial E/\partial \mathbf{k}$. We mean by $\partial E/\partial \mathbf{k}_n$ such a derivative when **k** is

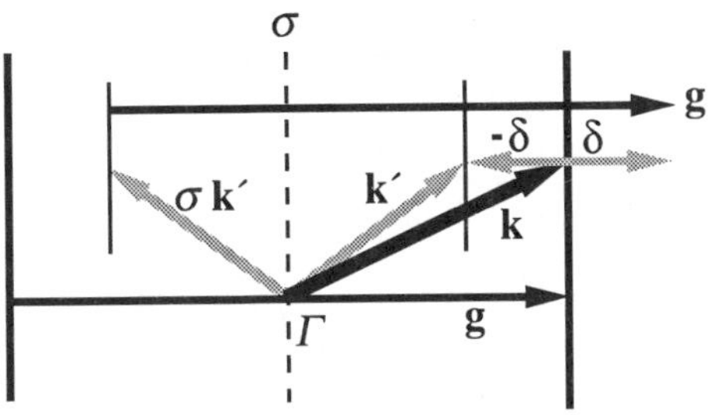

Fig. 8-2.2. Conditions for $\partial E/\partial \mathbf{k_n}$ on a Brillouin zone face. The external vertical lines represent the traces of two Brillouin zone faces separated by a vector $\mathbf{g}$ of the reciprocal lattice.

allowed to vary only in the direction $\mathbf{n}$ normal to a Brillouin zone face. We shall consider here a particular case in which there is a symmetry plane σ *in the crystal*, through the origin and parallel to the Brillouin zone face. (See Fig. 2.) Denote with $\sigma\mathbf{k}'$ the result of reflecting any vector $\mathbf{k}'$ through σ, as illustrated in the figure. Then, because σ is a symmetry operation of the crystal, the energy must be symmetrical with respect to it:

$$E(\mathbf{k}') = E(\sigma\mathbf{k}'). \tag{2}$$

For any arbitrary vector $\mathbf{k}''$, on the other hand, the energy must always be periodic in $\mathbf{k}$ space:

6-4.1
$$E(\mathbf{k}'') = E(\mathbf{k}'' + \mathbf{g}). \tag{3}$$

We can now compute the desired derivative, which is defined as follows, for an infinitesimal vector $\boldsymbol{\delta}$ parallel to $\mathbf{n}$ (see Fig. 2),

$$\frac{\partial E}{\partial \mathbf{k_n}} = \lim_{\boldsymbol{\delta}\to 0} \frac{E(\mathbf{k}+\boldsymbol{\delta}) - E(\mathbf{k}-\boldsymbol{\delta})}{2\boldsymbol{\delta}}. \tag{4}$$

From Fig. 2,

,2
$$E(\mathbf{k}-\boldsymbol{\delta}) = E(\mathbf{k}') = E(\sigma\mathbf{k}') \tag{5}$$

3
$$= E(\sigma\mathbf{k}' + \mathbf{g}) \tag{6}$$

F2
$$= E(\mathbf{k}+\boldsymbol{\delta}). \tag{7}$$

It follows from (7) that the fraction on (R4) must always vanish, whence we can have only two possible cases. The first is that the derivative exists, in which case it must vanish:

$$\frac{\partial E}{\partial \mathbf{k_n}} = 0. \tag{8}$$

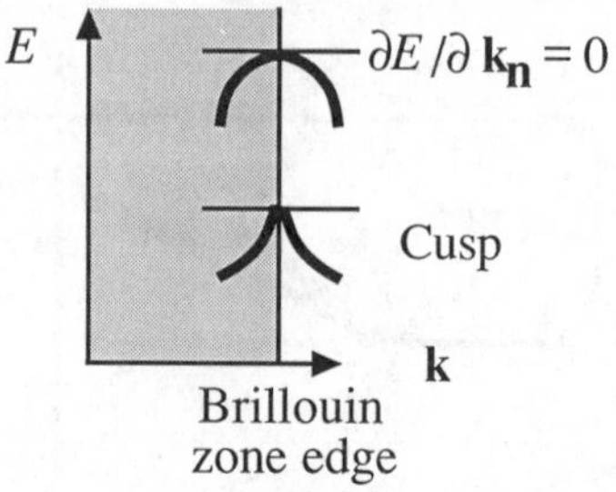

Fig. 8-2.3. Conditions for $\partial E/\partial \mathbf{k_n}$ on a Brillouin zone face. Cusps.

The second possibility is that the derivative does not exist (which means that the slopes on either side of the Brillouin zone face do not tend to the same limit), in which case there is a *cusp*, as illustrated in Fig. 3. (Notice that, from eqn 7, the energy must be symmetrical with respect to the Brillouin zone face, as shown in the figure.) Cusps, however, cannot be expected to appear (except trivially; see Problem 8.5) in the case of the electron bands which we are discussing for reasons which will be given in § 3. It should be noticed, however, that they do appear quite normally in phonon bands. (See the end of § **11**-4.)

It should be remembered that condition (8), or its alternative cusp condition, obtains for *all* points of a Brillouin zone face if and only if the symmetry plane σ exists. Of course, if continuity in the derivative is required, a cusp cannot exist and (8) will be the only valid condition. On this assumption, it is worth knowing that, even if σ does not exist, (8) is still valid, but only for some specific points of the Brillouin zone face and subject, possibly, to certain conditions. First, it is unconditionally valid for the centre of the Brillouin zone face. (Problem 8.3.) Secondly, it is valid over any line of the Brillouin zone face which is parallel to a binary axis. (Problem 8.1.) It is useful to apply these results on the Brillouin zone of the face-centred cubic lattice depicted in Fig. 1.1. Subject to the required continuity, condition (8) is valid for this Brillouin zone as follows: (i) Over the whole of the square faces, because the planes (100) are symmetry planes. (ii) Along the whole of the LW line of the hexagonal face, because, as shown in the figure, it is parallel to a binary axis that joins the mid-points of opposite cube edges. (iii) At the points X and L, because they respectively belong to the geometrical elements just described. All these properties have important consequences, as we shall see later on.

3 Properties of the $E(\mathbf{k})$ surfaces and curves. The gradient

As we already know, the electronic states in a crystal will be fully described by plotting their energies as a function of $\mathbf{k}$, $E(\mathbf{k})$, with $\mathbf{k}$ ranging over the

Brillouin zone. This can be done in two ways. One is by constructing isoenergetic surfaces over the whole of the Brillouin zone. Since this is hard work, one often tries to gain some idea of the way the energy goes by plotting it with **k** varying over specific directions in the reciprocal lattice, as we have done already in Fig. **6**-4.1. We want to discuss in this section some formal properties of these surfaces and curves, which largely depend on the results of §2. It is useful for this purpose to discuss in more detail the derivatives $\partial E/\partial \mathbf{k}$ therein considered. Because **k** is a vector in the reciprocal space, this derivative has three components and it is itself a vector:

$$\frac{\partial E}{\partial \mathbf{k}} = \left[\frac{\partial E}{\partial k_x}, \frac{\partial E}{\partial k_y}, \frac{\partial E}{\partial k_z}\right] =_{\text{def}} \operatorname{grad}_{\mathbf{k}} E. \tag{1}$$

We recognize on (R1) that this vector is a gradient, since the gradient of a function ϕ has components $\partial\phi/\partial x, \partial\phi/\partial y, \partial\phi/\partial z$, along the coordinate axes. The latter here are the coordinate axes k_x, k_y, k_z, of **k** space (along the directions $\mathbf{a}^{\#}$, $\mathbf{b}^{\#}$, $\mathbf{c}^{\#}$, respectively) and the suffix **k** of the gradient is used as a reminder that this gradient is worked out in reciprocal or **k** space, rather than in the direct one. It has otherwise the usual properties of a gradient vector, in particular that of being normal to the level surfaces of the corresponding function. This gradient has a direct physical meaning since, from (**1**-8.14) and (1) it is, except for a constant, the group velocity with which the electrons propagate along the crystal:

$$\mathbf{v}_g = \frac{1}{\hbar} \operatorname{grad}_{\mathbf{k}} E. \tag{2}$$

We know that the energy is proportional to $\mathbf{k}^2$ for free electrons, which means that it is a parabola in one dimension or a sphere in three. Also that, in the same case, the momentum **p** is $\hbar\mathbf{k}$. It follows immediately from Fig. 1 that the group velocity and the momentum are parallel for free electrons, but not

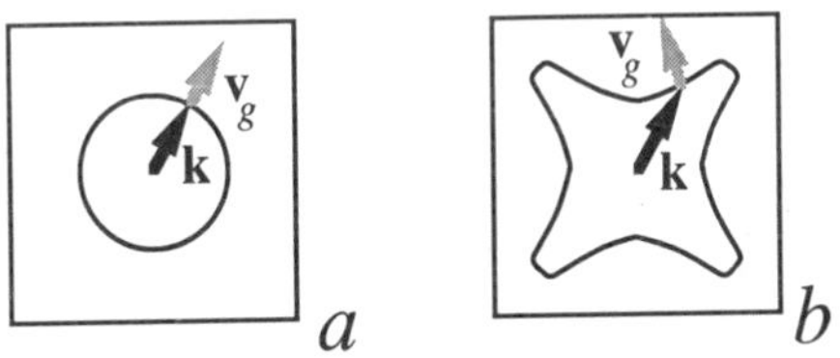

Fig. 8-3.1. Level surfaces for the energy. Both squares shown represent the Brillouin zone of a simple square lattice, and the curves inside them are level surfaces for which $E(\mathbf{k})$ has a constant value. The group velocity $\mathbf{v}_g$ is normal to these level surfaces, from eqn (2), since the gradient vector is always normal to the level surfaces of its corresponding function. A free-electron surface is depicted in *a* whereas *b* displays a level surface which may appear for a real non free-electron metal.

otherwise. (This is after all, why the momentum is used as distinct from the velocity!)

Condition (2.8) can now be rewritten as

$$(\text{grad}_{\mathbf{k}} E)_{\mathbf{n}} = 0, \tag{3}$$

which means that the component of the gradient of E normal to a Brillouin zone face must vanish, subject to the conditions required for (2.8) in § 2. This relation is very important because, when it is applicable, it determines distortions of the free-electron curves and surfaces near the Brillouin zone faces, as shown in Fig. 2.

We can now obtain a better insight into the Bragg reflection discussed in relation with (2.1). We can see in Fig. 2 that, as the propagation vector **k** of a wave approaches the Brillouin zone face the modulus of the group velocity corresponding to it must diminish until it reaches the value zero at precisely the Brillouin zone edge. Thus, the Brillouin zone face has a braking effect on the wave, the group velocity first decreasing, then becoming zero, then changing sign, as can readily be seen in the figure. This effect can easily be detected experimentally for an X-ray wave propagating along a crystal, as first done by Bragg.

We can also consider now the possible existence of cusps in the $E(\mathbf{k})$ curves or surfaces at the Brillouin zone faces which was discussed after eqn (2.8). These cusps can exist only when $\partial E/\partial \mathbf{k}$ is discontinuous there. We now recognize that this would require the group velocity to be discontinuous at the Brillouin zone surface. If these surfaces have any physical meaning at all, we cannot expect discontinuities in the group velocity to occur, so that cusps

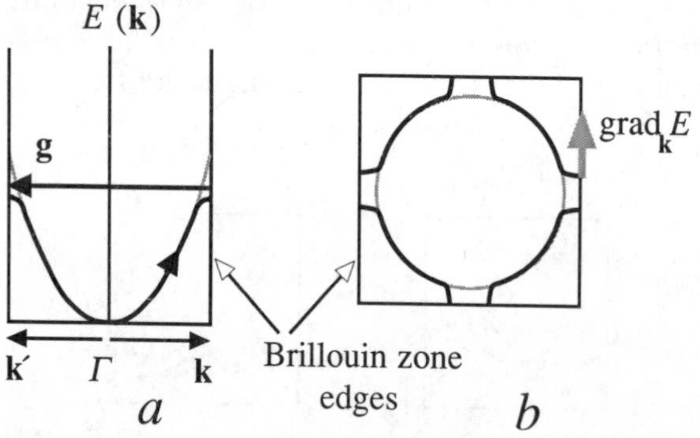

Fig. 8-3.2. Distortion of the free-electron $E(\mathbf{k})$ curves and surfaces near the Brillouin zone edges. It is assumed in both cases that symmetry planes exist in direct space that are parallel to the Brillouin zone faces represented, thus requiring the vanishing of the normal component of the gradient. The arrow in the curve shown in *a* indicates the change in the energy states when the wave vector is increased until it reaches the value **k**, when it is reflected into **k′** (compare with Fig. 2.1).

will exist only when unphysical Brillouin zone edges are introduced. (See Problem 8.5.) This is true only for electron bands: in the phonon bands, the energy of which is also given as a function of **k**, the group velocity is not significant so that the derivative can be discontinuous and cusps are actually common. (See § **11**-4.)

4 Energy bands: a one-dimensional example

We shall want to discuss later on various ways in which the energy bands can be plotted in **k** space and it will be useful for this purpose to consider a one-dimensional example, which will also serve as a revision of the methods discussed in Chapter **6**, as well as an introduction to some of the distinctive properties of metals, insulators and semiconductors. We illustrate in Fig. 1 a strictly one-dimensional linear chain and its Brillouin zone. The only symmetry operations in the space group, besides translations, are the identity E and the inversion $\mathbb{i}$ at the origin O. (Remember that in a strict one dimension, a reflection at the origin coincides with the inversion.) The point group is thus $\mathbf{C}_i$. We show three representative points in the Brillouin zone, Γ, Δ, and X, and we list their respective stars and groups of the **k** vectors in Table 1. Clearly, Γ is transformed into itself by the two operations of $\mathbf{C}_i$, so that its star consists of Γ alone and its group is $\mathbf{C}_i$. The vector Δ, on the other hand, changes into $-\Delta$ under $\mathbb{i}$, so that these two vectors are in the star. The

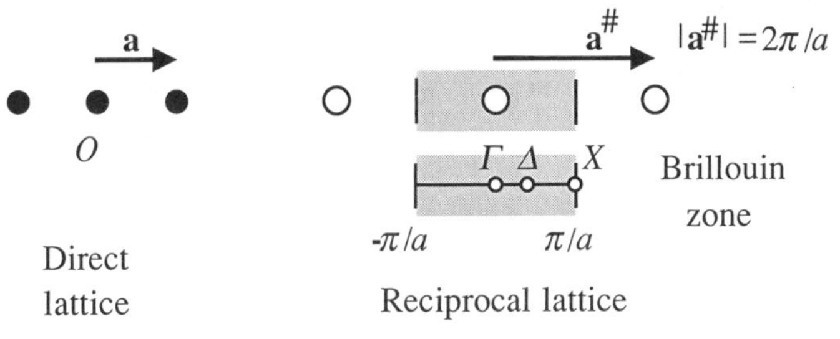

Fig. 8-4.1. A linear chain and its Brillouin zone

Table 8-4.1. Stars and groups of the **k** vector in the linear chain.

k vector	Star	Group of **k**
Γ	Γ	$\mathbf{C}_i$
Δ	$\Delta, -\Delta$	$\mathbf{C}_1$
X	X	$\mathbf{C}_i$

Table 8-4.2. The small representations for Γ and X: the irreducible representations of $\mathbf{C}_i$.

Rep.	E	i	Bases
A_g	1	1	s
A_u	1	-1	p

only operation which leaves Δ invariant is the identity, whence the group of Δ is $\mathbf{C}_1$ (the group whose only element is E). X is transformed into $-X$ by i, but X and $-X$ differ by $\mathbf{a}^{\#}$ and are thus equivalent. The inversion, therefore, leaves X invariant, so that its star contains only X, and its group consists of E and i, and it is again $\mathbf{C}_i$, as for Γ. Before we can apply these results we need the small representations, that is the irreducible representations of $\mathbf{C}_1$ and $\mathbf{C}_i$. The first of these groups, having only the identity, is trivial, and the representations of the second are given in Table 2. The representation A_g is *gerade*, that is symmetrical with respect to i, and it is spanned by s-type functions, which are symmetrical under inversion. The representation A_u is *ungerade* (antisymmetrical with respect to i) and it is spanned by p-type functions.

We know that the stars will give us the overall symmetry of $E(\mathbf{k})$ over the Brillouin zone. Thus, in particular, the star of Δ entails that $E(\Delta)$ and $E(-\Delta)$ are degenerate. The important fact is that no small representation is degenerate, (notice that this word is now used in a different sense!), which means that, for a given $\mathbf{k}$, no two Bloch functions can have the same energy, that is, that nowhere in the Brillouin zone can two bands touch or cross. The general form of the first two bands is shown in Fig. 2. This figure has been constructed by using free-electron parabolic curves modified at the Brillouin zone edges so as to satisfy the condition that their derivatives vanish at those points. (See Problem 8.4.) At Γ and X, from Table 2, there are two eigenvalues and in each case we have taken A_g to be the one lower in energy. We are assuming, in fact, that all states of the first band derive asymptotically from the $1s$ lowest level of the atom which composes the structure, whereas then the second band would derive from the $2p$ state, which is higher in energy. Naturally, bands of higher energy could be drawn corresponding to higher atomic states.

We have seen that if the linear lattice contains N primitive cells (N atoms), then the Brillouin zone also contains N permitted $\mathbf{k}$ values, to each of which there corresponds one energy level, although the Bloch functions for $\mathbf{k}$ and $-\mathbf{k}$ are degenerate, except at the centre Γ and at the edge X of the Brillouin zone. Each band will thus contain precisely N states (or $2N$ with spin). Notice that the energy levels form a quasi-continuous distribution, except that there is a *forbidden energy gap*, which is a vestigial trait left from the gaps between

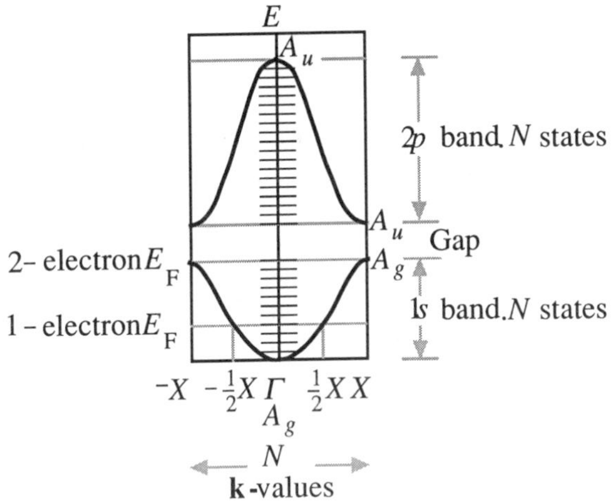

Fig. 8-4.2. The electronic energy bands for a linear lattice. Γ and X are singly degenerate. (Remember that $-X$ does not belong to the Brillouin zone.) $\frac{1}{2}X$ is a point of type Δ (compare with Fig. 1) and it is doubly degenerate.

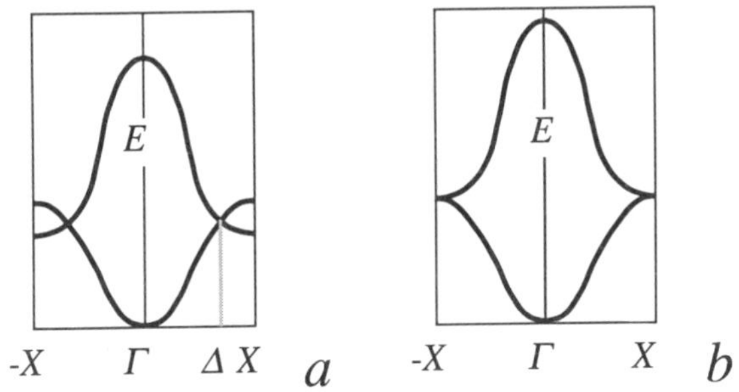

Fig. 8-4.3. Forbidden crossings and contacts in a linear lattice. In *a* the two Bloch functions at Δ for the first and second bands would be degenerate, whereas no degenerate small representation exists for this point of the Brillouin zone. Similarly for the point X at *b*.

the atomic levels that originate the bands. This energy gap is due to the fact that two bands cannot cross or touch, because of the one-dimensional nature of the small representations. If such contacts were permitted, as shown in Fig. 3, such gaps would disappear. The magnitude of the gap depends on the bending of the bands required by the condition that $\partial E/\partial \mathbf{k}$ must vanish at the Brillouin zone edge.

Filling the bands, conductors, semiconductors, and insulators

Consider a regular linear chain of N atoms (N large), each with one electron, and with bands as in Fig. 2. Each **k** value in the Brillouin zone will take two electrons (because of the spin), so that we must fill $\frac{1}{2}N$ values of **k**. Naturally, the lower energy levels must fill first, so that we must start from the centre of the Brillouin zone and at the first band, moving out until precisely $\frac{1}{2}N$ values of **k** have been occupied. Because the density of the grid of **k** values in the Brillouin zone is uniform, this means that **k** states are filled from the origin until precisely one half of the Brillouin zone is covered, which will immediately give the corresponding value of the Fermi energy E_F, as shown in the figure. For a material with two electrons per atom we have $2N$ electrons and thus the whole of the Brillouin zone is full. In the first case (a chain of hydrogen atoms, say), when an electric field is applied the electrons can take up the energy supplied by the field since, just as in Fig. **1**-4.1, vacant energy levels exist immediately above the Fermi energy, to which the electrons excited by the field can jump. Such a crystal will thus be a *conductor*, because the same mechanism will apply as in Fig. **1**-4.1. In the case of two electrons per atom (as in helium), the whole of the Brillouin zone is occupied, the Fermi energy is just below a forbidden gap (see Fig. 2) and the electrons, in order to take up the electrical field energy, would have to jump over the gap and go over into states in the second band. If the gap is larger than a fraction of an eV, the field would never provide enough energy to the electrons to allow them to surmount the gap, so that the material will not be able to conduct and will thus be an *insulator*. Naturally, the larger the gap, the better the insulator will be.

The conductivity of a metal (conductor) diminishes with the temperature, because higher temperatures excite vibrational modes of the crystal which decelerate the conduction electrons (§ **11**-3). Consider now the case when the band gap is much less than 1 eV, of the order of the thermal excitation energy of 0.03 eV and assume that the Brillouin zone is full. At absolute zero the upper band would be empty. As the temperature is raised, however, electrons will begin to percolate into the upper band and conduction will thus increase with the temperature. The upper, very partly filled band, is here responsible for the conduction process and it is called the *conduction band*. The material is called an *intrinsic semiconductor*, because the levels from which the conduction electrons are promoted belong to the band system of the material itself. It is possible, on the other hand to have a fairly wide band gap, say of the order of 1 eV and yet to have a different type of semiconductor, called an *extrinsic semiconductor*. In this case, the material must have some impurity which will create energy levels in the otherwise forbidden band gap (see § **14**-1). If these levels are now sufficiently near the bottom of the conduction band, temperature-dependent percolation and thus conductivity can take place, just as before.

5 Reduced and extended zone schemes: Brillouin zones of higher order

We have so far kept under cover an apparent contradiction in our work. On the one hand, in the free-electron approximation, the energy must be plotted as a function of **k** over the whole of the reciprocal space. On the other, we have repeatedly stressed that it is sufficient to restrict ourselves to the Brillouin zone, since the periodicity condition in the energy (eqn **6**-4.1) asserts that once $E(\mathbf{k})$ is known inside the Brillouin zone, it is then known over the whole of the reciprocal space. How do we reconcile these two views? We attempt this task in Fig. 1 in which energy bands are plotted for the linear lattice of Fig. 4.1, both over the whole of **k** space and over its first primitive cell or Brillouin zone. Careful attention must be paid to reading Fig. 1 in the correct way, which must be done in the successive steps described below.

1. Start with the axes E and **k** with their scales, the latter in units of $\mathbf{a}^{\#}$, and then draw the free-electron parabola.

2. Consider along the **k** axis the point pairs

$$(-\tfrac{1}{2}\mathbf{a}^{\#}, \tfrac{1}{2}\mathbf{a}^{\#}), \quad (-\mathbf{a}^{\#}, \mathbf{a}^{\#}), \quad (-\tfrac{3}{2}\mathbf{a}^{\#}, \tfrac{3}{2}\mathbf{a}^{\#}), \quad (-2\mathbf{a}^{\#}, 2\mathbf{a}^{\#}). \tag{1}$$

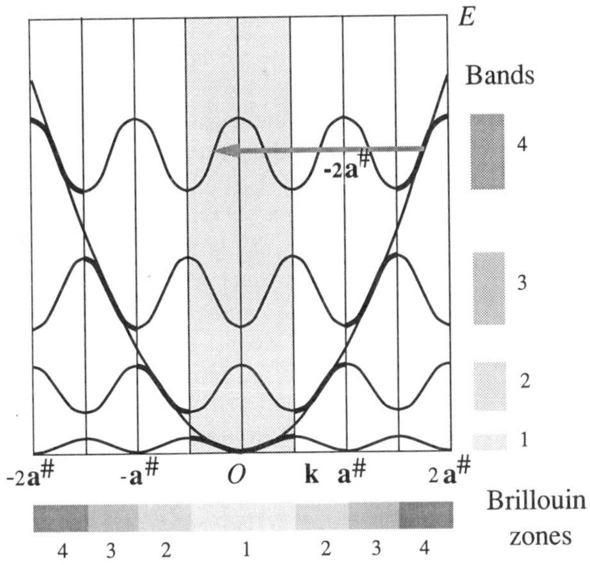

Fig. 8-5.1. The extended zone scheme. The sections of the **k** axis which belong to the same Brillouin zone are tinted in the same shade, and so are the corresponding bands. The centrally shaded region of the figure (reduced scheme) corresponds to the first Brillouin zone, from $-\tfrac{1}{2}\mathbf{a}^{\#}$ to $\tfrac{1}{2}\mathbf{a}^{\#}$.

The points in each of these four pairs are respectively separated by the following vectors:

$$\mathbf{a}^{\#}, \qquad 2\mathbf{a}^{\#}, \qquad 3\mathbf{a}^{\#}, \qquad 4\mathbf{a}^{\#}. \qquad (2)$$

The vectors in (2) are all vectors of the reciprocal lattice, **g**-like. Therefore:

3. Draw the vertical lines which correspond to each of the **k** pairs in (1); they satisfy the condition of Fig. 2.2 of being separated by a vector **g**, so that $\partial E/\partial \mathbf{k}$ must vanish on all these lines. Therefore:

4. The free-electron parabola must be modified when it crosses the just-mentioned vertical lines, as shown by the thick curves in the figure, so as to ensure the vanishing of the derivatives there.

5. Because the energy is periodic in **k**, (eqn **6**-4.1), the modified $E(\mathbf{k})$ curves are periodically extended over the whole of the **k** space. These periodic curves correspond to the first, second, third, fourth, and so on, bands, with energy states depicted on the right of the figure.

6. In order to correlate with our previous picture, in which the bands are depicted exclusively in the Brillouin zone, we retain only the shaded part of the figure, the rest of it being redundant owing to the periodicity condition for the energy bands. (Compare the first two bands thus obtained with Fig. 4.2.)

We have reconciled in this way our picture of almost free-electron bands with their description solely in the Brillouin zone. For reasons which will become clear in a moment, this Brillouin zone is now called the *first Brillouin zone* and the representation of the energy bands entirely within the first Brillouin zone is called the *reduced zone scheme.*

Because of its compactness, the reduced zone scheme is the most common way in which the results of band structure calculations are presented. It is sometimes useful, however, to follow a different approach. If we look again at Fig. 1, it is clear that, if we want to start from the free-electron picture and then correct it, either by some sensible hand waving, as we have done, or by means of some perturbation method of calculation, say, there is some merit in presenting the results in terms of curves such as the thick lines of the figure, since they follow the free-electron parabola as closely as possible. This means that, rather than representing the first, second, third, and so on, bands all on the same (first) Brillouin zone, we want to represent each band in a different piece of **k** space, thus covering the whole of this space, as free-electron curves do. Two points must be considered. First, each band corresponds, without spin, to precisely N states (the number of primitive cells in the crystal), which is the number of states in a primitive cell of the reciprocal lattice. Thus, each band will be plotted in a primitive cell, the first band in the first cell, the second in the second, and so on. The second point, which is clear from Fig. 1, is that, if we want to follow the free-electron approximation in which we start at the origin of **k** space with E vanishing, and we then increase our energy moving equally along the **k** and $-\mathbf{k}$ directions, then the successive primitive cells which we need for the successive bands must be *nested*, the second cell embracing the first symmetrically, and so on. These successive nested primitive cells in **k** space, symmetrically disposed around the origin, are called the successive Brillouin zones, or *Brillouin zones of higher order*, first, second, and so on. The scheme in which $E(\mathbf{k})$ is plotted as a singly-valued function in each Brillouin zone, with the n-th band in the n-th zone, is called the *extended zone scheme*. It corresponds in Fig. 1 to the depiction of $E(\mathbf{k})$ by the curves in thick lines. It is easy to pass from the extended to the reduced zone scheme: it is sufficient to shift the bands from the higher order zones to the first one by means of translations by appropriate **g** vectors, as illustrated in Fig. 1 for one branch of the fourth band.

There is yet a third way in which energy bands may be plotted, also illustrated in Fig. 1, which is by means of the *periodically repeated scheme*, in which $E(\mathbf{k})$ for a band is first given over the first Brillouin zone, which is then periodically repeated over the whole of **k** space. It should be stressed that Brillouin zones of higher order, that is nested primitive cells are *not* used in this scheme: it is the first cell which is periodically repeated. Although this procedure is highly redundant, it is advantageous for some purposes, as will be illustrated in § 7.

6 Construction of the Brillouin zones of higher order

The Brillouin zones are nested primitive cells centred at the origin of reciprocal space and each of them is thus a maximal set of translationally inequivalent points. They are therefore obtained by the method described in § **3**-2 in order to construct such sets. Given a reciprocal lattice direction passing through the origin, a neighbour of order n of the origin in that direction, separated from the origin by a vector **g**, must correspond to another such neighbour at the point $-\mathbf{g}$. The two planes which are the perpendicular bisectors of **g** and $-\mathbf{g}$ form a slab in **k** space such that any point outside the slab differs from some point inside it by the vector **g** and such that all points inside the slab are translationally inequivalent under **g**. Therefore, in order to form maximal sets of translationally inequivalent points, such slabs must be constructed and all points outside them eliminated.

As a result of this principle, the method for constructing the successive Brillouin zones is as follows. First, we draw the perpendicular bisector planes to the first, second, and higher order neighbours from the origin. Secondly, we start from the origin and we move away from it in all directions until a first set of planes is met: the region inside these planes is the first Brillouin zone. (Notice that not all bisector planes in such a set need belong to the same order of neighbours.) We then move away from the surface of the first zone until the next set of bisector planes is met: this is the second zone, nesting outside the first. In general, the $(n + 1)$th Brillouin zone is obtained by

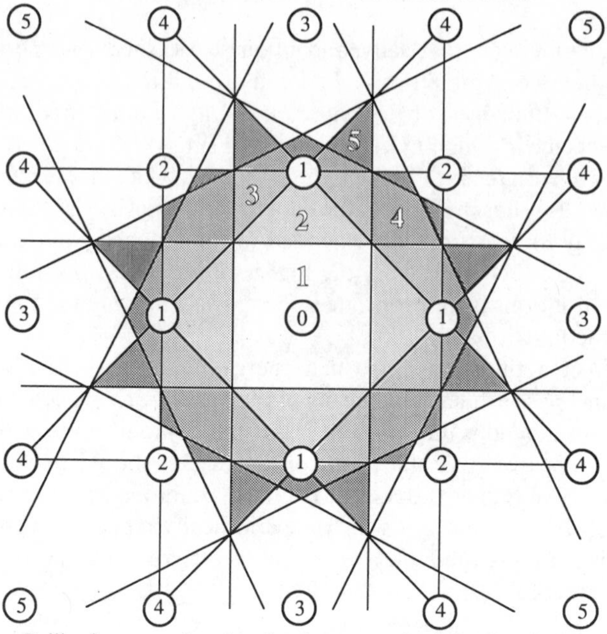

Fig. 8-6.1. Brillouin zones for the simple square lattice. The circles are reciprocal lattice points and the numerals inside them identify the order of the neighbours from the origin. Bisectors planes to neighbours up to only order four are shown. The first five Brillouin zones are identified with successively darker shading and their order *n* can be identified by the numeral *n* inserted in one of the sectors of the corresponding zone.

moving away from the surface of the *n*-th zone until the first new set of bisector planes are met. Because all these zones are maximal sets of translationally inequivalent points, they are all identically of the same volume and, like the first Brillouin zone, they all contain N values of $\mathbf{k}$, if N is the number of primitive cells in the crystal lattice.

This method of construction is illustrated in Fig. 1 for the reciprocal lattice of a simple square lattice. It is easy to see how, for example, the four pieces which make up the second Brillouin zone can be folded back, by means of translations by $\mathbf{g}$ vectors, into the first Brillouin zone. (Problem 8.6.)

7 The energy level surfaces and the Fermi surface

If we go back to Fig. 5.1, it is clear that the free-electron approximation in the extended zone scheme, coupled with the condition that $\partial E/\partial \mathbf{k}$ vanishes at the zone edges, determines the presence of energy discontinuities at the Brillouin zone face. The origin of these discontinuities is clearer, as we can see, in the extended scheme and we know that they are not a mere accident of our rough

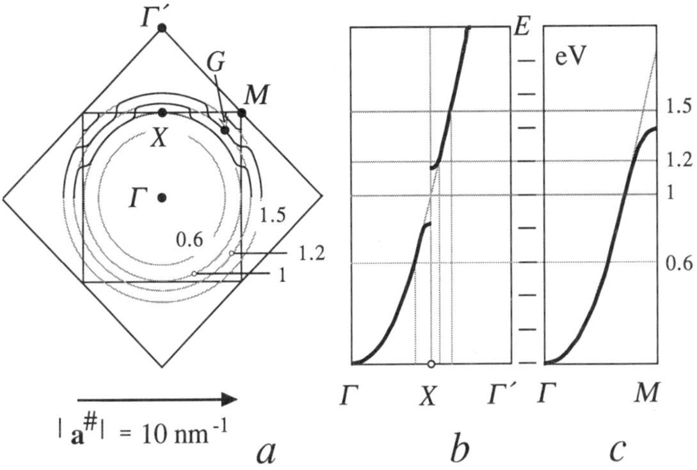

Fig. 8-7.1. Energy discontinuities for *s* bands at the edges of the Brillouin zone, and the Fermi surface. In order to simplify the construction of the figure it has been assumed that the modulus $|\mathbf{a}|$ of the lattice constant in direct space is $(2\pi/10)$ nm, which leads to the lattice constant in reciprocal space shown in *a*. This part of the figure shows the first two Brillouin zones in the extended scheme. The point labelled Γ' is a point of Γ type because it is the centre of a first Brillouin zone in the periodically repeated scheme. The values of the energy for the energy contours are in eV. The horizontal scale of Figs *b* and *c* is identical with that in *a*.

approximations, because the group-theoretical considerations of § 4 preclude their disappearance. The presence of energy discontinuities at the Brillouin zone surface is a very characteristic feature of band structures and we illustrate it in Fig. 1, where we draw energy surfaces (really, energy contours, since we work only in a two-dimensional case) for the simple square lattice, the Brillouin zones of which have been obtained in Fig. 6.1. In the free-electron approximation, as valid for the *s*-type bands which we are considering, the energy is a function of $|\mathbf{k}|^2$, given by (**1**-9.4), and these parabolae are plotted in Figs 1*b*, *c* along the directions $\Gamma X \Gamma'$ and ΓM of the Brillouin zone, sensible guesses being made as to the necessary corrections at the Brillouin zones edges in order to satisfy the vanishing of the derivatives therein. Clearly, the free-electron energy contours are circles, of which four are shown in Fig. 1*a* in increasing order of energy. As shown in the figure, we must use not only the first but also the second Brillouin zone in order to draw these full circles. At the Brillouin zone edges the circles have to be deformed to satisfy the vanishing of the energy derivatives. (A plane of symmetry exists in this lattice parallel to the Brillouin zone faces, as required in § 2.) Of course, we have done these deformations by sensible guesswork, with the guidance where necessary of the curves drawn in Figs 1*b*, *c* and, for clarity, they are only shown on the upper half of Fig. 1*a*. We can see that, along the edge XM, the contours for 1 eV from the first zone (first band) and 1.2 eV from the second zone (second band) meet at the same point and the same happens for the contours 1.2 eV (first band) and 1.5 eV (second band). This means that we have energy discontinuities at the points where these contours meet and, in the same manner, it is clear that energy discontinuities must exist for every point on the surface of this Brillouin zone. Only one of these discontinuities, of about 0.3 eV at X, is shown in Figs 1*b*, *c*. It must be appreciated that the existence of these discontinuities is contingent on the existence of a parallel reflection plane, so that every Brillouin zone face must be individually investigated from the point of view of discontinuities, although these are quite frequent.

The reader should notice that here, as a difference with § **6**-4, we are dealing with *s* bands for which there are no degeneracies at Γ or M (the small representations being in both cases of A_1 *type* in Table **6**-3.2). Thus the discontinuities at the Brillouin zone edge discussed here are valid only for this type of bands. (See Fig. **6**-4.1 where the two *p* bands touch at Γ and at M.)

We shall now consider the form of the Fermi surface in the same simple square lattice, always for the *s* bands, assuming that we have two electrons per atom. In this case, we must fill an area of **k** space precisely equal to that of the first Brillouin zone, which is $|\mathbf{a}^{\#}|^2$. In the free-electron approximation this is a circle of radius $|\mathbf{a}^{\#}|/\sqrt{\pi}$, that is $0.564\,|\mathbf{a}^{\#}|$, which is precisely the radius of the contour labelled 1.2 eV in the figure. We start filling states from the centre, moving out of it in all directions. When the point G is reached from Γ, however, we cannot go on filling states in this direction, because the states from G to M have energies larger than 1.2 eV, whereas moving up from X into the second zone there are plenty of states with energy lower than this. Thus, the corner around M remains empty, whereas the second zone (that is, the second band: we are in the extended zone scheme with one band per zone) begins to fill upwards from X until the contour 1.2 eV is reached again. If the number of states now filled, up to 1.2 eV, is precisely $2N$, this is the value of the Fermi energy. Otherwise, we would have to consider a slightly different contour for which this happens. Clearly, what we require is that the area above this contour in the first zone (empty states near M) must equal that of the occupied pockets around X in the second zone. For the sake of our example, however, it is sufficient to accept that 1.2 eV is the Fermi energy, as we have assumed in Figs 1*b*, *c*, where it can clearly be seen that the top of the first band at M is empty and that a pocket of electrons at X fills the second band. We reproduce in Fig. 2*a* the Fermi surface derived from Fig. 1*a*.

The first thing we want to do with this Fermi surface is to plot it in the reduced zone scheme, that is, entirely within the first Brillouin zone. Of course, the part of the Fermi surface belonging to the first band needs nothing doing and we merely repeat it in Fig. 2*b*, where we can clearly see the pockets of unoccupied states at M. In order to plot the part of the Fermi surface

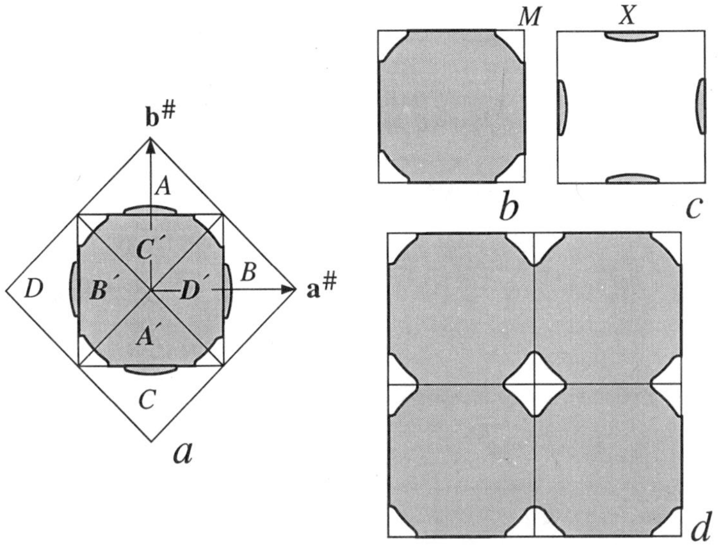

Fig. 8-7.2. The Fermi surface for a simple square lattice with two electrons per atom. *a*: Extended zone scheme. *b*: Fermi surface for the first band in the reduced zone scheme. *c*: Fermi surface for the second band in the reduced zone scheme. *d*: Fermi surface for the first band in the periodically repeated scheme. All the shaded areas correspond to occupied states (electrons).

belonging to the second band, we must first *fold back* the second Brillouin zone (and its contents) onto the first one. *Folding back* a Brillouin zone of order n into one of order $n-1$ means the following: since the first-named zone is nested outside the second one and since both of them are of the same area, each part of the outer zone can be translated by an appropriate **g** vector until it superimposes a part of the inner zone, and this process can be continued until the whole of the inner zone is covered by the outer one. In our case this is accomplished by translating the triangle labelled A by $-\mathbf{b}^{\#}$ into the one labelled A', B by $-\mathbf{a}^{\#}$ into B', and so on. We get in this way the Fermi surface of the second band shown in Fig. 2*c*.

It must also be understood that the first and second bands discussed here will correspond, say, to the 1*s* and 2*s* states of the atom of which the lattice is made up. They would thus be the 1*s* and 2*s* bands, respectively, whereas the bands in § **6**-4 would be of $2p_x$ and $2p_y$ types.

For reasons which will be explained a little later, it is very useful to represent the Fermi surface in the periodically repeated scheme, which we do in Fig. 2*d* for the first band. This figure is obtained by mere periodic repetition of Fig. 2*b*. We can proceed in the same manner for the second band, and the pictures for the two bands are shown in Fig. 3. Although this figure implies the periodic scheme, it is obtained by showing the first zone only, (as

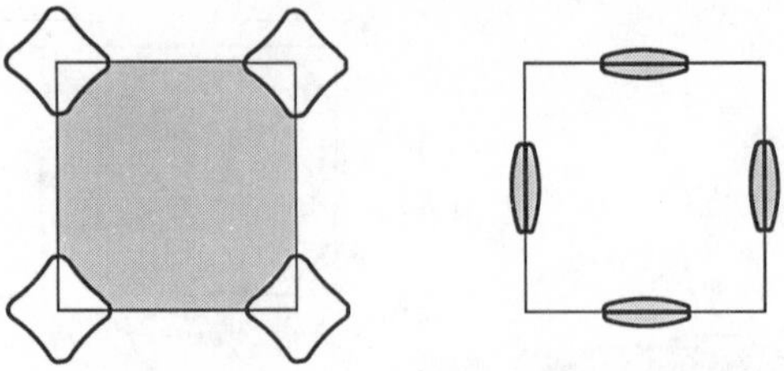

Fig. 8-7.3. The Fermi surface for a simple square lattice with two electrons per atom. All the shaded areas correspond to occupied states (electrons).

displayed right at the centre of Fig. 2*d*) with the Fermi surface completed by repetition on the neighbouring zones. The advantages of this approach, as compared with the description of the Fermi surface in the reduced schemes of Figs. 2*b*, *c*, will now be discussed.

When a magnetic field is applied to a piece of metal, the perturbation entailed by the field does not alter the energy of the electron states. This means that if an electron state is on the Fermi surface, it will continue to be on the Fermi surface after the field is applied. If the magnetic field is normal to the curves of Fig. 3, however, it can be proved that the electron state will change in such a way that its representative point will move clockwise or counter-clockwise (depending on the direction of the field) keeping on the Fermi surface, that is to the curves shown in the figure. Moreover, this displacement of the representative point of the electron state over the Fermi surface will be periodic. For this reason, such curves as shown in Fig. 3 (which in a real three dimensional case would be cross-sections of the Fermi surface) are called *orbits*. It must be realized, however, that the resulting motion in real space is quite complex, the orbits entailing not changes in the position of the actual electron but rather of the representative point of its state. The important result is that such orbits in a magnetic field (as cross-sections of real Fermi surfaces) are experimentally accessible: it is in fact even possible to measure their area by a method called the *de Haas–van Alphen effect*. (This is so only for *extremal* cross-sections of the Fermi surface.) Clearly, a much better depiction of these orbits is given in Fig. 3 in the periodic scheme, than in the reduced scheme of Figs. 2*b*, *c*. The experimental determination of these orbits produces much valuable information about the shape of the Fermi surface.

Holes, semiconductors, and semimetals

It should be noticed that the two types of orbits in Fig. 3 show a very substantial difference: the orbit in the first band surrounds an area of unoccupied states, whereas the orbit in the second band surrounds occupied states. It is common for this reason to refer to the areas at the corners of the first Brillouin zone here as being occupied by *holes*, whereas those in the second band are said to be occupied by *electrons*. The area (or volume in the three-dimensional case) occupied by electrons in the second band must be equal, for our divalent metal, to the area (or volume) occupied by holes in the first band, in order to ensure that the total number of electron states occupied equals that in the first Brillouin zone (that is $2N$, for N primitive cells in the direct lattice).

Notice that, in Fig. 1, if the free-electron picture were no good as an approximation for the divalent metal in question, then the energy gaps at the edge of the Brillouin zone might be so large that no state of the second band could be lower in energy than states in the first band. In this case, the whole of the first Brillouin zone would be full and the whole of the second zone would be empty. The electrons on the Fermi surface (which would in this case be the surface of the first Brillouin zone), would not be able to surmount the gaps, if they are large enough, in order to jump to unoccupied second band states. In this case, the material would not be able to become excited by the electric field and would thus be an *insulator*. If the gap were small enough, it could be a *semiconductor*. Clearly, considering a case such as that in Fig. 3, the larger the Fermi surface, the larger the number of electrons which may be excited by the electric field and thus the larger the electrical conductivity. (See § **11**-1.) Already, the hypothetical material shown has a pretty small Fermi surface and would thus be a poor conductor, whereas a monovalent metal would have a fairly large surface in the first Brillouin zone only (in the free-electron approximation a sphere half its volume) and would thus be a good conductor. If the Fermi surface, for the divalent metal of Fig. 2, were to become even smaller, the material would become a very poor conductor. Such materials with very small Fermi surfaces are called *semimetals*, of which arsenic and bismuth are examples.

Effective mass and holes

We have already referred to the unoccupied electron states at the top of the first band near M as *hole* states, and we shall now consider their properties, which are unusual. For simplicity of the discussion, we shall work in one conveniently chosen direction, so that the group velocity will now take the form

1-8.14
$$v_g = (\hbar)^{-1} \mathrm{d}E|\mathrm{d}k. \tag{1}$$

In a decent free-electron situation we would expect the group velocity associated with an electron state to increase as the energy increases, and it is clear that this is the case for $\mathrm{d}E/\mathrm{d}k$ near the bottom of the first band in Fig. 1*c*. It can be seen, however, that this band has an inflection point near the top of the band, after which the group velocity actually *decreases* as the energy increases, as if the particles in such states were subject to a negative acceleration a, a point which we shall now investigate:

,1 $$a = \frac{\mathrm{d}v_g}{\mathrm{d}t} = \frac{1}{\hbar}\frac{\mathrm{d}}{\mathrm{d}t}\frac{\mathrm{d}E}{\mathrm{d}k} = \frac{1}{\hbar}\frac{\mathrm{d}^2E}{\mathrm{d}k^2}\frac{\mathrm{d}k}{\mathrm{d}t}. \qquad (2)$$

It is well known, on the other hand, that the force f and the momentum p are related as follows,

,2 $$f = \frac{\mathrm{d}p}{\mathrm{d}t} = \hbar\frac{\mathrm{d}k}{\mathrm{d}t} = \hbar^2\left(\frac{\mathrm{d}^2E}{\mathrm{d}k^2}\right)^{-1} a. \qquad (3)$$

From (3), we can define an *effective mass* m^*:

3 $$m^* = \hbar^2\left(\frac{\mathrm{d}^2E}{\mathrm{d}k^2}\right)^{-1}. \qquad (4)$$

Quite clearly, for the electron states near M at the top of the first band, the effective mass is negative. Rather than dealing with such eccentric electron states, we shall now define some new 'particles', to be called *holes*, whose charge is the negative of the electron charge e (and therefore *positive*). We shall prove that their effective mass in the situation under discussion will be positive, for which purpose we must first show that the velocities v_e and v_h corresponding to the *same* state (point of the Brillouin zone) occupied by an electron or a hole, respectively, are equal. In order to do this, we write the *currents* j_e and j_h corresponding to the motion of the electron and the hole, respectively:

$$j_e = -|e|v_e, \qquad j_h = |e|v_h, \qquad (5)$$

where, for safety, we emphasize the change in sign of the charges. Let us now denote with Σv_i the sum over all the *electron* states over the *whole* of the Brillouin zone. We must accept that the current associated with such a sum must vanish, because of cancellations of positive and negative velocities:

5 $$-|e|\,\Sigma v_i = 0. \qquad (6)$$

The left-hand side of (6) can be split into two parts, by defining $\Sigma' v_i$ as the sum over all the electron states over the Brillouin zone, *except* the state v_e:

$$-|e|\,\Sigma' v_i - |e|v_e = 0 \quad \Rightarrow \quad -|e|\,\Sigma' v_i = |e|v_e. \qquad (7)$$

The current $-|e|\,\Sigma' v_i$, on the other hand, is precisely the current associated with having a full band with just one hole at the state v_e, which is given in (5) as $|e|v_h$. Comparison of this with (7) shows that the velocity of the hole v_h must equal the velocity v_e which an electron would have in the empty state corresponding to the hole.

We shall now prove that the effective mass of the hole must be the negative of the effective mass of the electron. Consider for this purpose the particles in an electric field $\mathscr{E}$. Clearly, the forces acting on an electron and a hole must be equal and opposite, which entails a change of sign in (L3) in changing from the electron state therein treated to the corresponding hole state. The velocities, in (2), on the other hand, remain unchanged, as just shown, so that a must also remain unchanged in (2) and (3). It follows from the new form of (3) that (4) must change sign for a hole state. Holes, thus, have effective masses of opposite sign to those of the electrons and, in cases such as the one discussed in this section, these effective masses are positive, which makes it easier to visualize their motion.

8 Problems

1. Prove that $\partial E/\partial \mathbf{k}_n$ vanishes for all $\mathbf{k}$ vectors which end on a line of a Brillouin zone face which is parallel to a binary axis.

   Hint. Use Fig. 2.2, on replacing σ by a binary axis C_2. Note that whereas $\mathbf{k}'$ in that figure can be any vector from Γ to a point outside the plane of the figure, $\mathbf{k}'$ must now lie on the latter plane, since, otherwise, $C_2\mathbf{k}'$ would not belong to the plane which contains $\mathbf{k}$ and $\mathbf{k}'$.
2. Prove that $\partial E/\partial \mathbf{k}_n$ vanishes whenever $\mathbf{k}$ ends at the centre of a Brillouin zone face.

   Hint. Take $\mathbf{k}$ equal to $\frac{1}{2}\mathbf{g}$ in Fig. 2.2 and use (**6**-5.6). No element of symmetry is thus required.
3. Prove that $\partial E/\partial \mathbf{k}$ vanishes for $\mathbf{k}$ equal to Γ, the centre of the Brillouin zone.
4. Prove that, for a one-dimensional Brillouin zone, $\partial E/\partial \mathbf{k}_n$ vanishes at the edge.

   Hint. Use a figure like Fig. 2.2 in one dimension and condition (**6**-5.6). No symmetry element is thus required.
5. Draw all the *uncorrected* free-electron bands of Fig. 5.1 in the periodically repeated scheme, and show that cusps appear at the Brillouin zone edges. Why is this possible?
6. Fold back the second, third and fourth Brillouin zones of Fig. 6.1 onto the first Brillouin zone.
7. Identify the sixth Brillouin zone in Fig. 6.1.
8. On introducing the free-electron expression for the energy in eqn (7.4) prove that the effective mass of free electrons coincides with the mass m of the electron.
9. Consider the three basic $\mathbf{k}$ vectors Γ, Δ, X of the Brillouin zone of a strictly linear chain with lattice constant a (Fig. 4.1) and show that the

bases and corresponding irreducible representations for the space group, ignoring inversion, are

$$k = \Gamma, \text{ basis } \langle \psi_{\mathbf{k}} |, \quad \text{matrix for } \{E|\mathbf{t}\}: \quad 1, \quad \forall \mathbf{t}. \tag{1}$$

$$k = \Delta, \text{ basis } \langle \psi_{\mathbf{k}}, \psi_{-\mathbf{k}} |, \quad \text{matrix for } \{E|\mathbf{t}\}: \quad \begin{bmatrix} e^{-i\mathbf{k}\cdot\mathbf{t}} & \\ & e^{i\mathbf{k}\cdot\mathbf{t}} \end{bmatrix}, \quad \forall \mathbf{t}. \tag{2}$$

$$k = X, \text{ basis } \langle \psi_{\mathbf{k}} |, \quad \text{matrix for } \{E|\mathbf{t}\}: \quad e^{-i\mathbf{k}\cdot\mathbf{t}}, \quad \forall \mathbf{t}. \tag{3}$$

What happens when the inversion is introduced?

10. Discuss the values of $\partial E/\partial \mathbf{k}$ in Figs **6**-4.1*b* and *c*, at the points Γ, Z, X, and M. (The derivative here is understood for $\mathbf{k}$ varying in the direction of the horizontal axis of the figure. It therefore gives directly the slopes of the curves shown in the figure.)

Further reading

See the references provided for Chapter **5**.

9

Bands and Fermi surfaces in metals and semiconductors

Although quantitative methods for the calculation of band structures will not be discussed until Chapter **10**, we shall give here a few important examples of their applications, since we already have the tools for understanding these results. This will provide better continuity with the methods so far studied for the description of bands. We shall first consider a *semiconductor*, silicon, for which at absolute zero the whole of the Brillouin zone is fully occupied. (As it is now standard practice in describing computed bands, the reduced band scheme will be exclusively used for this purpose.) After this, the *noble metal* copper will be treated and, since the Brillouin zone is only partly occupied here, the Fermi surface will also be discussed. The example of copper will allow us to understand also the qualitative features of the bands in *simple metals* (like the alkalis) and in *transition metals* (like nickel).

1 Silicon

Silicon, like all the Group IV semiconductors (C, Si, Ge) has the *diamond crystal pattern* as crystal structure, and we must first of all study this structure and its properties.

The silicon structure (diamond crystal pattern)

The silicon structure is depicted in Fig. 1, which shows quite clearly the tetrahedral configuration which is typical of the bonding in the Group IV elements. The four black atoms shown form a tetrahedron, represented as inset in a cube, and the three edges of this tetrahedron which meet at the origin O form the primitive vectors and thus span the primitive cell, which is not shown in the figure. It is clear, however, that the primitive cell will show five atoms, the four in black at the corners of a tetrahedron, and one in grey at the centre of it. Each of the atoms at the tetrahedron corners is shared by four primitive cells, so that this set of atoms contributes a total of one atom to the primitive cell. On adding to it the atom at the centre we arrive at a count of two atoms per primitive cell. We thus have a lattice with basis. We shall take

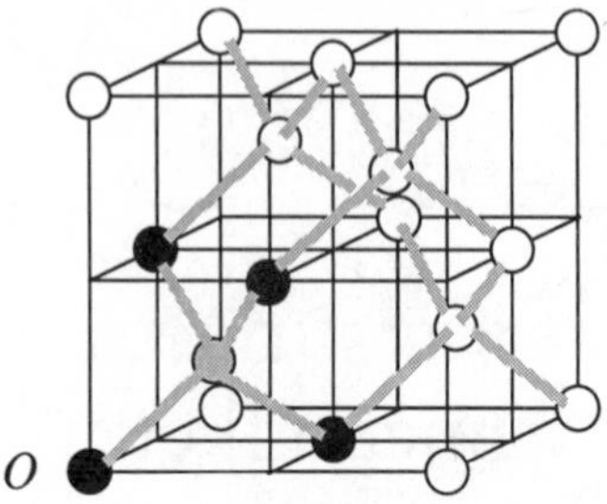

Fig. 9-1.1. The silicon structure (diamond crystal pattern). The point O is the origin of coordinates. This point and the atom in grey form the crystal basis. The unit vectors **a**, **b**, and **c** are along the cube edges as in Fig. **4**-7.1*a*.

this basis to be given by the atom at the origin O plus the grey atom at the centre of the tetrahedron.

We must next determine the Bravais lattice, since the Brillouin zone will entirely depend on it. In order to do this, we must realize that on translating the atom at O, (dragging with it also the second atom of the basis), until this atom occupies all the sites of a face-centred cubic lattice (compare with Fig. **4**-7.1*a*), then the whole of the diamond crystal pattern is generated. (It is easy to see that the translations just defined must, of course, correspond to the primitive vectors and that these agree with those shown for the face-centred cubic lattice in Fig. **4**-7.1*a*.) Precisely the same result would obviously be achieved by translating the second atom of the basis in the same manner, dragging with it the first atom. Thus, the diamond crystal pattern may be described as two interlocking face-centred cubic lattices, each corresponding to one of the two atoms of the basis. Either way, the Bravais lattice, as defined by the primitive translation vectors, is clearly a face-centred cubic lattice and this is our vital piece of information for the construction of the Brillouin zone.

The space group G of the silicon structure is non-symmorphic, because of the presence of the basis. Notice in Fig. 1, in fact, that the vector from the origin O to the atom shown in grey, which we shall call **v**, is *half* a lattice translation vector and, therefore, that it does not belong to the Bravais lattice. As an example of the appearance of this vector **v** in defining the operations of G, let us consider $\mathbb{i}$, the inversion at O: it carries the grey atom to a void site like the one shown at the centre of the cube drawn at the right-hand top-back side of Fig. 1. It is thus necessary, in order to ensure covering, to follow this operation with a translation by **v**, thus forming the operator $\{\mathbb{i}|\mathbf{v}\}$. It is not difficult to see geometrically that this operation is the inversion at the mid-point between the two atoms of the basis.

It is fairly easy to verify that the point group that leaves invariant the face-centred cubic Bravais lattice and its origin O is the full symmetry group of the cube, $\mathbf{O}_h$. Because of the presence of a basis in the structure, not all operations of $\mathbf{O}_h$ will be operations of G. We have seen, for example, that the inversion $\mathbb{i}$ appears only as $\{\mathbb{i}|\mathbf{v}\}$; in the same manner, the four-fold rotation axis becomes a screw axis, and so on. All the operations of $\mathbf{O}_h$, however, appear in G either on their own or associated with a fractional translation vector and thus $\mathbf{O}_h$ is the point group of G. (Therefore, because P coincides with the point group of the Bravais lattice, the space group G is holohedric.)

The orbital basis

The electron configuration of silicon in its ground state is $1s^2$, $2s^2, 2p^6, 3s^2, 3p^2$ but its normal valence state arises when one of the $3s$ electrons is promoted to a $3p$ level, thus giving the fourvalent configuration $1s^2, 2s^2, 2p^6, 3s, 3p^3$. (As it is well known, the $3s, 3p^3$ electrons hybridize to form tetrahedral sp^3 hybrids which are responsible for the tetrahedral configuration of the crystal structure in Fig. 1.) The states $1s^2, 2s^2, 2p^6$ have all large negative energies and they can be regarded as core states that do not take part in the band structure. Thus, we need only consider the four valence electrons $3s$, $3p^3$ centred on each of the two atoms of the crystal basis. What we are saying is that we expect that the various bands in the crystal will be given in terms of cell functions derived asymptotically from these orbitals. We call this set of orbitals the *orbital basis*. In order to depict this basis, we first choose the two sets of $3s$, $3p^3$ orbitals one inverted with respect to the other through the centre of inversion at the mid-point of the bond between the two atoms of the basis, as shown in Fig. 2*a*. What we have in mind here is the following. Remember that the cell function $u^j_{\mathbf{k}}(\mathbf{r})$ must be defined over the whole of the primitive cell, and this is why we have to display it over the two atoms of the basis. As a very rough approximation, we might expect that for some $\mathbf{k}$ vector (of very high symmetry) the orbitals in Fig. 2*a* might give a reasonable depiction of the cell function $u^j_{\mathbf{k}}(\mathbf{r})$, but this will certainly depend

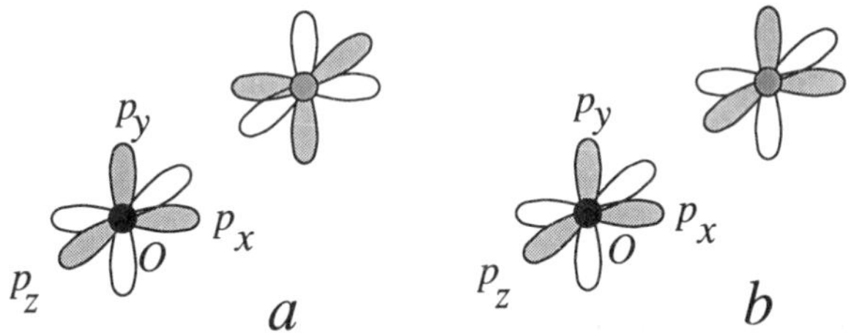

Fig. 9-1.2. The orbital basis of p orbitals on the lattice basis. The black and grey circles shown correspond respectively to the first and second atoms of the basis in Fig. 1. The fractional translation vector $\mathbf{v}$ is the vector between the two sites shown in the Figure and $\{\mathbb{i}|\mathbf{v}\}$ is the inversion at the mid-point of $\mathbf{v}$. The shaded lobes of the p orbitals are positive and the others are negative. The z direction is along $\mathbf{v}$ and the x, y directions are normal to it.

strongly on the value of **k** and on the band index j. Thus, for some different band and some particular value of **k**, a configuration such as the one displayed in Fig. 2*b* might be a rough approximation to the corresponding new value of $u^j_{\mathbf{k}}(\mathbf{r})$.

The main difference between Figs 2*a* and 2*b* is that there is a good overlap in the former between the two orbitals p_z directed along the bond between the two atoms, whereas in the latter there is a node between them, which corresponds to the nil value that the wave function must take in going from a positive to a negative lobe. (Notice that this relation is inverted when the p_x, p_y overlaps are considered, but these are so much smaller that their contribution to the bonding cannot upset the balance.) We thus conclude that the orbitals in Fig. 2*a* are *bonding* and those in Fig. 2*b* *antibonding*. The bond length between the two silicon atoms in the structure is 0.235 nm, whereas in diamond it is 0.154 nm. Thus, overlaps in silicon must be much smaller than in diamond so that the difference between the bonding and the antibonding states must be much smaller in silicon than in diamond. (We are ignoring here, of course, the different nature of the orbitals in these two materials and a host of other effects, but the result suggested by our argument is indeed supported by detailed calculations.) Although this result will possibly be valid for one value of **k** and for the two bands belonging to it which correspond to the bonding and antibonding cell functions discussed, the resulting effect happens to affect the positioning of whole bands and thus to determine that the energy gaps in diamond and silicon are large and small, respectively, therefore making diamond an insulator and silicon a semiconductor, as we shall show later on. It is somewhat awe-inspiring to reflect that it is this very simple effect that makes our silicon-based civilization possible!

The Brillouin zone and **k**-vector symmetries

The Bravais lattice of the diamond structure is face-centred cubic, so that the Brillouin zone is the one discussed in Fig. **8**-1.1 and shown here in Fig. 3. Great care, though, must be exercised in interpreting this figure: although the overall group of the polyhedron shown is $\mathbf{O}_h$, which is the point group of G, (the space group of the diamond structure of Fig. 1), not all operations of $\mathbf{O}_h$ are symmetry operations of G. We have seen, e.g., that $\mathbb{i}$, the inversion at O in Fig. 1, which is an operation of $\mathbf{O}_h$ is not in G but, rather, that it appears as $\{\mathbb{i}|\mathbf{v}\}$. Let us consider the null **k** vector Γ, which is obviously left invariant by all operations of $\mathbf{O}_h$. Its group, however, will not be $\mathbf{O}_h$ but rather a group in which $\mathbb{i}$ in $\mathbf{O}_h$ is replaced by $\{\mathbb{i}|\mathbf{v}\}$, and similarly for all the operations of $\mathbf{O}_h$ which contain the inversion. The group of Γ is nevertheless pretty similar to $\mathbf{O}_h$ and with a little care comparison with $\mathbf{O}_h$ is both possible and useful. Because of its importance in silicon, we are interested in the s and p orbitals. A table of the irreducible characters of $\mathbf{O}_h$ will show that the s orbitals belong to the irreducible representation A_{1g} (singly degenerate and totally symmetric under all operations). Likewise, the s orbitals in the group of Γ belong to

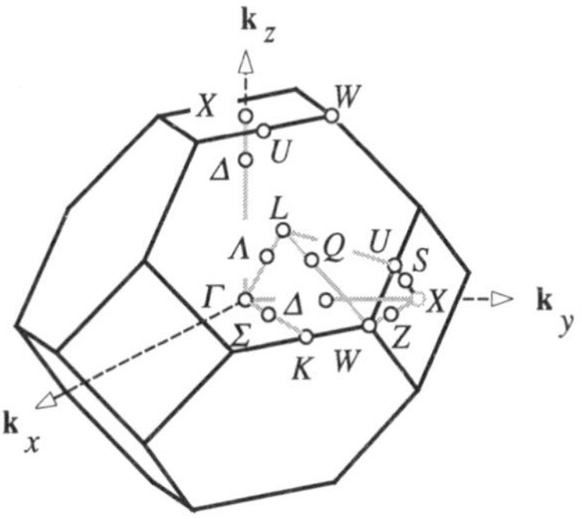

Fig. 9-1.3. The Brillouin zone for silicon. One **k** vector of each possible type of star is shown, except for the points Δ, X, U, and W, for which one other element of the star is given.

a singly degenerate, totally symmetric representation, conventionally called Γ_1. The three p orbitals p_x, p_y, p_z, belong to a triply degenerate representation of $\mathbf{O}_h$, called T_{1u}, the u indicating that this representation is antisymmetric with respect to $\mathbb{i}$. (The p orbitals change sign under inversion, that is, they are ungerade.) These orbitals, however, when set as shown in Fig. 2*a* are now *symmetric* with respect to $\{\mathbb{i}|\mathbf{v}\}$, the inversion at the mid-point between the two atoms of the basis and they belong to a triply degenerate representation called Γ'_{25}. One other important small representation of Γ is Γ_{15}, also triply degenerate but antisymmetric in $\{\mathbb{i}|\mathbf{v}\}$, which is spanned by the antibonding p orbitals shown in Fig. 2*b*.

We shall now consider a **k** value Δ, that is, any intermediate point on the line ΓX, which we take along the z axis. If there was only one atom per unit cell of the crystal, the group of this **k** vector would be $\mathbf{C}_{4v}$, the group of symmetry operations of a square pyramid. Although this group has to be modified in the diamond structure, because the vertical reflection planes are no longer in G as pure point group operations, the basic features of the representations of $\mathbf{C}_{4v}$ (see Table **6**-3.2) remain. There is the totally symmetric representation Δ_1, to which the s functions belong, and the three p orbitals split between two irreducible representations, one singly degenerate, spanned by p_z and called Δ'_2, and one doubly degenerate, spanned by p_x, p_y and called Δ_5. As regards X, all we need to know is that *all* its small representations are two-dimensional (which would never be the case for a symmorphic group!). These features of the small representations will be essential in order to understand the silicon band structure.

The band structure of silicon

We display in Fig. 4 a part of the band structure of silicon which will be sufficient in order to understand the basic features of this material. In reading

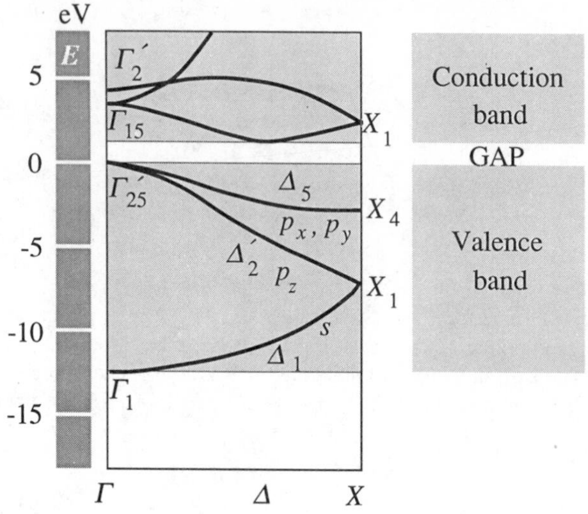

Fig. 9-1.4. The band structure of silicon along the ΓX direction (after Lodge, Altmann, Lapiccirella and Tomassini, 1984). All energy levels have been referred to the top of the valence band, Γ'_{25}.

such bands, the first thing one must do is to find out how the energy levels are filled. Silicon has two atoms each with four electrons (s, p_x, p_y, p_z) in the primitive cell, and we know that each band will accommodate two electrons per primitive cell. If a band is doubly degenerate, however, it will take four electrons, and so on. Thus, the singly degenerate bands Δ_1 and Δ'_2 will take each two electrons (s and p_z, respectively), four in total. The Δ_5 band is doubly degenerate and will take a further four electrons, raising the total to the eight electrons which we have in the primitive cell. Thus the first three bands in the figure will be fully occupied by the eight valence electrons and they are accordingly called the *valence band*. All the bands above the valence band are empty at 0 K and they are separated from the valence band by an energy gap, as shown, of 1.17 eV. For pure silicon at 0 K the electrons at the Fermi level, that is at the top of the valence band, cannot be excited by an electric field, because the energy provided by it is too small to surmount the energy gap. Thus conductivity through electron carriers will arise only if, in some way to be explained, the bands above the energy gap become occupied, for which reason these bands are called the *conduction band*. There are two ways in which the conduction band can become partly occupied. First, we can have thermal excitation of the electrons at the Fermi level. This, however, entails energies of only about 0.03 eV at room temperature, so that pretty high temperatures would be required even to surmount the small band gap of silicon. This type of conductivity is called *intrinsic semiconductivity* (see §8-4), but it is not very significant in a material like silicon, owing to the size of its band gap. On the other hand, impurities may create energy states right in the band gap (see § **14**-1) and give rise to *extrinsic semiconductivity* (§ **8**-4), which is the important effect in silicon. It should be noted that in these cases the Fermi energy lies in the band gap, which is an important feature of semiconductors.

There are two ways in which extrinsic semiconductivity can be manifested in silicon, which depend on the choice of impurities with which this material is doped. Phosphorus is the first element on the right of silicon in the periodic table and it therefore has one extra valence electron. Impurity states generated by phosphorus will thus be occupied by electrons provided by the impurity atom (*donor*) and these levels in the band gap are sufficiently near the bottom of the conduction band so as to permit the electrons to be promoted thermally to states in the conduction band. The carriers in this type of material are *electrons* in the conduction band. Such crystals are called *n-type semiconductors* (because the carriers are *negative*).

In a *p-type semiconductor* the impurity atoms are positively charged with respect to the matrix. If the material is doped with aluminium, this has only three valence electrons, (being on the left of silicon in the periodic table). Impurity levels are created, as always, in the band gap and near the top of the valence band. (How near, it will depend on the choice of atom; boron or gallium, with the same number of valence electrons as aluminium, may be used, to obtain levels sufficiently near the top of the valence band.) Because these levels have fewer electrons than the silicon structure sites, they are *acceptor* levels which attract electrons. These, naturally, will come from the nearest states, that is, from the top of the valence band, thus creating *hole* states. Conductivity carriers in this material will be holes instead of electrons, and thus are positively charged.

What makes silicon a useful semiconductor is the fact that the energy gap between the valence and the conduction bands is small enough. In diamond, instead, the band structure of which is qualitatively very similar to that of silicon, the gap is 5.4 eV and the material is in normal conditions an insulator, since the energy gap is too high either to be directly surmounted or to be bridged by impurity states. This difference between diamond and silicon can easily be understood from the figure. The determining feature from the point of view of the energy gap is the position of the level Γ_{15} above Γ'_{25}: if Γ_{15} were raised much higher the whole of the bottom of the conduction band would be raised in the same manner. Ar Γ, $\{\bar{\mathbb{1}}|\mathbf{v}\}$, that is the inversion at the mid-point between the two atoms of the basis, is a symmetry operation of the group of the **k** vector, so that the orbitals in the crystal basis (primitive cell) must be symmetrical (bonding) or antisymmetrical (antibonding) with respect to this operation, as shown in Figs 2*a* and 2*b* respectively. We have seen that these orbitals correspond to the small representations Γ'_{25} and Γ_{15} respectively. We have also seen that the difference in energy between these bonding and antibonding states must be much larger in diamond than in silicon, because of its short bond length and correspondingly large overlaps. This explains why the energy gap in diamond is so much larger than in silicon. Likewise, on comparing silicon (bond length 0.235 nm) with germanium (bond length 0.245 nm, see Problem 4.1) we should expect the band gap to diminish further in the latter material, which is in fact the case: the band gap is 0.74 eV, thus making intrinsic semiconduction more significant for germanium than for silicon.

One feature of the above description must be stressed, and it concerns the **k** vector X. We have seen that in this case *all* the small representations are two-dimensional, a feature which can only appear in non-symmorphic space groups. For symmorphic space groups, because the small representations must be irreducible representations of point groups, only a few multi-dimensional representations may exist at a given **k**, which means that at that particular **k** some bands will touch or cross for the levels which correspond to these representations. In the case of silicon, instead, the successive bands must always *stick together* in pairs at X. (Notice in Fig. 4 that this is still true for the second level at X, X_4, the bands p_x and p_y touching there, as they do all along the ΓX line in this band.) In some non-symmorphic space groups it is indeed possible for this sticking together of bands to occur over all points of a Brillouin zone face, which thus will not be a surface of energy discontinuities (see § **10**-1).

Guessing band trends

It is possible to provide plausible arguments, in the spirit of § **6**-4, to explain the form of the s, p_z, and p_x, p_y bands in Fig. 4. As in that section, the key to the work is the eigenvalue equation (**6**-1.2) of the Bloch functions, whereby, on translation by $\{E|\mathbf{t}\}$ they are multiplied by $\exp(-\mathrm{i}\mathbf{k}\cdot\mathbf{t})$. Thus, for $\Gamma(\mathbf{k}=\mathbf{0})$ the factor is unity for all $\{E|\mathbf{t}\}$, indicating that the Bloch function is totally periodic. Consider now how the Bloch function with **k** equal to X (that is $\frac{1}{2}\mathbf{c}^{\#}$) behaves under a displacement by **c** along the z axis. It must be multiplied by the factor

$$\exp(-\mathrm{i}\mathbf{k}\cdot\mathbf{t}) = \exp(-\tfrac{1}{2}\mathrm{i}\mathbf{c}^{\#}\cdot\mathbf{c}) = \exp(-\tfrac{1}{2}\mathrm{i}2\pi) = -1. \qquad (1)$$

We now assume that around each atom of the basis the wave function is reasonably approximated by the s, p_z, and p_x, p_y orbitals for the corresponding bands, at least for **k** vectors of high symmetry. We thus show in Fig. 5 how they vary along the **c** axis for Γ (when they are periodic) and for X (when, from eqn 1, they change sign from one cell to the next). We should really represent in the figure the two atoms of the basis in the primitive cell, but the same picture is repeated for the second atom. We must use again the rule of thumb whereby the larger the number of nodes the higher the energy of the wave function. (In fact, in the free-electron picture, the energy is proportional to $\mathbf{k}^2$ and the larger the wave number **k** is, and thus the energy, the larger

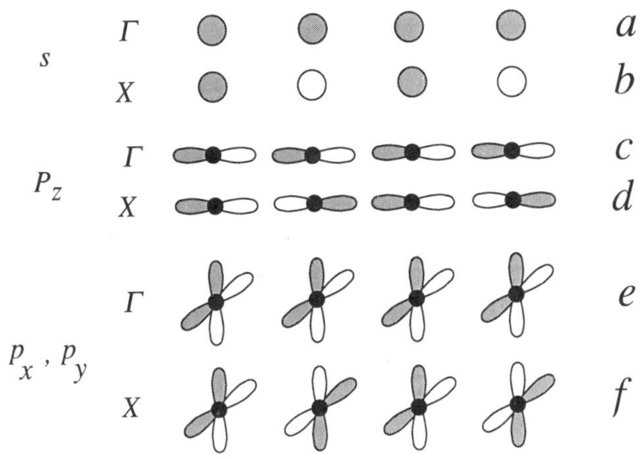

Fig. 9-1.5. Behaviour of the Bloch functions for the s, p_z, and p_x, p_y bands of silicon along the z axis, which is parallel to **c** as described in Fig. 1.1. The interatomic distance shown is equal to $|\mathbf{c}|$ in all cases. Only one of the two atoms of the basis is shown in each cell.

must be the number of nodes in the wave.) On using this rule in comparing Figs 5*a* and *b*, we see that the energy should increase from Γ to X for the s band, as is indeed the case in Fig. 4. For the p_z band we find, in Figs 5*c* and *d*, that, at Γ, there are nodes at the atomic sites as well as at the mid-points of the bonds, whereas at X only the atomic nodes remain. Thus, X should have lower energy than Γ, as indeed found in Fig. 4. Finally, for the p_x, p_y band, the overlap between cell and cell must be much smaller than in the p_z band, because the orbitals are now transverse to the **c** direction (which is the direction of propagation for this **k** vector). Thus, the change of phase from Fig. 5*e* to Fig. 5*f* should not affect the energy much, so that this band should be rather flat, as shown in Fig. 4. (Counting nodes in the figure, however, would indicate X to have higher energy than Γ, contrary to the correct result, but you must remember that the other atom of the basis should also be considered, which accounts for the difference.) It is also useful to recognize that the reason why p_x, p_y are degenerate is that the C_4 rotation around the z axis (rotation by $\frac{1}{2}\pi$) transforms one into the other, whereas there is no symmetry operation which will transform these orbitals into p_z. This degeneracy, however, must disappear for a **k** vector away from the ΓX line, since such a **k** vector would break the rotational symmetry around the **c** axis.

2 Bands and Fermi surfaces in copper

Copper is a noble metal with the electronic structure $1s^2, 2s^2, 2p^6, 3s^2, 3p^6, 3d^{10}, 4s$. As it is characteristic in the noble metals, the $3d^{10}$ shell is sufficiently below the 4*s* level not to interact with it very much, but it is not low enough in energy for it to be considered a core level and thus irrelevant in the formation of the metallic bands. These must therefore be constructed from all the eleven $3d^{10}$, 4*s* electrons, although we must expect that the 4*s* electrons will contribute the dominant features of the band system. Copper is a face-centred cubic metal, so that its Brillouin zone is that of Fig. 1.3, but things are now much simpler than in silicon, since there is no basis in the primitive cell so that the space group is symmorphic. The bands for copper are shown in Fig. 1 along several directions in the Brillouin zone. In order to discuss the bands, we shall concentrate our attention on the ΓX direction, as we did for silicon. If we first look at the Δ_1 band, as completed by the free-electron band drawn with a grey line, (see Problem 4.4) we observe a broad band, which we can immediately identify as corresponding to the 4*s* level, since the Γ_1 level is spanned by *s* orbitals and thus, just as in silicon, the Δ_1 band is of *s* type. This *s* band is of some 12 eV in width. The narrow bands which are superimposed on the 4*s* band, from about 4 to about 7.5 eV above its bottom, are the 3*d* bands. There are four of them but, because Δ_5 is doubly degenerate (as in silicon, see § 1), they carry ten electrons, as they should. It is clear that there is a strong interaction between the 3*d* and the 4*s* bands, the latter retaining its individuality only at the bottom and at the top of the band.

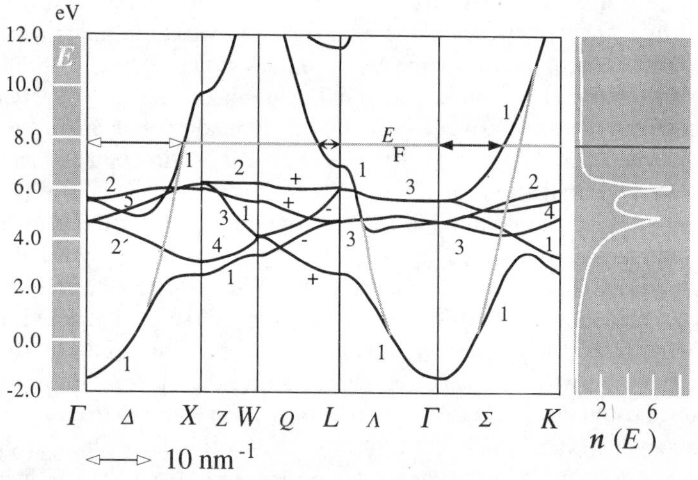

Fig. 9-2.1. The band structure and density of states for copper, adapted from Papaconstantopoulos (1986). The grey lines indicate the free-electron bands. In order to simplify the figure, the irreducible representations corresponding to each band, for the general points along the symmetry lines shown (which are given in smaller size along the horizontal scale), are denoted only by the suffices which should be added to the symbol of the corresponding point. Thus the first numeral 1 on the left indicates Δ_1 and so on. The density of states $n(E)$ is given in electron states per eV.

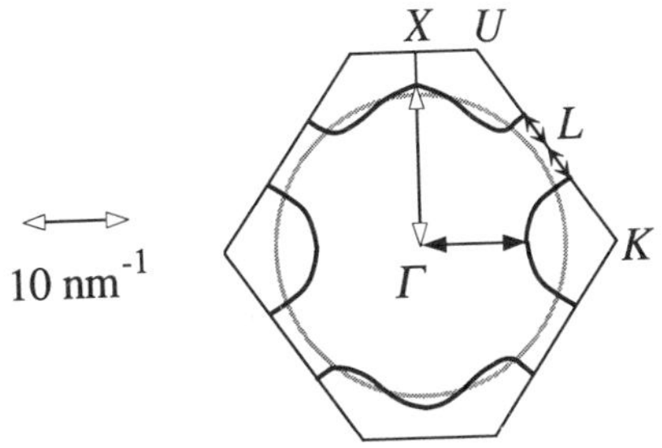

Fig. 9-2.2. A section of the Fermi surface of copper, as obtained from the energy bands in Fig. 1. The free-electron spherical Fermi surface is shown in grey.

The 4s band must be only half full, since it carries only one electron, and this roughly determines the position of the Fermi level. As shown in the figure, the 4s band is full from Γ to X, from Γ to K, and from L to W only along those parts of these lines as are shown along the E_F line by means of segments marked with arrowheads at both ends.

Some of the more important features of the Fermi surface of copper can be recognized at once from the bands in Fig. 1, and we do this in Fig. 2, where the section $\Gamma X U K$ of the Brillouin zone of Fig. 1.3 is shown on using exactly the same scale as in Fig. 1. (See Problem 4.2.) It is very easy to calculate the radius of the free-electron sphere that contains exactly one electron, since its volume must be one half the volume of the Brillouin zone (see Problem 4.2). Because of the anisotropy of the Brillouin zone the free-electron Fermi surface almost touches the Brillouin zone hexagonal face at L, which is rather surprising, since it is easy to imagine that a sphere with half the volume of the zone would be much smaller, whereas its radius is actually 90% of the ΓL length. The shape of the Fermi surface for the copper bands is obtained as follows. Along ΓX and ΓK the occupied region of the bands is given by the doubly-arrowed segments of Fig. 1, which are exactly copied along the appropriate directions of Fig. 2. They indicate, of course, the edge of the Fermi surface along those directions. It is important to see what happens along UL since we know (see Problem 4.5) that $\partial E/\partial \mathbf{k}$ must vanish along this line, so that the free-electron curve should be distorted here so as to give a vanishing slope. Unfortunately, UL has not been computed, but from various symmetry considerations one knows that this line must be pretty much the same as WL (see Fig. 1.3), which is occupied from L to W, as shown by the corresponding double arrow in Fig. 1. We thus find that, as expected, the Fermi sphere of Fig. 2 is distorted near L forming a neck, which can be clearly seen in the Fermi surface shown in Fig. 3. It should be noticed, however, that the bulk of the Fermi surface is not very much distorted from the free-electron Fermi sphere.

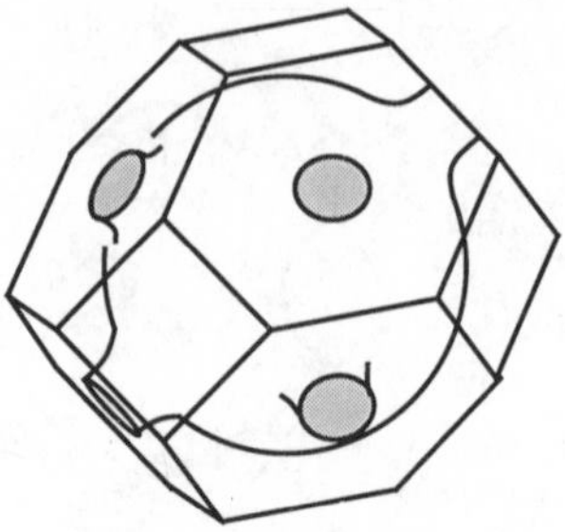

Fig. 9-2.3. The Fermi surface of copper in the Brillouin zone of Fig. 1.3.

It must be appreciated that despite the interaction of the *d* and *s* bands evident in Fig. 1, the Fermi surface of copper is almost entirely determined by the behaviour of the *s* band. Thus most of the major properties of copper result from the *s* band and they show comparatively little influence of the *d* electrons. We can see how this is the case by considering the density of states $n(E)$ (see § **1**-5) computed from the bands in Fig. 1 and shown on the right of this figure. It is important to recognize that the peak in this figure comes from the *d* bands which are precisely in this region of the energy scale: their density of states must be much larger than that of the *s* band for the following reason. From Fig. 1, if N is the number of primitive cells in the crystal structure, the *s* band has N states, not counting spin, spread over a range of energy of some 12 eV, whereas the *d* bands have $5N$ states over some 3 to 4 eV. We see that the Fermi energy E_F appears well after the peak of the density of states and, therefore, that the density of states at the Fermi surface is low and characteristic of an *s* rather than a *d* band.

3 Noble and transition metals

The density of states in copper from Fig. 2.1 can be considered as resulting from the superposition of a broad *s* band of low density and a narrow *d* band of high density. What characterizes a noble metal is that the *d* band is well below the Fermi energy and therefore that the density of states at the Fermi energy is low. In a *transition metal* like nickel or tungsten, the *d* levels in the free atom are much nearer in energy to the outer *s* level. As a result, the *d* bands in Fig. 2.1 move up. We illustrate this situation in Fig. 1, where we see that for a transition metal the *d* band moves to the right (higher energies) in such a way that the Fermi surface now cuts across it. Not only this produces a high density of states at the Fermi energy but it also means that some of the *d* states are now empty, which has important consequences on the magnetic properties of these materials, because a partly empty *d* band will

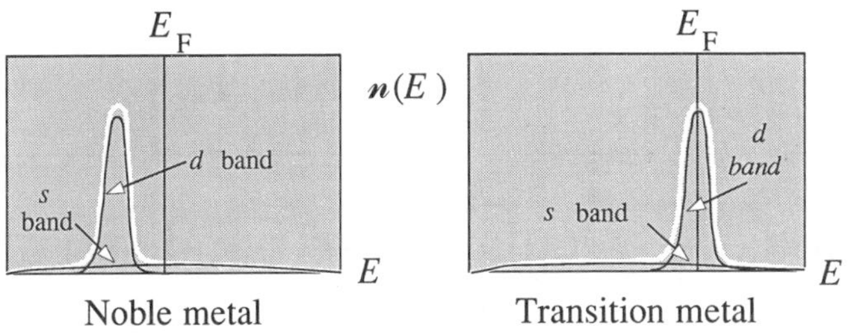

Fig. 9-3.1. Density of states for noble and transition metals. The total density of states is given by the white lines. The states above the Fermi energy E_F are unoccupied.

lead to net free spins. Notice that, of course, the density of states of a noble metal is a fairly idealized version of that for copper in Fig. 2.1.

4 Problems

1. Prove that, in the diamond crystal pattern of Fig. 1.1, the bond length d is given by the relation $d = 0.433\,a$ in terms of the lattice constant a. The lattice constant for C, Si, and Ge equals 0.357 nm, 0.543 nm, and 0.566 nm respectively. Calculate the corresponding bond lengths.
2. The lattice constant of the face-centred cubic unit cell in copper is 0.361 nm. Draw Fig. 2.2 from first principles in a scale of, say, 2 mm to 1 nm^{-1}. Find the radius of the free-electron Fermi surface for one electron per atom and plot it in the figure.
3. Find the components in units of the lattice constant of the reciprocal lattice of the following points of the Brillouin zone for copper (see Problem 2): Γ, X, W, L, K. Find the lengths in nm^{-1} of the lines ΓX, XW, WL, ΓL, ΓK. Compare your results with the figure drawn in Problem 2. Find the ratio between the radius of the free-electron Fermi surface (from Problem 2) to ΓL.
4. Find the free-electron energies in copper (see Problem 2) for the points X, L, K. Compare your results with Fig. 2.1. Compare the free-electron Fermi energy of copper obtained in Problem **1**-9.5 with the value of E_F in Fig. 2.1.
5. Show that $\partial E/\partial \mathbf{k}_n$ vanishes along the UL line in the face-centred cubic Brillouin zone of Fig. 1.3.

Further reading

See Rosenberg (1988), Ashcroft and Mermin (1976), Blakemore (1985), and Kittel (1986).

10

Methods of calculation of band structures

In parallel with the development of modern computers in the last twenty or thirty years, several methods for the calculation of the energy eigenvalues $E(\mathbf{k})$ in crystals have been perfected, and computer programs now exist which make the calculation of band structures a fairly routine matter. Some results of such calculations have already been discussed in the last chapter. No attempt, however, will be made in this chapter to give an account of the accurate methods of numerical calculation now available. The discussion will concentrate instead on the methods that provide some simple, although approximate, insights into the main features of band structures.

One of the major problems in band structure calculations is the determination of the crystal potential field $V(\mathbf{r})$. In principle, one can start from calculated atomic potentials defined in convenient regions around each crystal site. Various effects have then to be considered. To start with, the electron cloud associated with the metal electrons throughout the crystal screens the atomic-like potentials mentioned in various degrees depending on the nature of the material. If this screening is very strong the final potential will be a nearly free-electron one because the charges of the atomic cores will be largely neutralized, but if the screening is weak the final potential will be largely atomic-like.

It should be appreciated that, unfortunately, the screening exerted by an electron distribution depends on its calculated wave function which should be the end result of all this work. It is thus often necessary to use *self-consistent field methods* in which one starts with estimated wave functions which allow the screening and thus the potential field to be computed. Once new wave functions are calculated with this field a new potential is calculated and the process is iterated until the improved wave functions agree with those used in working out the potential determined in the previous step of the iteration. Screening is of course a manifestation of electron–electron interactions and it must be appreciated that these are not merely of a semi-classical or coulombic type but that allowance has to be made for exchange interactions, for which appropriate correction terms have to be added to the potential. Various detailed prescriptions are provided for determining all these effects and can be obtained from the references at the end of this chapter.

1 The nearly free-electron method

We have made frequent use of the free-electron approximation in a qualitative fashion, and we shall now put it into a more quantitative form. We shall assume that the kinetic energy term $\mathbb{T}$ is much larger than the potential energy $\mathbb{V}$ in the Hamiltonian $\mathbb{H}$,

$$\mathbb{H} = \mathbb{T} + \mathbb{V}, \tag{1}$$

so that the eigenfunctions of the operator

$$\mathbb{T} = \mathbb{p}^2/2m \tag{2}$$

are a good starting point. From (**4**-10.3), the eigenfunctions of $\mathbb{T}$ are the plane waves

4-10.2
$$|\,\mathbf{k}\,\rangle = V^{-1/2} \exp{(\mathrm{i}\mathbf{k}\cdot\mathbf{r})}. \tag{3}$$

Notice here that, as a difference with (**4**-10.2), we now normalize the plane waves over the crystal volume and not over that of the unit cell. This is important, since we shall want to express the crystal eigenfunctions in terms of them. (See Problem **5**-7.4.) That these functions are eigenfunctions of the kinetic energy is easily proved, bearing in mind the change stated:

4-10.3
$$\mathbb{T}\,|\,\mathbf{k}\,\rangle = \frac{1}{2m}\mathbb{p}^2\,|\,\mathbf{k}\,\rangle = \frac{\hbar^2\mathbf{k}^2}{2m}\,|\,\mathbf{k}\,\rangle = E_{\mathbf{k}}\,|\,\mathbf{k}\,\rangle, \tag{4}$$

where we write $E_{\mathbf{k}}$ for the *free-electron* energy. The next thing we want to do is to write the eigenfunction $\psi(\mathbf{r})$ of the full Hamiltonian as a linear combination of n of the plane waves (3):

$$\psi(\mathbf{r}) = \sum_{\mathbf{k}} c_{\mathbf{k}}\,|\,\mathbf{k}\rangle. \tag{5}$$

This technique is well known in elementary quantum mechanics under the name of the *linear variational method.* In this method, whenever the wave function is written, as in (5), as a linear combination of some known functions, then the coefficients $c_{\mathbf{k}}$ are varied until the eigenvalues ε of the total Hamiltonian,

$$\mathbb{H}\,\psi(\mathbf{r}) = \varepsilon\,\psi(\mathbf{r}), \tag{6}$$

are minimized. It is shown that a necessary condition arises for this to be the case:

$$\det{(H_{\mathbf{kk}'} - \varepsilon\, S_{\mathbf{kk}'})} = 0. \tag{7}$$

Here, $\mathbf{k}$ and $\mathbf{k}'$ range over the n values of $\mathbf{k}$ in the summation (5) and

$$H_{\mathbf{kk}'} = \langle\mathbf{k}\,|\,\mathbb{H}\,|\,\mathbf{k}'\rangle, \tag{8}$$

$$S_{\mathbf{kk}'} = \langle\mathbf{k}\,|\,\mathbf{k}'\rangle. \tag{9}$$

The angular brackets here must be understood, as in (**4**-9.3), as an integral between the complex conjugate of $|\mathbf{k}\rangle$ in (3) and $|\mathbf{k}'\rangle$ as defined by (3), but with the important difference that this integral is now *over the crystal volume* V and not over the unit cell. $H_{\mathbf{kk}'}$ and $S_{\mathbf{kk}'}$ are called the *matrix element* and the *overlap integral*, respectively. Because of the range of $\mathbf{k}$ and $\mathbf{k}'$ in (7), the matrix in it is of dimension n by n, whence the so-called *secular determinant* (7) is an n by n *algebraic equation* in ε, which introduced in a computer will quickly yield n roots, that is, the first n eigenvalues in (6).

All this looks pretty good, but, alas, the method described will be useless unless we are quite a bit more clever. This is so because one of the most useful results of group theory is that integrals like (8) or (9) vanish unless the functions which they contain on their right and left belong to the same irreducible representations. Plane waves are particular cases of the Bloch functions (**5**-6.9) and, therefore, span irreducible representations of the translation group T, $|\mathbf{k}\rangle$ being a basis of the representation ${}_{\mathbf{k}}\hat{T}$. It can thus readily be seen that if n different functions $|\mathbf{k}\rangle$ are used in (5), with n corresponding energies $E_{\mathbf{k}}$, then the sad result of the whole work will be to obtain again the original n states, each with precisely the same energy $E_{\mathbf{k}}$ as before. (See Problem 5.2.) This result is, in any case understandable, since if all the matrix elements and overlaps vanish, except the diagonal ones $H_{\mathbf{kk}}$ and $S_{\mathbf{kk}}$, there is no mixing of the n functions $|\mathbf{k}\rangle$ in (7) and then the expansion (5) does not make sense. We thus realize that we must ensure that the n functions which appear in (5) must all belong to the same irreducible representation of T. We know that the only way in which $|\mathbf{k}\rangle$ and $|\mathbf{k}'\rangle$ will belong to the same irreducible representation of T will be when $\mathbf{k}'$ equals $\mathbf{k}+\mathbf{g}$, as follows from (**5**-3.4). That is, all the values of $\mathbf{k}$ in (5) must be equivalent to a single $\mathbf{k}$, which means that we have to have plane waves of the form $|\mathbf{k}+\mathbf{g}\rangle$, for a constant $\mathbf{k}$ and for the vector $\mathbf{g}$ of the reciprocal lattice ranging over n values (to be conveniently chosen):

$$\psi_{\mathbf{k}}(\mathbf{r}) = \sum_{\mathbf{g}} c_{\mathbf{g}} |\mathbf{k}+\mathbf{g}\rangle. \tag{10}$$

It is extremely gratifying, of course, that we can now identify the $\mathbf{k}$ value of the function on (L10), which we could not do before in eqn (5).

The secular determinant (7) will be written as follows from the expansion (10):

$$\det(H_{\mathbf{gg}'} - \varepsilon\, S_{\mathbf{gg}'}) = 0, \tag{11}$$

with

8 $$H_{\mathbf{gg}'} = \langle \mathbf{k}+\mathbf{g} \,|\, \mathbb{H} \,|\, \mathbf{k}+\mathbf{g}'\rangle. \tag{12}$$

9 $$S_{\mathbf{gg}'} = \langle \mathbf{k}+\mathbf{g} \,|\, \mathbf{k}+\mathbf{g}'\rangle. \tag{13}$$

It is easy to see in (13) that the exponentials in $-\mathbf{k}$ on the left (negative because of the implied conjugation) and in $\mathbf{k}$ on the right cancel, so that

,4-9.9 $$S_{\mathbf{g}\mathbf{g}'} = \langle \mathbf{k}+\mathbf{g} \,|\, \mathbf{k}+\mathbf{g}' \rangle = \langle \mathbf{g} \,|\, \mathbf{g}' \rangle = \delta(\mathbf{g}, \mathbf{g}'). \tag{14}$$

On the other hand,

1|12 $$H_{\mathbf{g}\mathbf{g}'} = \langle \mathbf{k}+\mathbf{g} \,|\, \mathbb{T} + \mathbb{V} \,|\, \mathbf{k}+\mathbf{g}' \rangle \tag{15}$$

$$= \langle \mathbf{k}+\mathbf{g} \,|\, \mathbb{T} \,|\, \mathbf{k}+\mathbf{g}' \rangle + \langle \mathbf{k}+\mathbf{g} \,|\, \mathbb{V} \,|\, \mathbf{k}+\mathbf{g}' \rangle. \tag{16}$$

We do two things on (R16). One is to apply (4):

4 $$\mathbb{T} \,|\, \mathbf{k}+\mathbf{g}' \rangle = E_{\mathbf{k}+\mathbf{g}'} \,|\, \mathbf{k}+\mathbf{g}' \rangle. \tag{17}$$

(Remember that the E's are the *free-electron* eigenvalues in eqn 4, so that $E_{\mathbf{k}+\mathbf{g}'}$ is not the same as $E_{\mathbf{k}}$!) The other is to take into account that the potential energy operator depends only on the position vector $\mathbf{r}$, and that it is thus nothing else than the potential function $V(\mathbf{r})$, which we can expand over the crystal in Fourier series,

4-9.25 $$V(\mathbf{r}) = \sum_{\mathbf{g}''} \mathscr{V}_{\mathbf{g}''} \,|\, \mathbf{g}'' \rangle, \tag{18}$$

where the coefficients $\mathscr{V}_{\mathbf{g}''}$ are the structure factors (**4**-9.26). We must be careful when introducing (18) into (R16). From (3), the right-hand side of the second integral in (16) will contain two exponential factors corresponding to $|\mathbf{g}''\rangle$ and $|\mathbf{k}+\mathbf{g}'\rangle$. On adding up exponents, these will give the plane wave $|\mathbf{k}+\mathbf{g}'+\mathbf{g}''\rangle$, except that, whereas the original product contains two factors $V^{-1/2}$, the symbol $|\mathbf{k}+\mathbf{g}'+\mathbf{g}''\rangle$ carries only one, as given in (3). It is therefore necessary to include explicitly the missing extra factor $V^{-1/2}$:

18,17|16 $$H_{\mathbf{g}\mathbf{g}'} = E_{\mathbf{k}+\mathbf{g}'} \langle \mathbf{k}+\mathbf{g} \,|\, \mathbf{k}+\mathbf{g}' \rangle$$

$$+ V^{-1/2} \sum_{\mathbf{g}''} \mathscr{V}_{\mathbf{g}''} \langle \mathbf{k}+\mathbf{g} \,|\, \mathbf{k}+\mathbf{g}'+\mathbf{g}'' \rangle. \tag{19}$$

14|19 $$= E_{\mathbf{k}+\mathbf{g}'} \delta(\mathbf{g}, \mathbf{g}') + V^{-1/2} \sum_{\mathbf{g}''} \mathscr{V}_{\mathbf{g}''} \delta(\mathbf{g}, \mathbf{g}'+\mathbf{g}''). \tag{20}$$

Equations (14) and (20), in principle, solve our problem, since all that remains is to feed them into the secular determinant (11). We shall be concerned, however, with a more simplified and interesting use of these equation.

The band gap

In order to get an approximate expression for the band gap, we shall simplify drastically the expansion (10), by taking only two values of $\mathbf{g}$ in it, namely $\mathbf{0}$ and one other vector of the reciprocal lattice, which we shall still denote as

g. The secular determinant (11) will therefore contain only four elements, with suffices **00**, **0g**, **g0**, and **gg**, respectively:

14 $$S_{\mathbf{00}} = S_{\mathbf{gg}} = 1 \qquad S_{\mathbf{0g}} = S_{\mathbf{g0}} = 0. \tag{21}$$

20 $$H_{\mathbf{00}} = E_{\mathbf{k}} + V^{-1/2}\mathscr{V}_{\mathbf{0}}. \tag{22}$$

20 $$H_{\mathbf{0g}} = V^{-1/2}\sum_{\mathbf{g}''}\mathscr{V}_{\mathbf{g}''}\,\delta(\mathbf{0}, \mathbf{g}' + \mathbf{g}'') = V^{-1/2}\mathscr{V}_{-\mathbf{g}} = V^{-1/2}\mathscr{V}^*_{\mathbf{g}}. \tag{23}$$

In the last step here we have expressed $\mathscr{V}_{-\mathbf{g}}$ by taking into account (**4**-9.26) and (**4**-9.19). We now continue with the remaining matrix elements:

20 $$H_{\mathbf{g0}} = V^{-1/2}\mathscr{V}_{\mathbf{g}}. \tag{24}$$

20 $$H_{\mathbf{gg}} = E_{\mathbf{k}+\mathbf{g}} + V^{-1/2}\sum_{\mathbf{g}''}\mathscr{V}_{\mathbf{g}''}\,\delta(\mathbf{g}, \mathbf{g} + \mathbf{g}'') = E_{\mathbf{k}+\mathbf{g}} + V^{-1/2}\mathscr{V}_{\mathbf{0}}. \tag{25}$$

The secular determinant can now easily be constructed:

11 $$\begin{vmatrix} E_{\mathbf{k}} + V^{-1/2}\mathscr{V}_{\mathbf{0}} - \varepsilon & V^{-1/2}\mathscr{V}^*_{\mathbf{g}} \\ V^{-1/2}\mathscr{V}_{\mathbf{g}} & E_{\mathbf{k}+\mathbf{g}} + V^{-1/2}\mathscr{V}_{\mathbf{0}} - \varepsilon \end{vmatrix} = 0. \tag{26}$$

It is convenient to shift the energy scale for ε so as to incorporate in it the constant term $V^{-1/2}\mathscr{V}_{\mathbf{0}}$:

$$\mathscr{E} =_{\text{def}} \varepsilon - V^{-1/2}\mathscr{V}_{\mathbf{0}}. \tag{27}$$

27|26 $$\begin{vmatrix} E_{\mathbf{k}} - \mathscr{E} & V^{-1/2}\mathscr{V}^*_{\mathbf{g}} \\ V^{-1/2}\mathscr{V}_{\mathbf{g}} & E_{\mathbf{k}+\mathbf{g}} - \mathscr{E} \end{vmatrix} = 0. \tag{28}$$

28 $$E_{\mathbf{k}}E_{\mathbf{k}+\mathbf{g}} - \mathscr{E}(E_{\mathbf{k}} + E_{\mathbf{k}+\mathbf{g}}) + \mathscr{E}^2 - V^{-1}|\mathscr{V}_{\mathbf{g}}|^2 = 0. \tag{29}$$

On solving this quadratic in $\mathscr{E}$, we find:

29 $$\mathscr{E}_{\pm} = \tfrac{1}{2}(E_{\mathbf{k}} + E_{\mathbf{k}+\mathbf{g}}) \pm \tfrac{1}{2}\{(E_{\mathbf{k}} + E_{\mathbf{k}+\mathbf{g}})^2 - 4(E_{\mathbf{k}}E_{\mathbf{k}+\mathbf{g}} - V^{-1}|\mathscr{V}_{\mathbf{g}}|^2)\}^{1/2}. \tag{30}$$

We started with two free-electron states of energies $E_{\mathbf{k}}$ and $E_{\mathbf{k}+\mathbf{g}}$, respectively, and we have ended with two improved energy states $\mathscr{E}_+$ and $\mathscr{E}_-$. The difference in energy $\Delta\mathscr{E}$ between these two states is:

30 $$\Delta\mathscr{E} = \{(E_{\mathbf{k}} + E_{\mathbf{k}+\mathbf{g}})^2 - 4E_{\mathbf{k}}E_{\mathbf{k}+\mathbf{g}} + 4V^{-1}|\mathscr{V}_{\mathbf{g}}|^2\}^{1/2} \tag{31}$$

$$= \{(E_{\mathbf{k}} - E_{\mathbf{k}+\mathbf{g}})^2 + 4V^{-1}|\mathscr{V}_{\mathbf{g}}|^2\}^{1/2}. \tag{32}$$

If $E_{\mathbf{k}} - E_{\mathbf{k}+\mathbf{g}}$ is large, $\Delta\mathscr{E}$ will approximately be equal to $E_{\mathbf{k}} - E_{\mathbf{k}+\mathbf{g}}$, which means that there is no change in energy when mixing the two plane waves $|\mathbf{k}\rangle$ and $|\mathbf{k}+\mathbf{g}\rangle$. It can be seen in Fig. 1 that this will be the case whenever **k** is far away from the Brillouin zone edge. As we approach the latter, $E_{\mathbf{k}} - E_{\mathbf{k}+\mathbf{g}}$ becomes small and the correction term in (32) becomes significant. Finally, at

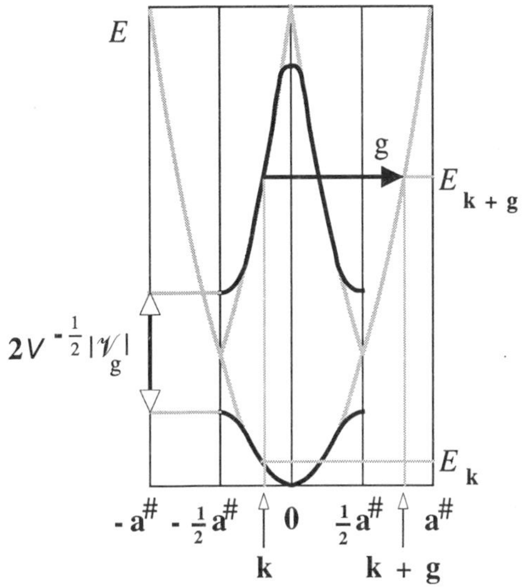

Fig. 10-1.1. The energy gap at the Brillouin zone edge. The first and second free-electron bands are depicted in the extended zone scheme. The first and second Brillouin zones of a linear chain of lattice constant a are given. The branches of the parabola in the second zone are folded back onto the first one by translations of $-\mathbf{a}^{\#}$ and $\mathbf{a}^{\#}$ for the right and left branches respectively. The states with vectors $\mathbf{k}$ and $\mathbf{k}+\mathbf{g}$ (where $\mathbf{g}$ equals $\mathbf{a}^{\#}$) are mixed in the linear variational method. $E_{\mathbf{k}}$ and $E_{\mathbf{k}+\mathbf{g}}$ are energies in the first and second free-electron bands respectively.

the Brillouin zone edge itself, $E_{\mathbf{k}}$ equals $E_{\mathbf{k}+\mathbf{g}}$, so that

32
$$\Delta\mathscr{E} = 2\,V^{-1/2}\,|\mathscr{V}_{\mathbf{g}}|. \tag{33}$$

As it can be seen from Fig. 1, this is the value of the energy gap at the edge of the Brillouin zone. It is easy to verify (Problem 5.3) that the same correction applies at the centre of the Brillouin zone for the gap between the second and third zones at $\mathbf{k}=\mathbf{0}$. Notice that in all this work we give $\mathbf{g}$ its lowest possible value ($\mathbf{a}^{\#}$), that is we deal with two contiguous bands. If this were not so, $E_{\mathbf{k}} - E_{\mathbf{k}+\mathbf{g}}$ would throughout the Brillouin zone be large, whence, from (32), the energy correction would be negligible.

We have already mentioned in § **9**-1 that in some non-symmorphic space groups Brillouin zone faces exist over which no energy discontinuity may occur. We can now discuss this very important fact in a different way. An example is the hexagonal close-packed structure, depicted in Fig. **4**-9.1. Once the reciprocal vectors $\mathbf{a}^{\#}$, $\mathbf{b}^{\#}$, $\mathbf{c}^{\#}$, are constructed, the top and bottom faces of the Brillouin zone will be separated by $\mathbf{c}^{\#}$, that is by $[001]^{\#}$, and we have seen in (**4**-9.29) that $\mathscr{V}_{\mathbf{g}}$ vanishes for this vector. Thus the first and second bands must *stick together* on these faces, and the same must be the case for all successive pairs of bands. As a result, energy discontinuities that normally exist at the Brillouin zone surfaces do not appear on these faces. This has a variety of consequences since the energy gaps, as we have seen, affect electrical conductivity and they are also influential in determining structural properties (see Chapter **12**).

Band gap classification

In the discussion of some effects which will be studied later on, such as surface states, it is useful to have a rough classification of the band gaps in two different types. Consider a simple lattice, in which $\mathscr{V}_{\mathbf{g}}$ in (33) coincides with $V_{\mathbf{g}}$:

4-9.19 $$V_{\mathbf{g}} = \langle \mathbf{g} | V(\mathbf{r}) \rangle = V^{-1/2} \int_V \exp(-\mathrm{i}\mathbf{g}\cdot\mathbf{r})\, V(\mathbf{r})\, \mathrm{d}v. \tag{34}$$

34|33 $$\Delta\mathscr{E} = 2V^{-1} \left| \int_V \exp(-\mathrm{i}\mathbf{g}\cdot\mathbf{r})\, V(\mathbf{r})\, \mathrm{d}v \right|. \tag{35}$$

Let us now consider a one-dimensional crystal of length L and period $\mathbf{a}$. We take $\mathbf{g}$ to be the first vector of the reciprocal lattice (width of the Brillouin zone) $\mathbf{a}^{\#}$. In order to perform the integration in the correct units, the component x of $\mathbf{r}$, like L itself, must be measured *in units of length* (nanometers or Å) and not in the usual units of $\mathbf{a}$. That is, $\mathbf{r}$ must be written as $x\,\mathbf{a}/|\mathbf{a}|$. Therefore,

35 $$\Delta\mathscr{E} = 2L^{-1} \left| \int_0^L \exp\left(-\frac{2\pi \mathrm{i} x}{a}\right) V(x)\, \mathrm{d}x \right|. \tag{36}$$

For a simple one-dimensional chain the potential $V(x)$ must be gerade over the unit cell and thus, because of periodicity, even over the length L of the crystal. The sine, instead, is odd, whence only the cosine term survives in (36):

36 $$\Delta\mathscr{E} = 2L^{-1} \int_0^L \cos\left(\frac{2\pi x}{a}\right) V(x)\, \mathrm{d}x. \tag{37}$$

It follows from (37) that $\Delta\mathscr{E}$ is a function of the lattice parameter a, so that it must vanish in general for one particular value of it. We have seen that all the energy states in a band will go asymptotically (that is for $a \to \infty$) into the same atomic state, as depicted in Fig. 2*a*. Here, however, $\Delta\mathscr{E}$ never vanishes as a function of a, so that the general situation must be as shown in Fig. 2*b*, which permits the classification of the energy gaps into *type* I and *type* II, corresponding to large or small interaction, respectively, between the bands.

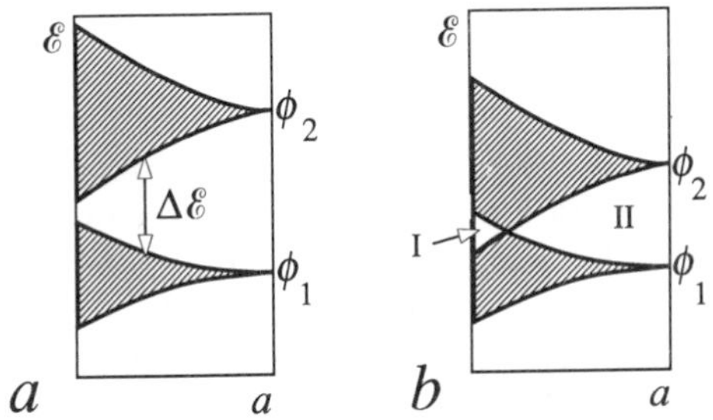

Fig. 10-1.2. Classification of band gaps. The hatched areas indicate the region occupied by the bands for each value of the lattice constant a, the horizontal scale going from zero to infinity. The energy scale $\mathscr{E}$ is defined in eqn (27). The functions ϕ_1 and ϕ_2 denote two successive atomic states.

(In silicon crystal, e.g., the gap can be considered to be of type II whereas in calcium metal it would be of type I.)

2 The tight-binding method

We have seen that the Bloch functions go over into plane waves for vanishing potentials, which supports the nearly free-electron approximation of § 1. At large interatomic distances, on the other hand, Bloch functions go over into atomic orbitals, so that it is natural in such a case to try to expand them as linear combinations of such orbitals. Let us denote with the symbols $\phi(1)$, $\phi(2), \ldots, \phi(m), \ldots$, an atomic orbital (such as the 3s orbital of silicon), at each of the atomic sites. For simplicity, we shall denote the orbital $\phi(m)$ with the symbol $|m\rangle$, so that our linear variational functions can be written as

$$\psi(\mathbf{r}) = \sum_m c_m \, |m\rangle. \tag{1}$$

Just as before, the condition for the coefficients c_m to minimize the energy is

$$\det(H_{mm'} - \varepsilon \, S_{mm'}) = 0, \tag{2}$$

with

$$H_{mm'} = \int_V \phi(m)^* \, \mathbb{H} \, \phi(m') \, \mathrm{d}v =_{\mathrm{def}} \langle m \,|\, \mathbb{H} \,|\, m' \rangle, \tag{3}$$

$$S_{mm'} = \int_V \phi(m)^* \, \phi(m') \, \mathrm{d}v =_{\mathrm{def}} \langle m \,|\, m' \rangle. \tag{4}$$

Even if we assume that we take only one orbital per site, since we have N lattice sites, the secular determinant will be of dimension $N \times N$ and thus too large to handle, even in a computer. Enormous simplification, however, may be achieved by carrying out the work in two steps. In the first we transform the functions $|m\rangle$ in (1) into linear combinations of them, $\varphi^i(\mathbf{r})$, adapted to the irreducible representations, denoted with the superscript i, of the space group G. In the second step, just as in (1), linear combinations are written in terms of the functions $\varphi^i(\mathbf{r})$. When this is done, the matrix elements in (2) will be of the form $H_{ii'}$, $S_{ii'}$, and because of the famous orthogonality property they will vanish unless i and i' are the same irreducible representation. This will cut down very considerably the size of the secular determinant.

Because, as we have seen, all the irreducible representations of G can be obtained from those of T, the first step in the symmetrization of the orbitals $|m\rangle$ is to symmetrize them with respect to T. We do this by means of the projection operator (**2**-5.4): in order to symmetrize with respect to an irreducible representation ${}^i\hat{G}$ we operate on a function ϕ with each operation g of G, multiply by the complex conjugate of the character of g in ${}^i\hat{G}$, and add up over all g. Applied to the translation group, the representations of which are labelled by the $\mathbf{k}$ vectors, ${}_\mathbf{k}\hat{T}$, we must form, for one of the atomic orbitals $\phi(n)$ or $|n\rangle$, say, the combinations

$$\sum_m \chi(\{E\,|\,\mathbf{t}_m\}\,|\,{}_\mathbf{k}\hat{T})^* \,\{E\,|\,\mathbf{t}_m\}\,|n\rangle. \tag{5}$$

We know a few things here. First, the representations ${}_\mathbf{k}\hat{T}$ are all one-dimensional, so that their characters coincide with the value $\exp(-\,\mathrm{i}\mathbf{k}\cdot\mathbf{t}_m)$ given by (**5**-2.9) for the representatives. Secondly, we can choose $|n\rangle$ as we please: if we take it to be the orbital at the origin, then the effect of $\{E\,|\,\mathbf{t}_m\}$ on this orbital is to take it to the site m, that is, to transform it into $|m\rangle$. Thirdly, the linear combination (5) will be a basis of ${}_\mathbf{k}\hat{T}$ and, therefore, a Bloch function $\psi_\mathbf{k}(\mathbf{r})$. On putting all these bits and pieces together, we have:

$$\psi_\mathbf{k}(\mathbf{r}) = \sum_m \exp(\mathrm{i}\mathbf{k}\cdot\mathbf{t}_m)\,|m\rangle. \tag{6}$$

It is usual to call *Bloch sums* the Bloch functions thus constructed. It is quite easy to verify directly that they are indeed the correct eigenfunctions of the translations (Problem 5.4) and that they are orthogonal for different $\mathbf{k}$ (Problem 5.5).

In principle, we would have to symmetrize further the Bloch sums in order to adapt them to the irreducible representations of G. Suppose, however, that we were to form linear combinations

$$\sum_\mathbf{k} c_\mathbf{k}\,\psi_\mathbf{k}(\mathbf{r}), \tag{7}$$

with coefficients chosen variationally so that a secular determinant like (2) would be formed. Now, because of the orthogonality of the Bloch functions for different **k**, all the $H_{\mathbf{kk}'}$ and $S_{\mathbf{kk}'}$ will be zero unless **k** and **k**′ are equal. The only irreducible representations of G for which there is more than one **k** in the same irreducible representation are those for which the small representation is two or three-dimensional. So, *except for such cases*, the secular determinant is fully diagonalized and the energy eigenvalues will be given by

2, 3, 4 $$\varepsilon_{\mathbf{k}} = H_{\mathbf{kk}}/S_{\mathbf{kk}} = \langle \psi_{\mathbf{k}} | \mathbb{H} | \psi_{\mathbf{k}} \rangle / \langle \psi_{\mathbf{k}} | \psi_{\mathbf{k}} \rangle. \tag{8}$$

It is important to remember, however, that we are assuming in the above that we have only one orbital per atomic site. If, like in silicon, we have to consider, say, orbitals s, p_x, p_y, p_z, at each atomic site, then each orbital will produce its own Bloch sum $\psi_{\mathbf{k}}(\mathbf{r})$ and, therefore, there will always be more than one Bloch function for each **k**, so that further symmetrization will be necessary.

An example of the use of Bloch sums and of the tight-binding method is given below and further examples are discussed in Chapter **14**. The use of Bloch sums in lattices with bases is illustrated in Problem **14**-5.8.

An example: the linear chain

We want to apply the tight-binding method to the example of the linear chain discussed in §**8**-4. We know that no representation of dimension higher than unity exists in this case, so that all that we have to do is to form the Bloch sums (6) and then introduce them into (8). As discussed in §**5**-1, for the linear chain of N atoms the translation operations are given in terms of the translation vectors

$$\mathbf{t}_m = m\mathbf{a}, \qquad m = 1, 2, \ldots, N. \tag{9}$$

The representations are labelled by the **k** vectors,

5-4.3 $$\mathbf{k} = \kappa \mathbf{a}^{\#}/N, \qquad \kappa = 1, 2, \ldots, N. \tag{10}$$

In all the work that follows we shall assume for convenience that N is even. Thus, because of the periodicity in N of κ, we can use the following, more symmetrical interval for it:

$$\kappa = -\tfrac{1}{2}N + 1, -\tfrac{1}{2}N + 2, \ldots, \tfrac{1}{2}N. \tag{11}$$

The Bloch sums can now be written immediately from (6), on remembering that $\mathbf{a} \cdot \mathbf{a}^{\#}$ equals 2π:

9, 10 | 6 $$\psi_{\mathbf{k}}(\mathbf{r}) = \sum_m \exp(2\pi \mathrm{i} \kappa m/N) \, | m \rangle. \tag{12}$$

We shall introduce some approximations and simplifications in order to form the eigenvalues (8):

$$\langle m' \,|\, m \rangle = \delta(m\, m'). \tag{13}$$

$$\langle m' \,|\, \mathbb{H} \,|\, m \rangle = E_0, \qquad m' = m, \tag{14}$$

$$= \beta, \qquad m' = m \pm 1. \tag{15}$$

$$= 0, \qquad \text{otherwise}. \tag{16}$$

Equation (13) entails the use of normalized atomic orbitals $|\, m \rangle$ plus the rather drastic approximation of neglecting all overlaps (somewhat justified however by the built-in assumption that the interatomic distances are large). Equation (14) requires recognition that the atomic orbitals $|\, m \rangle$ are eigenfunctions of $\mathbb{H}$ with eigenvalue E_0:

$$\mathbb{H} \,|\, m \rangle = E_0 \,|\, m \rangle. \tag{17}$$

17, 13 $$\langle m \,|\, \mathbb{H} \,|\, m \rangle = E_0 \langle m \,|\, m \rangle = E_0, \tag{18}$$

which agrees with (14). Equations (15) and (16) entail neglecting all except first-neighbour interactions. The parameter β is called the *exchange* (or *hopping*) *integral* and it is found in general to be negative.

In order to deal with eqn (8) we first form its denominator:

12 $$\langle \psi_{\mathbf{k}} \,|\, \psi_{\mathbf{k}} \rangle = \sum_{m'm} \exp(-2\pi i \kappa m'/N) \exp(2\pi i \kappa m/N) \langle m' \,|\, m \rangle \tag{19}$$

13|19 $$= \sum_m \exp(-2\pi i \kappa m/N) \exp(2\pi i \kappa m/N) = N. \tag{20}$$

Therefore, in (8),

$$\varepsilon_{\mathbf{k}} = N^{-1} \langle \psi_{\mathbf{k}} \,|\, \mathbb{H} \,|\, \psi_{\mathbf{k}} \rangle \tag{21}$$

12|21 $$= N^{-1} \sum_{m'm} \exp(-2\pi i \kappa m'/N) \exp(2\pi i \kappa m/N) \langle m' | \mathbb{H} | m \rangle. \tag{22}$$

On applying now eqns (14), (15), and (16) to (22), we obtain,

$$\varepsilon_{\mathbf{k}} = N^{-1} N E_0 + N^{-1} \sum_m [\exp\{-2\pi i \kappa (m+1)/N\} \exp(2\pi i \kappa m/N) + \exp\{-2\pi i \kappa (m-1)/N\} \exp(2\pi i \kappa m/N)]\beta \tag{23}$$

$$= E_0 + N^{-1}\beta \sum_m \{\exp(-2\pi i \kappa/N) + \exp(2\pi i \kappa/N)\} \tag{24}$$

$$= E_0 + N^{-1}\beta \sum_m 2\cos(2\pi\kappa/N) \tag{25}$$

$$= E_0 + 2\beta\cos(2\pi\kappa/N), \qquad \kappa = -\tfrac{1}{2}N + 1, \ldots, \tfrac{1}{2}N. \tag{26}$$

This is the final result, but in order to have a more concrete example we shall take N equal to six, in which case κ will go from -2 to $+3$. We know, by considering the stars in Fig. 7-3.1, that the κ pairs 1, -1, and 2, -2 are degenerate, whereas the levels for κ equal to 0 and 3 are non degenerate. It is thus necessary to work out (26) only for κ equal to 0, 1, 2, and 3, which we do in Table 1. In interpreting this table, it must be remembered that β is negative, so that the lowest state is that for κ equal to zero. Since we have six electrons, only the first two levels in the table will be occupied, and the Fermi level will be the energy of the state with κ equal to 1. Many readers will recognize the values of ε_κ as the energy levels of the π electrons in benzene, which is indeed a particular case of our example. These energy levels are displayed in Fig. 1 in the Brillouin zone of Fig. 7-3.1. When used in chemical applications (as in the finite-size linear chain) the tight-binding method goes under the names of the *linear combination of atomic orbitals* (LCAO) or the *Hückel approximation.* (This latter name is more specifically associated with approximations 13 to 16.)

Table 10-2.1. Energy levels for a linear chain of six atoms.

$N = 6.$

κ	$\cos(\pi\kappa/3)$	ε_κ	Degeneracy
0	$\cos 0$	$E_0 + 2\beta$	single
1	$\cos(\pi/3)$	$E_0 + \beta$	double
2	$\cos(2\pi/3)$	$E_0 - \beta$	double
3	$\cos \pi$	$E_0 - 2\beta$	single

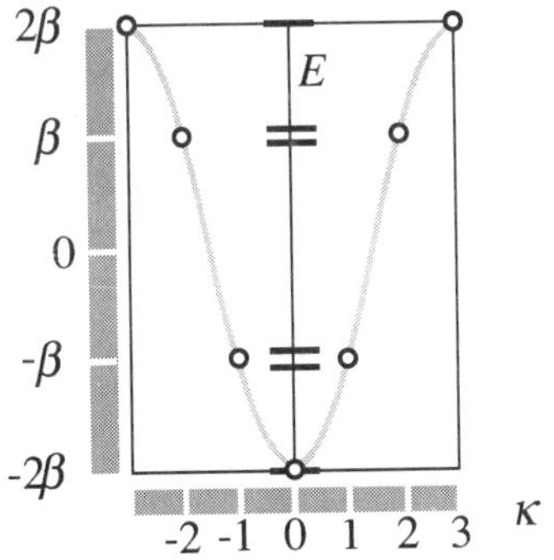

Fig. 10-2.1. The energy levels of Table 1 displayed in the Brillouin zone of Fig. 7-3.1. The zero of energy has been taken at E_0. The grey curve, representing a band, is meaningless in this discrete example but would become significant as the number of atoms increases. This figure can be taken to represent the 2p 'band' in benzene (cf. Fig. 2-6.2).

3 The orthogonalized plane waves method (OPW)

We are mainly interested in the OPW method as an introduction to the study of pseudopotentials, which is a subject of growing importance in the treatment of solids and molecular systems. The OPW method starts from the following observation. Consider the expansion (1.10), in a slightly changed notation,

$$|\psi_{\mathbf{k}}\rangle = \sum_{\mathbf{g}} c_{\mathbf{g}} \,|\mathbf{k}+\mathbf{g}\rangle, \tag{1}$$

which gives the crystal wave functions as linear combinations of plane waves. Such an expansion can behave like a wave packet (**1**-8.12) but this will only be the case when very large wave numbers, that is very short wave lengths, are involved in (R1). Such localization is required in order to represent the crystal wave function near the nuclei, when it has to coincide largely with the core functions. This means, conversely, that because of the presence of the core functions, one is forced to go up to very large values of **g** in the expansion (1) which is, of course, very time consuming and not very satisfactory because, in any case, the core functions are often of minor interest as regards the band structure. Both the OPW and the pseudopotential methods are attempts to get rid of these troublesome core functions as far as possible. In transition metals, however, the separation between core and valence states is not always acceptable, which is of course a limitation of the OPW method.

We must first of all define our notation. The letter j will designate a crystal site and the letter s a core state. Correspondingly,

$$|sj\rangle = \psi_s(\mathbf{r}-\mathbf{r}_j) \tag{2}$$

denotes a core function of energy E_s at the site j. (Notice that this function depends on a position vector measured from the site j.) In order to simplify the notation, we shall use the letter σ to denote the double index sj:

$$|\sigma\rangle = |sj\rangle = \psi_s(\mathbf{r}-\mathbf{r}_j), \qquad E_\sigma \equiv E_s. \tag{3}$$

Core states which differ either in s or in j are orthogonal, because different eigenfunctions on the same atom are orthogonal or, if they are on different atoms, because the overlap between core states is very small. Therefore,

$$\langle s'j'|sj\rangle = \delta(s'\,s)\,\delta(j'\,j), \tag{4}$$

which, from (3) we abbreviate as

$$\langle \sigma'|\sigma\rangle = \delta(\sigma'\,\sigma). \tag{5}$$

Orthogonalization to the core

The fundamental idea of the OPW method is that a plane wave $|\mathbf{k}\rangle$ may be orthogonalized to the core states by the addition of a conveniently chosen correction term:

$$|\mathbf{k}\rangle - \sum_\sigma |\sigma\rangle \langle\sigma|\mathbf{k}\rangle =_{\text{def}} |\phi_\mathbf{k}\rangle, \tag{6}$$

thus defining a new function $|\phi_\mathbf{k}\rangle$ which we shall now prove is, in fact, orthogonal to any core state $|\sigma'\rangle$:

6 $$\langle\sigma'|\mathbf{k}\rangle = \langle\sigma'|\mathbf{k}\rangle - \sum_\sigma \langle\sigma'|\sigma\rangle \langle\sigma|\mathbf{k}\rangle \tag{7}$$

5|7 $$= \langle\sigma'|\mathbf{k}\rangle - \sum_\sigma \delta(\sigma'\,\sigma) \langle\sigma|\mathbf{k}\rangle \tag{8}$$

8 $$= \langle\sigma'|\mathbf{k}\rangle - \langle\sigma'|\mathbf{k}\rangle = 0. \tag{9}$$

It is convenient to simplify the notation in (6) by defining the *projection operator*

$$\mathbb{P} =_{\text{def}} \sum_\sigma |\sigma\rangle \langle\sigma|, \tag{10}$$

in terms of which the orthogonalized plane waves $|\phi_\mathbf{k}\rangle$ are:

$$|\phi_\mathbf{k}\rangle = |\mathbf{k}\rangle - \mathbb{P}|\mathbf{k}\rangle = (1 - \mathbb{P})|\mathbf{k}\rangle. \tag{11}$$

It is useful to take another look at the correction term in eqn (6). This is a linear combination of the orthogonal core states $|\sigma\rangle$ with coefficients $\langle\sigma|\mathbf{k}\rangle$. On comparing with (**4**-9.17) and (**4**-9.19) it is easy to see that this term is the generalized Fourier series expansion of the plane wave $|\mathbf{k}\rangle$ over the orthogonal core states. (A *generalized Fourier series* is an expansion over any set of orthogonal functions, whereas the Fourier series of Chapter **4** were expansions over plane waves, which are of course orthogonal.)

The crystal eigenfunctions

In analogy with (1), we shall expand the crystal wave function $|\psi_\mathbf{k}\rangle$ as a linear combination of the OPW's in (11):

$$|\psi_\mathbf{k}\rangle = \sum_\mathbf{g} c_\mathbf{g} |\phi_{\mathbf{k}+\mathbf{g}}\rangle. \tag{12}$$

(Notice that, as in eqn 1, the coefficient $c_\mathbf{g}$ should also carry a suffix **k**, since it must also depend on the function being expanded.) From (11),

11|12 $$|\psi_\mathbf{k}\rangle = \sum_\mathbf{g} c_\mathbf{g} (1 - \mathbb{P}) |\mathbf{k} + \mathbf{g}\rangle. \tag{13}$$

The main virtue of the OPW method, as compared with the plane-wave expansion (1), is that, because the cores have been removed by orthogonalization, one does not have to go to very high values of **g** in (12) or (13), so that much fewer terms are required in the plane-wave expansion (1.10). The work, otherwise, goes on very much as with this expansion, that is, a secular determinant is formed from which the energy levels are derived.

4 Pseudopotential method

The starting point for the discussion of the pseudopotential method is the OPW expansion (3.13) of the crystal eigenfunction $|\psi_\mathbf{k}\rangle$. These will be eigenfunctions of the crystal Hamiltonian $\mathbb{H}$, with energy $E_\mathbf{k}$. (Notice that, as a difference with § 1, these are *not* free-electron energies: all the E's in this section will be either crystal or core-state eigenvalues.)

$$\mathbb{H}\,|\psi_\mathbf{k}\rangle = E_\mathbf{k}\,|\psi_\mathbf{k}\rangle. \tag{1}$$

The main idea of the work that follows is to replace the wave function $|\psi_\mathbf{k}\rangle$ in (1) by a form easier to handle than (3.13). This new form will be called a *pseudo wave function* and, correspondingly, the Hamiltonian $\mathbb{H}$ in (1) will also have to be replaced by an appropriately constructed *pseudo Hamiltonian.* The pseudo wave function $|\varphi_\mathbf{k}\rangle$ is constructed from (3.13) by removing from it the cumbersome operator $1 - \mathbb{P}$:

$$|\varphi_\mathbf{k}\rangle =_{\text{def}} \sum_\mathbf{g} c_\mathbf{g}\,|\mathbf{k}+\mathbf{g}\rangle. \tag{2}$$

2|3.13 $$|\psi_\mathbf{k}\rangle = (1 - \mathbb{P})\,|\varphi_\mathbf{k}\rangle. \tag{3}$$

The eigenvalue equation (1) takes now the following form,

3|1 $$\mathbb{H}\,(1 - \mathbb{P})\,|\varphi_\mathbf{k}\rangle = E_\mathbf{k}(1 - \mathbb{P})\,|\varphi_\mathbf{k}\rangle. \tag{4}$$

4 $$(\mathbb{H} - \mathbb{H}\mathbb{P} + E_\mathbf{k}\mathbb{P})|\varphi_\mathbf{k}\rangle = E_\mathbf{k}\,|\varphi_\mathbf{k}\rangle. \tag{5}$$

It is now clear that the terms in the round bracket on (L5) can be taken to be a modified Hamiltonian of which the pseudo wave function $|\varphi_\mathbf{k}\rangle$ is the eigenfunction. We thus define the pseudo Hamiltonian $\mathcal{H}$ as

$$\mathcal{H} = \mathbb{H} - \mathbb{H}\mathbb{P} + E_\mathbf{k}\mathbb{P} = \mathbb{H} + (E_\mathbf{k} - \mathbb{H})\mathbb{P}. \tag{6}$$

We shall write the Hamiltonian $\mathbb{H}$ in (6), when convenient, as a sum of the kinetic energy $\mathbb{T}$ and the potential energy $\mathbb{V}$:

$$\mathcal{H} = \mathbb{T} + \mathbb{V} + (E_\mathbf{k} - \mathbb{H})\mathbb{P} =_{\text{def}} \mathbb{T} + \mathbb{W}. \tag{7}$$

It is natural to regard $\mathbb{W}$ here as the modified potential or pseudopotential

$$\mathbb{W} = \mathbb{V} + (E_\mathbf{k} - \mathbb{H})\mathbb{P}. \tag{8}$$

The fundamental assumption of the pseudopotential method is this: whereas the crystal potential $\mathbb{V}$ is attractive and therefore negative (and generally large), the term $(E_\mathbf{k} - \mathbb{H})\mathbb{P}$ is positive *and it largely cancels* $\mathbb{V}$, so that the new eigenvalue equation

6|5 $$\mathcal{H}|\varphi_{\mathbf{k}}\rangle = E_{\mathbf{k}}|\varphi_{\mathbf{k}}\rangle, \tag{9}$$

7|9 $$(\mathbb{T} + \mathbb{W})|\varphi_{\mathbf{k}}\rangle = E_{\mathbf{k}}|\varphi_{\mathbf{k}}\rangle, \tag{10}$$

is an almost free-electron equation. We shall now verify our assertion that $(E_{\mathbf{k}} - \mathbb{H})\mathbb{P}$ is positive:

3.10 $$(E_{\mathbf{k}} - \mathbb{H})\mathbb{P} = \sum_{\sigma}(E_{\mathbf{k}} - \mathbb{H})|\sigma\rangle\langle\sigma|. \tag{11}$$

Remember that $|\sigma\rangle$ is a core state of energy E_σ (which is large and negative), so that

$$\mathbb{H}|\sigma\rangle = E_\sigma|\sigma\rangle. \tag{12}$$

12|11 $$(E_{\mathbf{k}} - \mathbb{H})\mathbb{P} = \sum_{\sigma}(E_{\mathbf{k}} - E_\sigma)|\sigma\rangle\langle\sigma|. \tag{13}$$

Since the band state eigenvalues $E_{\mathbf{k}}$ must be much smaller in absolute value than the core eigenvalues E_σ, (R13) must be positive. Whether it actually compensates the negative potential $\mathbb{V}$ in (8), which was the second prong of our fundamental assumption, remains a matter of conjecture, although in many practical cases it happens to be largely true.

It is important to realize that, assuming that electron exchange interactions are not explicitly treated, the crystal potential $\mathbb{V}$ depends only on $\mathbf{r}$, $\mathbb{V}(\mathbf{r})$, and it is thus called a *local potential.* The pseudopotential $\mathbb{W}$ in (8), on the other hand, is not a local potential, since it contains an operator $\mathbb{P}$. Another complication with the pseudopotential is that it depends on the very energy eigenvalues $E_{\mathbf{k}}$ which one wants to calculate. This problem, however, is not too difficult to circumvent, for $E_{\mathbf{k}}$ in (8) or, what is the same, in (13), appears as added to the much larger term E_σ, whence an approximate value for it can be taken without substantially affecting the accuracy of the results. It is a sensible approximation to replace all values of $E_{\mathbf{k}}$ in the band by the single value of the Fermi energy E_{F}. This will be, however, a very good approximation near E_{F}, but not so good near the bottom of the bands. Many physical properties, however, depend only on the shape of the bands near the Fermi surface, so that the approximation suggested is often acceptable. If this is the case, one can then use an almost free-electron approximation in which the pseudo wave function $|\varphi_{\mathbf{k}}\rangle$ is expanded in terms of plane waves $|\mathbf{k} + \mathbf{g}\rangle$ as in (3.1). The great advantage now is that this expansion should require only a few values of $\mathbf{g}$, since the pseudo wave function is free-electron like. On introducing this expansion into (9) the energy eigenvalues $E_{\mathbf{k}}$ will follow.

The Austin–Heine–Sham pseudopotential

On combining (6) with (11), the pseudo Hamiltonian $\mathcal{H}$ takes the form

$$\mathcal{H} = \mathbb{H} + (E_{\mathbf{k}} - \mathbb{H})\sum_{\sigma}|\sigma\rangle\langle\sigma|. \tag{14}$$

Austin, Heine, and Sham established that, to a very good approximation, $\mathcal{H}$ can be replaced by another pseudo Hamiltonian $\mathscr{H}$ given by

$$\mathscr{H} = \mathbb{H} + \sum_{\sigma} |\sigma\rangle \langle f(\sigma)|, \tag{15}$$

where $f(\sigma)$ is some conveniently chosen function which, in practice, can be taken to be independent of the crystal site j and which can therefore be considered as a function of the core states only, $f(s)$. This pseudo Hamiltonian is clearly much simpler to use than (14) and in order to see how it provides a good approximation to the eigenvalue $E_{\mathbf{k}}$ of the true Hamiltonian in (1), let us assume that the eigenvalue equation for $\mathscr{H}$ is solved, with the pseudo wave functions $|\varphi_{\mathbf{k}}\rangle$:

$$\mathscr{H} |\varphi_{\mathbf{k}}\rangle = E'_{\mathbf{k}} |\varphi_{\mathbf{k}}\rangle. \tag{16}$$

We would have to use for this purpose a nearly free-electron approximation and we must now prove that $E'_{\mathbf{k}}$ in (16) is a good approximation to $E_{\mathbf{k}}$ in (1):

15|16 $$\mathbb{H} |\varphi_{\mathbf{k}}\rangle + \sum_{\sigma} |\sigma\rangle \langle f(\sigma) | \varphi_{\mathbf{k}}\rangle = E'_{\mathbf{k}} |\varphi_{\mathbf{k}}\rangle. \tag{17}$$

We now act on both sides of this equation with the true wave function $\langle \psi_{\mathbf{k}}|$ of (1). (This means that we multiply with $\psi^*_{\mathbf{k}}$ and integrate over the crystal volume.) We obtain

17 $$\langle \psi_{\mathbf{k}} | \mathbb{H} | \varphi_{\mathbf{k}}\rangle + \sum_{\sigma} \langle \psi_{\mathbf{k}} | \sigma\rangle \langle f(\sigma) | \varphi_{\mathbf{k}}\rangle = E'_{\mathbf{k}} \langle \psi_{\mathbf{k}} | \varphi_{\mathbf{k}}\rangle. \tag{18}$$

Because $\mathbb{H}$ is Hermitian, the first bracket in (18) has the same value when $\mathbb{H}$ operates on $\psi_{\mathbf{k}}$, in which case, from (1), it multiplies it with $E_{\mathbf{k}}$. We can also assume that the overlap $\langle \psi_{\mathbf{k}} | \sigma\rangle$ will be very small, so that the term which contains it in (18) may be neglected:

18 $$E_{\mathbf{k}} \langle \psi_{\mathbf{k}} | \varphi_{\mathbf{k}}\rangle = E'_{\mathbf{k}} \langle \psi_{\mathbf{k}} | \varphi_{\mathbf{k}}\rangle. \tag{19}$$

This means that, unless the pseudo wave function has been chosen so badly that it is orthogonal to the true wave function, that is, that $\langle \psi_{\mathbf{k}} | \varphi_{\mathbf{k}}\rangle$ vanishes, then $E'_{\mathbf{k}}$ must equal $E_{\mathbf{k}}$. Notice that, in principle, this result is valid for any function $f(\sigma)$, although, should $\langle f(\sigma) | \varphi_{\mathbf{k}}\rangle$ be very large, then the neglect of the summation over σ in (18) would not be a very good approximation. The art of choosing adequate pseudopotentials, however, has been much developed and many successful calculations have been performed by this method.

We shall now close our discussion of computational methods, although we have said nothing about some of the most accurate and successful methods of calculation. These are basically number-crunching methods designed to solve the Schrödinger equation in the crystal as accurately as possible. There are basically three such methods, the cellular method, the augmented plane wave method (APW) and the Kohn–Korringa–Rostoker method (KKR). Various excellent programs using these methods are now available which provide extremely accurate eigenvalues.

5 Problems

1. Define general matrix elements and overlap integrals for eigenfunctions of the translations $\psi_{\mathbf{k}}(\mathbf{r})$ (see eqn **5**-6.2), of which the plane waves $|\,\mathbf{k}\rangle$ are a special case (see eqn **5**-6.9):

   1.8, 1.9 $$H_{\mathbf{kk}'} = \langle \psi_{\mathbf{k}}(\mathbf{r})\,|\,\mathbb{H}\,|\,\psi_{\mathbf{k}'}(\mathbf{r})\rangle, \qquad S_{\mathbf{kk}'} = \langle \psi_{\mathbf{k}}(\mathbf{r})\,|\,\psi_{\mathbf{k}'}(\mathbf{r})\rangle, \tag{1}$$

   and prove that they vanish unless the functions $\psi_{\mathbf{k}}(\mathbf{r})$ and $\psi_{\mathbf{k}'}(\mathbf{r})$ belong to the same representation of the translation group. Apply this result to the functions $|\,\mathbf{k}\rangle$.
2. Prove that, if $E_{\mathbf{k}}$ is the free-electron energy of the plane waves $|\,\mathbf{k}\rangle$ in eqn (1.5), then the eigenvalues ε in (1.7) take values $\varepsilon_{\mathbf{k}}$ equal to $E_{\mathbf{k}}$.
3. Prove that, under the nearly free-electron approximation, eqn (1.33) is also valid for $\mathbf{k} = \mathbf{0}$.
4. Verify directly that the Bloch sums (2.6) are eigenfunctions of the translations $\{E\,|\,\mathbf{t}_n\}$ with the correct eigenvalues $\exp(-\,\mathrm{i}\mathbf{k}\cdot\mathbf{t}_n)$.
5. Prove, on using the result of Problem 1, that the functions $\psi_{\mathbf{k}}(\mathbf{r})$ of (2.6) are orthogonal for different values of $\mathbf{k}$.

Further reading

A modern review of the tight-binding method is given in Bullett (1980). The OPW method was introduced by Herring (1940). Pseudopotentials were invented independently and almost simultaneously by Antončík (1959) and Phillips and Kleinmann (1959). Substantial contributions by Bassani and Celli (1961) and Austin, Heine, and Sham (1962) then followed. A wealth of information on pseudopotentials is contained in the book by Harrison (1966) as well as in the reviews by Heine (1970), Cohen and Heine (1970), and Heine and Weaire (1970). Many of the topics of this chapter, including pseudopotentials, are reviewed in Ziman (1971). A very useful introduction to pseudopotentials and their applications is given in Cottrell (1988). As regards the number-crunching methods: see Loucks (1967) and Dimmock (1971) for the augmented plane wave method, Altmann (1974) for the cellular method, and Weinberger (1990) for the KKR method. Dimmock (1971) contains also a simple account of the KKR method. See Papaconstantopoulos (1986) for a fairly complete compilation of results for metals.

11

Phonons and conductivity

We have already used more than once the idea that electrical conductivity is due to the fact that the electric field promotes electrons which are near the Fermi level to unoccupied levels above it. We have then assumed that the size of the Fermi surface is a determining factor in conductivity, a point which we shall confirm and refine in § 1. Once this is done we shall see that, because the electric field provides a constant acceleration to the Fermi electron sea, a mechanism must be found which prevents this process from continuing indefinitely, since otherwise a steady state of conduction by the electrical carriers would never be achieved. We must therefore discover in what way the electrons may be scattered and thus slowed down. There are two ways in which this may happen. One is through electron–electron collisions, but this, of course, cannot affect the energy of the electron sea as a whole, whereas it is this energy which we must ensure does not run up to infinity. A second possible mechanism is by collisions between the electrons and the crystal structure. We shall see, though, that a perfect, that is, a rigid crystal cannot exchange energy with the electron sea. It is only a vibrating crystal that can do this, as we shall discover when we study crystal vibrations, which will lead us to the important concept of the *phonon*.

1 Conductivity and the Fermi surface

An electric field $\mathscr{E}$ acting on an electron of charge e (negative), applies a force on it equal to $e\mathscr{E}$, which from elementary mechanics must be equal to the rate of change of the electron momentum $\hbar\mathbf{k}$:

$$\hbar\dot{\mathbf{k}} = e\mathscr{E}. \tag{1}$$

It must be appreciated that the above argument is strictly correct only for free electrons, although it can be proved (see Ziman 1960, p. 95) that eqn (1) is still valid when an electron is moving in the periodic crystal field. In such a case, however, the electron will not always move in the direction of the electrical field $\mathscr{E}$, that is the electron velocity, and thus the current density $\mathbf{J}$, is not in general parallel to $\mathscr{E}$, which means that the conductivity $\boldsymbol{\sigma}$, given by

$$\mathbf{J} =_{\text{def}} \boldsymbol{\sigma} \cdot \mathscr{E}, \tag{2}$$

cannot be a scalar or a vector quantity but rather a tensor, which transforms a vector into another vector, not in general parallel to the first one.

We cannot expect the force in (1) to act on the electron for an indefinitely long time, since then the electron velocity will become infinite. Obviously, any electron must collide from time to time with some other electron, thus losing kinetic energy. The mean time between two such collisions is called the *relaxation time* τ, and the average increment $\mathbf{dk}$ of $\mathbf{k}$ during this time can be taken to be $\tau\ \mathbf{dk}/\mathrm{d}t$. Thus, from (1),

$$\mathbf{dk} = \tau e \mathscr{E}/\hbar \tag{3}$$

The result of this increment on $\mathbf{k}$ is that the Fermi surface 'moves' as illustrated in Fig. 1. The number of new states occupied arising from the element of area $\mathrm{d}\mathbf{A}$ of the Fermi surface, is the volume of the cylinder of height $|\mathbf{dk}|$ whose base is the projection of $|\mathrm{d}\mathbf{A}|$ normal to $\mathbf{dk}$, that is, $\mathbf{dk}\cdot\mathrm{d}\mathbf{A}$. The number of electrons in these states is this volume multiplied by the density of states at the Fermi surface $n(E_\mathrm{F})$, so that the charge carried by these electrons is

$$\text{charge} = e n(E_\mathrm{F})\, \mathrm{d}\mathbf{A}\cdot\mathbf{dk} = \{e n(E_\mathrm{F})\tau e/\hbar\}\, \mathscr{E}\cdot\mathrm{d}\mathbf{A}. \tag{4}$$

The total current density $\mathbf{J}$ will equal this charge times the velocity $\mathbf{v}$, integrated over the Fermi surface:

$$\mathbf{J} = (\tau e^2/\hbar)\, n(E_\mathrm{F}) \int_{\mathrm{FS}} \mathbf{v}\mathscr{E}\cdot\mathrm{d}\mathbf{A}. \tag{5}$$

It follows at once from (2) and (5) that the conductivity tensor $\boldsymbol{\sigma}$ is

$$\boldsymbol{\sigma} = (\tau e^2/\hbar)\, n(E_\mathrm{F}) \int_{\mathrm{FS}} \mathbf{v}\, \mathrm{d}\mathbf{A}. \tag{6}$$

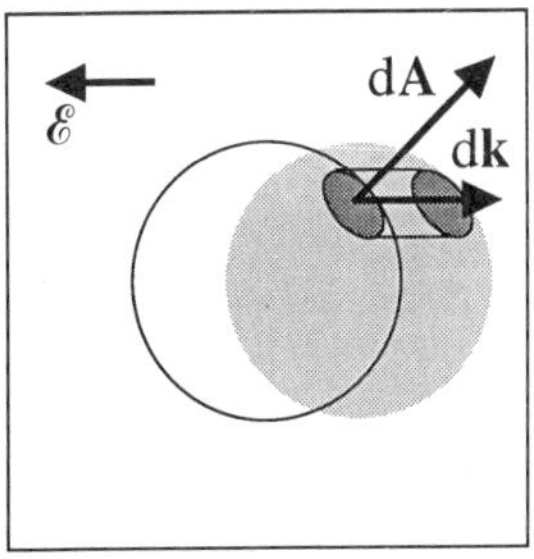

Fig. 11-1.1. Electron conductivity. The square represents the Brillouin zone and the white and grey circles are the traces of the Fermi surface without and with the electric field respectively. (Compare with Fig. **1**-4.1.)

Since the integral is over the whole of the Fermi surface, it follows that the smaller the Fermi surface the smaller the conductivity, as assumed in § **8**-7.

It is clear from Fig. 1 that the electric field, while promoting some electrons above the Fermi surface depletes other electron states. We must find a mechanism that replenishes these states: otherwise the current will grow up to infinity and no steady state will ever be reached. We shall find that such a mechanism is provided by the interaction between the electronic states and the vibrations of the crystal structure, which will now be studied.

2 Crystal normal modes

Before we can deal properly with crystal vibrations we have to revise some of the standard work on harmonic oscillators. In one dimension, a particle of mass μ and coordinate x will perform harmonic oscillations whenever it is acted on by a restoring force $-\kappa x$, where κ is the oscillator force constant. The potential of the particle, referred to zero at the origin ($x = 0$), will be $\frac{1}{2}\kappa x^2$, and its kinetic energy $p^2/2\mu$. Therefore, the quantum mechanical Hamiltonian $\mathbb{H}(x)$ will be

$$\mathbb{H}(x) = -\frac{\hbar^2}{2\mu}\frac{\mathrm{d}^2}{\mathrm{d}x^2} + \tfrac{1}{2}\kappa x^2. \tag{1}$$

The solutions of the corresponding Schrödinger equation

$$\mathbb{H}(x)\,\psi_n(x) = E_n\,\psi_n(x), \tag{2}$$

are well known:

$$E_n = \hbar\omega(n + \tfrac{1}{2}), \qquad n = 0, 1, 2, \ldots, \qquad \omega = (\kappa/\mu)^{1/2}, \tag{3}$$

and the wave functions $\psi_n(x)$ are the so-called Hermite functions, which express the harmonic motion of the particle with a frequency ω.

The situation will of course be far more complicated in a crystal because we shall have a large number of *coupled* oscillators. This means that the potential energy will contain terms which depend not only on each of the coordinates alone but also on pairs, triples, etc, of them, and which are the terms that establish the coupling between the oscillators. These are the awkward terms which make the work difficult. Let us, however, do a bit of pipe-dreaming and assume that it were possible to transform away the coordinates x_i ($i = 1, 2, \ldots$) into some new coordinates ξ_i with the miraculous property that, in these coordinates, no interaction terms exist, that is, that all the terms which appear in the Hamiltonian now depend each on a single coordinate and never, say, on a product of coordinates. This means that the Hamiltonian might now be expressed as a sum of harmonic oscillator

Hamiltonians, each depending on a single coordinate:

$$\mathbb{H} = \sum_i \mathbb{H}(\xi_i). \tag{4}$$

The result of Problem **1**-9.1 will show in this case that the energy of the system must be the sum of the individual energies of each of the oscillators on (R4), (each given by eqn 3), and that the wave function will be the product of the eigenfunctions:

$$E = \sum_i \hbar\omega_i(n_i + \tfrac{1}{2}), \qquad \Psi = \prod_i \psi_{n_i}(\xi_i). \tag{5}$$

Such a transformation would, of course, solve most neatly the problem of the coupled oscillators and we shall see that it is indeed possible to find a transformation that uncouples the crystal oscillators in this manner, in which case the coordinates ξ_i for which the coupling disappears are called the *normal coordinates.* Before we look for the transformation that generates such coordinates we must, however, write the crystal vibrational Hamiltonian in full.

Kinetic and potential energies in a crystal

We shall assume for simplicity (but see §4) that we have a symmorphic space group without a basis. All the crystal sites are therefore denoted by vectors *of* the lattice **t**. We can safely assume that all the crystal sites correspond to positions of equilibrium of the crystal and we shall be interested in *small oscillations* about these positions of equilibrium. An instantaneous picture of the crystal will thus look like Fig. 1, the small vectors in it denoting in each case the position of the atoms which, in equilibrium, are at their tails. As

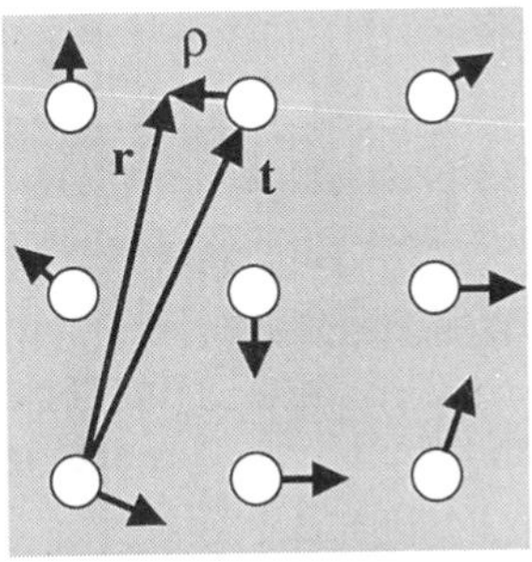

Fig. 11-2.1. Displacement vectors. The white circles represent the positions of equilibrium of the atoms of the crystal structure.

shown in the figure, the position vector **r** of each atom can be written as follows,

$$\mathbf{r} = \mathbf{t} + \boldsymbol{\rho}, \tag{6}$$

where $\boldsymbol{\rho}$, the *displacement vector* of an atom from its position of equilibrium, is such that $|\boldsymbol{\rho}|$ is always much smaller than $|\mathbf{t}|$. We shall assume that the crystal contains N atoms and that this number is large enough so that periodic boundary conditions can be used, leading to a finite translation group T of order N, as in §**5**-1. In what follows, we shall distinguish lattice vectors with symbols such as **t** and **t**′, and the displacement vector $\boldsymbol{\rho}$ corresponding to the site **t** will be labelled as $\boldsymbol{\rho}_{\mathbf{t}}$. The cartesian components of such vectors will be labelled with indices such as i, j. The range of these indices will therefore be

$$\mathbf{t}, \mathbf{t}' : \forall T, \qquad |T| = N; \qquad i, j = 1, 2, 3, \tag{7}$$

and all summations involving these labels will be taken over this range.

The potential energy of the crystal will be written as a function of the components of all the displacement vectors and, for the position of equilibrium of the crystal atoms, it can be taken to be zero and it must be a minimum:

$$V(\rho_{\mathbf{t}1}, \rho_{\mathbf{t}2}, \ldots, \rho_{\mathbf{t}'1}, \ldots) = 0, \qquad \text{for} \quad \rho_{\mathbf{t}1}, \rho_{\mathbf{t}2}, \ldots, \rho_{\mathbf{t}'1}, \ldots = 0. \tag{8}$$

$$\partial V / \partial \rho_{\mathbf{t}i} = 0, \qquad \text{for} \quad \rho_{\mathbf{t}i} = 0. \tag{9}$$

(In eqn 8, further translation vectors **t**″, etc. must be added until the whole of T is included.) The kinetic energy $\boldsymbol{T}$, on taking $\mu_{\mathbf{t}}$ as the mass of the atom at the site **t**, is given by

$$\boldsymbol{T} = \tfrac{1}{2} \sum_{\mathbf{t}i} \mu_{\mathbf{t}} \dot{\rho}_{\mathbf{t}i}^2, \tag{10}$$

whereas the potential energy V can be expanded in Taylor series around the position of equilibrium, remembering that all the $\boldsymbol{\rho}$'s are infinitesimals of first order:

$$V = V(0) + \sum_{\mathbf{t}i} \left(\frac{\partial V}{\partial \rho_{\mathbf{t}i}} \right)_{\boldsymbol{\rho}_{\mathbf{t}} = 0} \rho_{\mathbf{t}i} + \tfrac{1}{2} \sum_{\mathbf{t}i, \mathbf{t}'j} \left(\frac{\partial^2 V}{\partial \rho_{\mathbf{t}i}\, \partial \rho_{\mathbf{t}'j}} \right)_{\boldsymbol{\rho}_{\mathbf{t}} = \boldsymbol{\rho}_{\mathbf{t}'} = 0} \rho_{\mathbf{t}i}\, \rho_{\mathbf{t}'j} + \ldots . \tag{11}$$

The first term here is V as calculated in (8) and it vanishes. The second term vanishes from the equilibrium conditions (9). The coefficient in the third term will be written as $b_{\mathbf{t}i,\mathbf{t}'j}$ and the higher order terms will be neglected in the so-called *harmonic approximation.* Thus

$$V = \tfrac{1}{2} \sum_{\mathbf{t}i, \mathbf{t}'i} b_{\mathbf{t}i,\mathbf{t}'j}\, \rho_{\mathbf{t}i}\, \rho_{\mathbf{t}'j}. \tag{12}$$

It will be helpful to get rid of the masses in the kinetic energy in (10) by making the simple coordinate substitution into *reduced coordinates*

$$\mathbf{s}_{\mathbf{t}} = (\mu_{\mathbf{t}})^{1/2}\, \boldsymbol{\rho}_{\mathbf{t}}, \tag{13}$$

so that, with some new coefficients in (12),

$$\boldsymbol{T} = \tfrac{1}{2}\sum_{\mathbf{t}i} \dot{s}_{\mathbf{t}i}^{2}, \qquad V = \tfrac{1}{2}\sum_{\mathbf{t}i,\mathbf{t}'j} B_{\mathbf{t}i,\mathbf{t}'j}\, s_{\mathbf{t}i}\, s_{\mathbf{t}'j}. \tag{14}$$

These are the equations in which we have to introduce an ingenious coordinate transformation to get rid of the cross-terms in V and in order to do this with the minimum amount of work and trouble we need a change of notation. As we shall soon see, the variables which will replace the $\mathbf{s}_{\mathbf{t}}$ will turn out to be complex. We must in such a case ensure that the kinetic energy remains always real, whereas w^2 is not so for a complex number w. The standard trick in such cases is to replace w^2 by w^*w, which is always real. In any case, if we change the real vectors $\mathbf{s}_{\mathbf{t}}$ in (13) by $\mathbf{s}_{\mathbf{t}}^*$, absolutely nothing changes and we shall do this for the sake of future work:

$$\boldsymbol{T} = \tfrac{1}{2}\sum_{\mathbf{t}i} \dot{s}_{\mathbf{t}i}^{*}\, \dot{s}_{\mathbf{t}i}, \qquad V = \tfrac{1}{2}\sum_{\mathbf{t}i,\mathbf{t}'j} s_{\mathbf{t}i}^{*}\, B_{\mathbf{t}i,\mathbf{t}'j}\, s_{\mathbf{t}'j}. \tag{15}$$

Translations and their effect on $\boldsymbol{T}$ and V

What we would like now to do in (15) is to transform away the N vectors $\mathbf{s}_{\mathbf{t}}$ into some new vectors, say $\mathbf{z}_{\mathbf{q}}$, (also N in number, of course) so that the cross terms in V disappear. (The meaning of the $\mathbf{q}$ vectors here used to label the N vectors $\mathbf{z}$ will soon emerge.) In order to do this job it is useful to reflect a little about the behaviour of the kinetic and potential energies under translations. Obviously, $\boldsymbol{T}$ and V must be invariant under translations, since no symmetry operation can affect an energy term. The condition required, however, is stronger than this: each term in V in (15) must be invariant under translations. This is so because the invariance of V must still hold when all the values of the coordinate components $s_{\mathbf{t}i}, s_{\mathbf{t}'j}$ in the expansion of V in (15) vanish except for those in one single term and likewise the same invariance property for the potential energy terms must obtain after transformation of the $\mathbf{s}_{\mathbf{t}}$ into the $\mathbf{z}_{\mathbf{q}}$. Suppose that the vectors $\mathbf{z}_{\mathbf{q}}$ have been chosen to be eigenvectors of the translations $\{E\,|\,\mathbf{t}\}$, now simplified with the symbol t,

$$t\, \mathbf{z}_{\mathbf{q}} = \varepsilon_{\mathbf{q}}\, \mathbf{z}_{\mathbf{q}} \qquad \Rightarrow \qquad t\, \mathbf{z}_{\mathbf{q}}^{*} = \varepsilon_{\mathbf{q}}^{*}\, \mathbf{z}_{\mathbf{q}}^{*}, \tag{16}$$

with the additional constraint that the eigenvalues $\varepsilon_{\mathbf{q}}$ be *unimodular*

$$\varepsilon_{\mathbf{q}}\, \varepsilon_{\mathbf{q}}^{*} = 1, \tag{17}$$

a condition which can always be easily satisfied by appropriate multiplication of the equations in (16) with a constant.

Let us now see what happens to the kinetic and potential energies in their new form (see Problem 5.3 as regards the conservation of the kinetic energy and notice the new coefficients $\boldsymbol{B}$ in the new expression for the potential),

$$\boldsymbol{T} = \tfrac{1}{2}\sum_{\mathbf{q}i} \dot{z}^*_{\mathbf{q}i}\dot{z}_{\mathbf{q}i}, \qquad V = \tfrac{1}{2}\sum_{\mathbf{q}i,\mathbf{q}'j} z^*_{\mathbf{q}}\, \boldsymbol{B}_{\mathbf{q}i,\mathbf{q}'j} z_{\mathbf{q}'j}. \tag{18}$$

Quite clearly, from (16), the general term in V becomes multiplied, after a translation, by a factor equal to $\varepsilon^*_{\mathbf{q}}\varepsilon_{\mathbf{q}'}$. In order to guarantee invariance this factor must equal unity. It follows at once from (17) that $\varepsilon^*_{\mathbf{q}}\varepsilon_{\mathbf{q}'}$ equals $\varepsilon_{\mathbf{q}'}/\varepsilon_{\mathbf{q}}$, which can never be unity unless $\mathbf{q}$ equals $\mathbf{q}'$. This means that the coefficients $\boldsymbol{B}_{\mathbf{q}i,\mathbf{q}'j}$ must vanish whenever $\mathbf{q}$ and $\mathbf{q}'$ differ since, otherwise, the corresponding terms would not be invariant. We have now got rid of all our cross terms, except in the case when $\mathbf{q}$ equals $\mathbf{q}'$, when nine terms corresponding to the ranges of i and j survive. We have almost fulfilled our pipe-dream of eqn (4), except that rather than being left with N uncoupled one-dimensional harmonic oscillators, we have N uncoupled three-dimensional oscillators, corresponding to the N values of $\mathbf{q}$ and the three values of i and of j.

The eigenvectors of the translations. Normal coordinates

To put the above programme of work into operation we need the eigenvectors of the translations which, in fact, we already know how to obtain, since we constructed them in (**10**-2.6) as Bloch sums (see also Problem **10**-5.4). We shall revise briefly, however, in a notation more adapted to our present needs, the work required. All we have to do is to use the projection operator (**2**-5.4) in order to adapt functions (or, if necessary, vectors) to one of the irreducible representations of the translation group T. We know that these irreducible representations are all one-dimensional and N in number, and that they are labelled by each of the N permitted $\mathbf{k}$ vectors in the first Brillouin zone. In order not to confuse our labelling of the electronic states (labelled with the $\mathbf{k}$ vector) with that for the vibrational states, we shall label the latter with a vector $\mathbf{q}$ which otherwise has all the formal properties of a $\mathbf{k}$ vector. Thus, the character of a translation t by $\mathbf{t}$ in the irreducible representation ${}_{\mathbf{q}}\hat{T}$ is $\exp(-i\mathbf{q}\cdot\mathbf{t})$ from (**5**-2.9), $\mathbf{t}$ being the vector (of the lattice) corresponding to the translation t. The translation eigenvector $\mathbf{z}_{\mathbf{q}}$ corresponding to the representation ${}_{\mathbf{q}}\hat{T}$ is therefore

2-5.4
$$\mathbf{z}_{\mathbf{q}} = N^{-1/2}\sum_{\mathbf{t}} \chi(t\,|\,{}_{\mathbf{q}}\hat{T})^*\, t\,\mathbf{s}_0 = N^{-1/2}\sum_{\mathbf{t}} \exp(i\mathbf{q}\cdot\mathbf{t})\,\mathbf{s}_{\mathbf{t}}. \tag{19}$$

(We use here $\mathbf{s}_0$ as the generator for the projection operator, recognizing that its transform under t is $\mathbf{s}_{\mathbf{t}}$, as appropriate labelling in Fig. 2.1 will show. Notice also that we have added a normalization factor $N^{-1/2}$ which, of course, does not affect the symmetry. See Problem 5.3 as regards the importance of this factor.) It is simple to verify that these vectors are indeed eigenvectors of the translations (see Problem 5.1):

$$t\,\mathbf{z_q}(\mathbf{r}) = \exp(-\,\mathrm{i}\mathbf{q}\cdot\mathbf{t})\,\mathbf{z_q}(\mathbf{r}), \tag{20}$$

where we recognize that, as a difference with the $\mathbf{s_t}$, the $\mathbf{z_q}$ are not associated with a single site but they rather extend over the whole crystal and thus depend on the position vector **r**. Equation (20), however, shows how they vary from cell to cell, as we shall see later.

It follows from our previous discussion (eqn 5) that when the coordinates $\mathbf{z_q}$ are introduced instead of the displacement vectors $\mathbf{s_t}$ the N harmonic oscillators (N being the number of atoms in the crystal) become uncoupled and that to each coordinate $\mathbf{z_q}$ there corresponds one separate three-dimensional oscillator with frequency $\omega_{\mathbf{q}}$. The coordinates $\mathbf{z_q}$ which effect this separation of the total vibrational motion of the system into that of N uncoupled oscillators are called the *normal coordinates.* Notice that we start with N (vector) coordinates $\mathbf{s_t}$ and that we obtain exactly the same number of normal coordinates $\mathbf{z_q}$. Each of these coordinates varies harmonically in time but it must be clearly understood that this motion does not describe the motion of a single atom in the crystal but, rather, of all atoms in it. This motion is called a *normal mode of vibration* or a *normal mode.*

A few examples will show how this works. Take a normal coordinate for which **q** equals a vector of the reciprocal lattice **g**. In this case, the coefficient on (R20) equals unity (see eqn **5**-3.3). Eqn (20) tells us now that the value of the normal coordinate at the cell labelled by **t** (which is L20) equals that in the cell at the origin. That is, the displacement vectors are the same over all the crystal cells: this motion is a *translation* of the crystal as a whole and it is therefore not properly a vibration (see Fig. 2*a*). The case just considered corresponds of course to the centre Γ (**q** equal to zero) of the Brillouin zone. In order to have another example, consider a one-dimensional chain, and take q equal to X, a point at the edge of the Brillouin zone. (See § **8**-4.) It is easy to

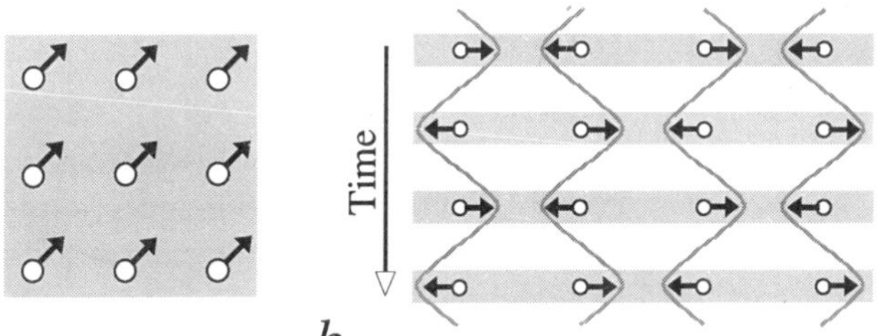

a *b*

Fig. 11-2.2. Normal modes of vibration. The normal mode in *a* corresponds to the centre Γ of the Brillouin zone of a simple square lattice. In *b* the normal mode at the edge X of the Brillouin zone of a simple linear chain is shown. The various rows in this picture show the time dependence of the motion of the atoms for a small selection of values of the time. For all other values the harmonic curves shown must be followed.

verify now that the exponent on (R20) equals -1, (see **6**-4.2 and the discussion below it), which means that the displacement vector is out of phase in going from one cell to the next, as depicted in the top line of Fig. 2*b*, where the successive lines show the evolution of the motion in time, the harmonic function depicted corresponding to the particular frequency of this normal mode.

Thermal excitation of the crystal will result in the excitation of a normal coordinate $\mathbf{z}_\mathbf{q}$ by jumps in quanta $\hbar\omega_\mathbf{q}$ (see eqn 5) or in the excitation of the whole system to some other normal mode. Such excitations are most often visualized by imagining that the various vibrational states are occupied by 'particles' called *phonons*, of energy $\hbar\omega_\mathbf{q}$, so that excitation of the system corresponds to phonons jumping from one state to another, in the same way as electrons do when the Fermi sea is excited, for example, by an electric field. (More properly, a 'jump' from the state $\mathbf{q}$ to the state $\mathbf{q}'$ entails the *annihilation* of a phonon in the state $\mathbf{q}$ and the *creation* of a phonon in the state $\mathbf{q}'$.) When using such a picture, it must be remembered that there is no Pauli principle for the phonons: all the normal coordinates, for instance, might be found in the lowest vibrational state, as will be the case at 0 K. Phonons must thus be regarded as *bosons* and not as fermions, as electrons are.

3 Electron–phonon scattering

We shall see that we now have a mechanism for replenishing some of the states in the Brillouin zone which are left empty when the electrons are excited by an electric field. (See the end of § 1.) Consider in Fig. 1*a* an electron excited to a state $\mathbf{k}$ just above the Fermi surface, as was shown in Fig. 1.1. The interaction of this electron with the vibrating crystal may result in the normal frequency $\omega_\mathbf{q}$ being excited. Whereas before the collision of the electron with the crystal structure we have an electron with wave vector $\mathbf{k}$ (slightly above the Fermi surface), after the collision the wave vector of the electron is $\mathbf{k}'$

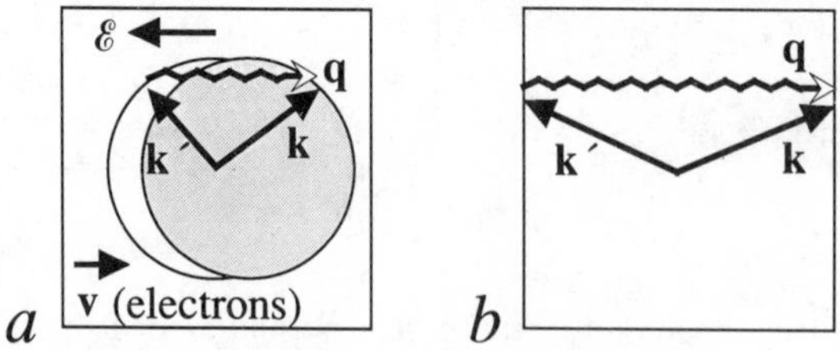

Fig. 11-3.1. Electron–phonon scattering. The squares represent the Brillouin zone of a simple square lattice. Case *a* is the normal type of scattering, whereas *b* illustrates the Bragg scattering.

(slightly below the Fermi surface), and a phonon appears with wave vector $\mathbf{q}$. As shown in the figure,

$$\mathbf{k} = \mathbf{k}' + \mathbf{q}. \tag{1}$$

Momentum and energy must be conserved in this process:

$$\hbar\mathbf{k} = \hbar\mathbf{k}' + \hbar\mathbf{q}, \qquad E(\mathbf{k}) = E(\mathbf{k}') + E(\mathbf{q}). \tag{2}$$

The energy of the scattered electron, $E(\mathbf{k}')$, is thus lower than the energy $E(\mathbf{k})$ of the excited electron. This is the mechanism required in order to reduce the electron energy and thus prevent the excitation provided by the field from running up to infinity. It is by a balance between the excitation of the electron by the field and its scattering by the vibrating crystal structure that a steady state is reached in the electrical conduction process. Clearly, as the crystal temperature is increased, scattering by the vibrating crystal structure must also increase, and thus the conductivity of a metal will diminish with temperature.

Scattering by a rigid crystal: Bragg scattering

In a rigid crystal structure none of the vibrational states can be excited, that is the crystal must be found at the state Γ corresponding to the centre of the Brillouin zone, for which $\mathbf{q}$ equals zero. We know that this wave vector is equivalent to any vector $\mathbf{g}$ of the reciprocal lattice, and we have seen that the normal modes that correspond to such vectors are translations of the crystal structure as a whole, for which their vibrational energy can be taken as zero (but see below). If we have

1 $\mathbf{k} = \mathbf{k}' + \mathbf{g}, \qquad \hbar\mathbf{q} = \hbar\mathbf{g} = \mathbf{0}, \qquad (3)$

then

2 $\mathbf{k} = \mathbf{k}', \qquad E(\mathbf{k}) = E(\mathbf{k}'). \qquad (4)$

This means that there is no scattering, in the sense that there is no energy exchange between the electron and the crystal structure. The conditions in (3) and (4) are a particular case of the von Laue conditions of §**8**-2 and this process is called the *Bragg scattering*. (Fig. 1*b*.)

We should, however, refine our interpretation of this process. Although it is quite legitimate at Γ to assign a zero momentum to the corresponding phonon, the momentum $\hbar\mathbf{g}$ is the momentum of the crystal translating as a whole and thus finite. We must assume, however, as it happens in experimental situations, that the crystal is clamped and thus prevented from moving. The momentum $\hbar\mathbf{g}$ is therefore cancelled by an equal and opposite momentum produced by the crystal support. When this momentum is added to the momentum conservation equation on the left of (2), the momentum $\hbar\mathbf{g}$ is cancelled, as given in (3). In either view of this process, the important thing is that no energy exchange takes place between the electron and the crystal structure, which is thus unable to scatter the electron into a lower energy state, as it was shown in Fig. 1*a*, and as it is necessary in order to reach the steady state during conduction.

The phonon drag

Let us now go back to Fig. 1*a*. It is clear that when an electron with momentum $\hbar\mathbf{k}$ is scattered by the phonon, the momentum $\hbar\mathbf{q}$ is given up by the electron to the phonon system, with $\mathbf{q}$ in the same direction as $\mathbf{k}$. That is, at the same time as we have a current density (see § 1) $-e\,\mathbf{v}$, we have a current of phonons dragged by the electrons in the same direction as that of their motion. The existence of this *phonon drag* is sensible, since we know that a thermal current is always associated with an electrical one: wires do get hot when conducting electricity! This process, however, must be more complex than so far discussed, since the phonon drag that we have found is always in the direction of electron propagation and therefore, unless a mechanism for partially reversing it can be found, we would never have a steady state. This mechanism was found by Peierls and it is called the *umklapp process*, illustrated in Fig. 2. For simplicity of the interpretation, we use the periodically repeated zone scheme (§ **8**-5). We start with an electron in a state $\mathbf{k}$, as in Fig. 1*a*, scattered by a phonon into $\mathbf{k}'$, exactly as in Fig. 1*a*. At this stage, we can add by equivalence a vector $\mathbf{g}$ to $\mathbf{k}'$, (which means that we allow the electron to collide with the 'rigid' crystal, i.e. to act on the crystal support and receive from it an equal and opposite reaction, in other words to experience a Bragg reflection: see eqns 3 and 4). We see that

F2
$$\mathbf{k} = \mathbf{k}' + \mathbf{g} + \mathbf{q}, \tag{5}$$

with $\mathbf{q}$ now in a direction opposite to $\mathbf{k}$. We can thus have both backward as well as forward drag and the balance of these two processes can allow a steady state to be reached.

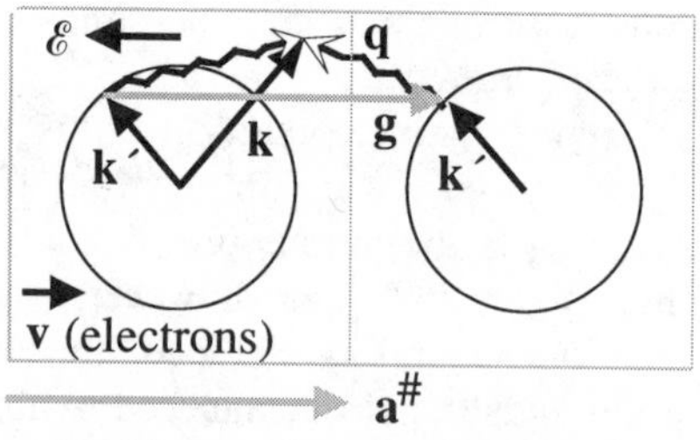

Fig. 11-3.2. The umklapp process. The Brillouin zones of a simple square lattice of lattice constant **a** are illustrated. The periodically repeated scheme is used in which $\mathbf{k}'$ and $\mathbf{k}' + \mathbf{g}$ are the same vector, as illustrated. The vector $\mathbf{g}$ of the reciprocal lattice equals $\mathbf{a}^{\#}$.

4 Lattice with basis. The silicon phonon spectrum

The eigenfunctions of the vibrational Hamiltonian, as those of the electronic one, must be classified by the irreducible representations of the space group and thus by means of the **k** vectors (now designated as **q** vectors) in the Brillouin zone, their stars, the groups of the **q** vectors, and their small representations, all as done in Chapter **6**. We shall illustrate how this works by jumping into the deep end and taking as an example a lattice with basis, that of silicon, the Brillouin zone of which is shown in Fig. **9**-1.3. As in the case of the bands for the electron states, shown in Fig. **9**-1.4, we illustrate in Fig. 1 the phonon bands along the ΓX direction. First of all, we notice that at the centre of the Brillouin zone, Γ, there are two levels, Γ_{15} and Γ'_{25} in order of energy. From our discussion in §**9**-1 they are both triply degenerate and antisymmetric and symmetric, respectively, with respect to the inversion at the mid-point between the two atoms of the basis. We illustrate such vibrational modes in Fig. 2, in a very partial manner, since the basis must be extended by translational symmetry, account being taken of the **q** vector,

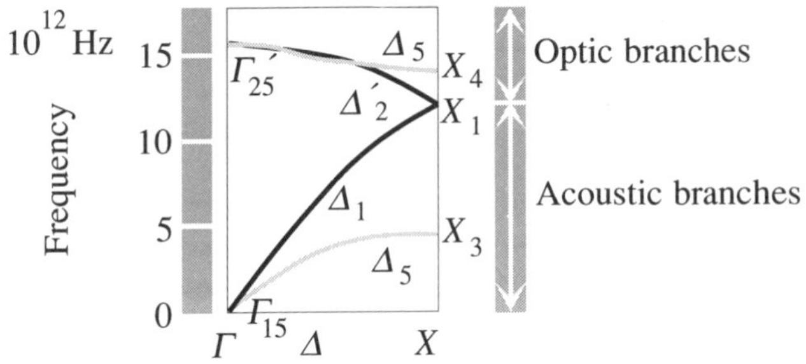

Fig. 11-4.1. The phonon bands for silicon along the ΓX direction, after Altmann, Lapiccirella, Lodge, and Tomassini (1982). The grey and black curves are transverse and longitudinal normal modes, respectively.

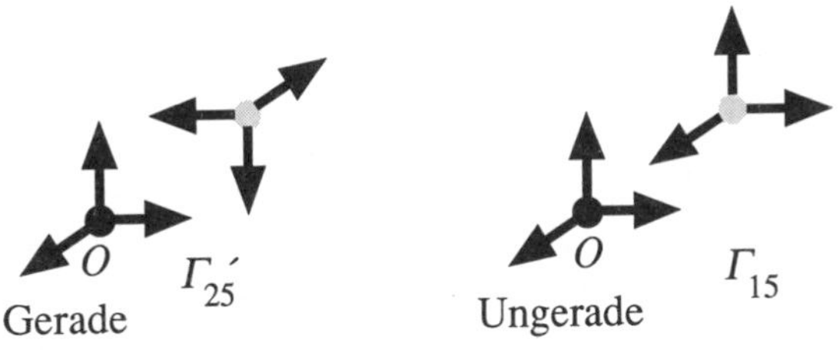

Fig. 11-4.2. Two normal vibrations for silicon at Γ, illustrated for the basis defined in Fig. **9**-1.1. Compare this figure with the orbital basis shown in Fig. **9**-1.2.

which determines the relation between one cell and the next. It is clear in Fig. 2, however, that the two atoms of the basis, which correspond to the shortest interatomic distance in the crystal, are in phase for Γ_{15} and out of phase for Γ'_{25}. Quite clearly, specially if one considers the motion of the atoms in the direction of the bond, the out-of-phase motion must correspond to a large force constant and thus to a large frequency, and in fact the difference between these two levels is very large. As for the electronic bands, as soon as we move away from Γ we lose the inversion symmetry, but, as in that case, the nature of the cell motion is propagated along the band by continuity and the two lower bands, starting from Γ_{15}, correspond to in-phase motion within the basis and thus to lower frequencies, and they are called the *acoustic bands* or *acoustic branches*, since their frequencies approach those of sound waves as $\mathbf{q} \to \mathbf{0}$. The bands arising from Γ'_{25} correspond to out-of-phase motion within the basis, and thus to high frequencies, and they are called the *optic bands* or *optic branches*, since they are found in the near infrared. Naturally, in a lattice without basis only the acoustic branches are found. It is useful to remark that the difference between the acoustic and optic branches is quite similar to that between the bonding and the antibonding bands in the electronic spectrum of silicon.

We have found in §**9**-1 that the level Δ_5 is doubly degenerate, spanned in the electronic levels by the functions p_x, p_y, normal to the direction of the basis. The corresponding normal mode must thus be normal to this direction, with two degenerate but independent vibrations at right angles. Thus, the corresponding normal modes are *transverse oscillations*. Δ_1 is singly degenerate and corresponds instead to a *longitudinal oscillation*. If we look at the optic branches, Δ'_2 is singly degenerate and thus longitudinal, while the other, Δ_5, is transverse. As in §**9**-1, we can understand why, for example, the acoustic branches go up from Γ to X, since at X each cell is out of phase with its near neighbour. (The effect of this relation cannot be as marked as that of the two atoms of the basis being out of phase, which determines the acoustic-optic separation, since the interatomic distance involved is much larger, 0.54 nm being the cell constant, against 0.235 nm for the bond length of the basis.)

One final remark about the phonon spectrum in Fig. 1. It is clear that the slope of the two lower bands at Γ does not vanish, and thus that we have a cusp at this point (see §**8**-2). The reason for this situation is as follows. In the case of the electronic bands we had the general guidance of the asymptotic case of free electrons, for which the bands are parabolae. The free-electron approximation is normally quite good at Γ and thus the parabola is significant, and it gives us the vanishing derivative. The asymptotic picture which one uses for the normal modes is instead that of the vibrations of a continuous medium for which it can be proved that, at low frequency, the frequency is linear in $\mathbf{q}$, as it can indeed be seen in the figure. It is this linear behaviour which is responsible for the cusp at Γ.

5 Problems

1. Verify that the vector functions $\mathbf{z}_{\mathbf{q}}(\mathbf{r})$ of (2.19) are eigenfunctions of the translations as required by eqn (2.20).
2. Prove from (2.19) that the displacement vectors are written as follows in terms of the normal coordinates:

$$\mathbf{s}_{\mathbf{t}} = N^{-1/2} \sum_{\mathbf{q}} \chi(\mathbf{t} \,|\, {}_{\mathbf{q}}\hat{T})^* \, \mathbf{z}_{\mathbf{q}}, \qquad s_{\mathbf{t}i} = N^{-1/2} \sum_{\mathbf{q}} \chi(\mathbf{t} \,|\, {}_{\mathbf{q}}\hat{T})^* \, z_{\mathbf{q}i}. \qquad (1)$$

3. On introducing the transformation (1) into the kinetic energy $\boldsymbol{T}$ in (2.15), show directly that the latter is left invariant, as assumed in (2.18). (Notice the importance of the normalization factor $N^{-1/2}$ in this work.)
4. Classify the normal modes for a linear chain without basis of period a in terms of the Brillouin zone of Fig. **8**-4.1. Show that for $\mathbf{q} = \Gamma$ the motion is a translation of the chain as a whole and that for $q = \pi/a$, the edge of the Brillouin zone, a higher energy normal mode arises. Sketch plausible bands. Do the same for a lattice with basis, identifying the acoustic and optic branches.

Further reading

Our discussion on conductivity in § 1 follows that of Ziman (1963). A good, albeit rather formal, discussion of the group theory of crystal vibrations is given by Streitwolf (1971). Rosenberg (1988) provides a delightful introduction to all these topics, including umklapp processes. A thorough treatment of crystal vibrations is given by Cochran (1973). All about phonons (at an advanced level) can be found in Ziman (1960).

12

Phase stability: Brillouin zone effects

Copper is face-centred cubic but, when zinc is added to it, this phase eventually becomes unstable and it changes over to the body-centred cubic structure. It is possible to explain such changes of phase by considering the change in the Fermi energy which takes place when the Fermi surface expands (as it must, for increasing electron concentrations) until it touches a face of the Brillouin zone. We shall study in the first few sections of this chapter the rules which determine such phase transformations in alloys. We shall then consider another effect which is due to the properties of the Fermi energy at the Brillouin zone edges, and which requires that regular one-dimensional chains be unstable with respect to transformations into structures with alternating short and long bonds. This effect is called the *Peierls instability*.

1 Effect of the Brillouin zone faces on the Fermi energy of alloys

We saw in § **9**-2 that the electronic properties of copper can largely be understood in terms of a single 4*s* electron per atom, except that perturbations on this picture occur near the Brillouin zone faces. When zinc is added to copper to form brass alloys, zinc atoms randomly replace copper atoms in the face-centred cubic lattice of copper and, since zinc has two valence electrons per atom, the electron concentration (or density) $\mathcal{N}$ increases. (This was defined in § **1**-5 as the electron concentration per unit volume but, in what follows, it will be more convenient to replace it by $\mathscr{C}$, the electron concentration per atom.)

A rough but simple description of the alloying process is provided by the *rigid-band model* in which the electronic structure of the alloy is described by assuming that the pure-copper band structure is valid throughout the concentration range of the alloy. In the same way that the Fermi surface of copper is obtained by filling these bands until an electronic concentration $\mathscr{C}$ is reached equal to one electron per atom, now more states are filled in the *same* bands until the required electron concentration $\mathscr{C} > 1$ of the alloy is reached. We further assume that free-electron theory can be used throughout, except that, when a face of the Brillouin zone is reached, changes must be effected as in Fig. **8**-7.1, which will result in energy discontinuities. If we thus look at the cross-section of the copper free-electron Fermi surface in Fig. **9**-2.2, the free-electron Fermi surface ($\mathscr{C}$ equal to 1) will expand without deformation until it touches the hexagonal face of the Brillouin zone at the electron concentration $\mathscr{C}'$ equal to 1.36. (See Problem 4.1.) Because of the discontinuities at the Brillouin zone face, any further increase in $\mathscr{C}$ will require an extra increase in the energy. If we look at the relation between the free-electron energy surfaces and the shape of the Brillouin zone, it is easy to prove that in the body-centred structure the surface which contains a concentration $\mathscr{C}$ equal to 1.36 is still well inside the Brillouin zone. Indeed, in this structure, the free-electron spheres do not touch the Brillouin zone face until $\mathscr{C}$ reaches the value 1.48, so that it is possible to increase $\mathscr{C}$ above 1.36 without having to surmount any energy discontinuity. It is therefore argued that for $\mathscr{C}$ larger than 1.36 the body-centred structure will be more stable than the face-centred one. The basic idea is that a phase change will take place every time the Fermi sphere would touch the Brillouin zone face and that a new structure will then appear in which this sphere is still well inside the Brillouin zone.

This simple theory, proposed by Harry Jones in 1934, was amazingly successful in explaining the pioneering results of Hume-Rothery as regards the structures of the brass-alloy phases. As early as 1926, Hume-Rothery had found that the determining factor in the structure of these alloys was the electron concentration $\mathscr{C}$. He found four different phases, of increasing zinc content, which are called α, β, γ, ε, respectively, and found that each phase was stable until a given concentration $\mathscr{C}$, at which a change of phase occurred into the phase with the next higher concentration. These concentrations $\mathscr{C}$ are compared in Table 1 with the concentrations $\mathscr{C}'$ calculated by Jones for the free-electron spheres which touch the Brillouin zone face and the agreement is, as one would modestly say, as good as it can be expected. We now know, alas, that we do not have any right at all to expect any such agreement.

Table 12-1.1. Phase changes in brass alloys.

$\mathscr{C}$ is the electron concentration in electrons per atom at which a phase begins to change into the one listed in the next line. $\mathscr{C}'$ is the calculated electron concentration at which the free-electron Fermi sphere touches for the first time a Brillouin zone face.

	Phase	$\mathscr{C}$	$\mathscr{C}'$
α	Face-centred cubic	1.38	1.36
β	Body-centred cubic	1.50	1.48
γ	Complex	1.62	1.54
ε	Hexagonal close-packed	1.75	1.69

If we look back at Fig. 9-2.2, we see that even at $\mathscr{C}$ equal to unity (pure copper) the Fermi surface already touches the Brillouin zone face! Despite this set-back for the Jones theory, the experimental fact remains that electron concentrations are significant in determining the structures of these alloys. And although the theory cannot be taken seriously in all its details, it is still true that when the Fermi surface touches the Brillouin zone, energy discontinuities appear, which are worthwhile looking into as *possible* sources of phase changes. Further discussion of the present theoretical status of the Hume-Rothery rules may be found in Cottrell (1988).

2 Phase change in a linear chain: the Peierls instability

Whereas the discussion in the last section was somewhat abortive, the effect which we shall now discuss, and its explanation, have a most important place in modern work on solids, because there are numerous materials for which a quasi one-dimensional model is valid and these have practical applications, mainly as organic conductors and the like. Thus, although the model which we shall discuss is very simple, it is nevertheless physically very significant. We wish to consider a regular one-dimensional chain of period $\mathbf{a}$ (Fig. 1) in which we have only one electron per atom. This regular structure could be *reconstructed* by forming short and long alternating bonds, as shown in the lower part of the figure. (In practice each alternate atom would move by $\pm\frac{1}{2}\boldsymbol{\delta}$, but the equivalent description used in the figure will simplify the work.) We want to find out which of these two structures is the stable one. This is significant for the following reason: in the regular linear chain with one electron per atom only half the Brillouin zone is full and we have a conductor whereas, as we shall see, this is not so for the alternating chain. Thus, the present problem is important because materials involving linear or quasi-linear conduction are often expected to be of interest in modern technology. Peierls showed in 1955 that the regular metallic chain is unstable and that it reconstructs into an alternating chain, an effect which is called the *Peierls instability* and which is readily understood from Fig. 2.

The lattice constant for the regular linear chain is a, whence $|\mathbf{a}^{\#}|$ equals $2\pi/a$. The first Brillouin zone, therefore, ranges from $-\pi/a$ to π/a, as shown

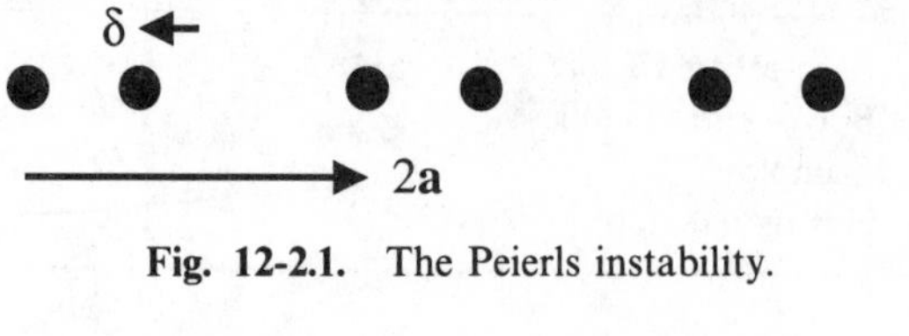

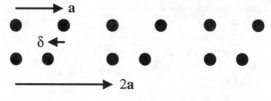

Fig. 12-2.1. The Peierls instability.

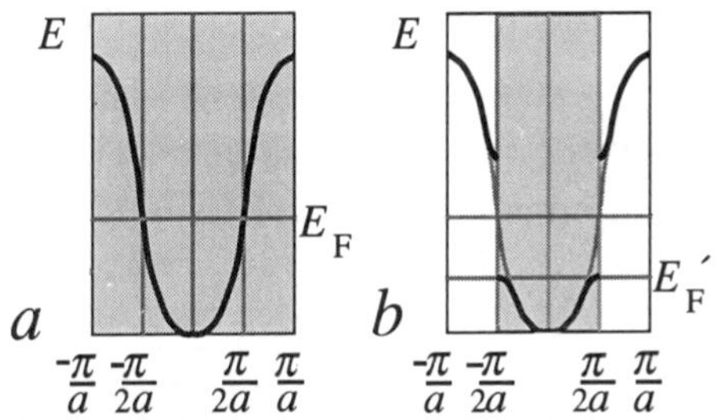

Fig. 12-2.2. Bands and Fermi energies for the regular linear chain (*a*) and the reconstructed linear chain (*b*) of Fig. 1. The shaded areas indicate the first Brillouin zone corresponding to each case.

in Fig. 2*a*. Since we have only one electron per atom, this zone is half full, from $-\pi/2a$ to $\pi/2a$, which determines the Fermi energy E_F shown in the figure. (Here and hereafter we always assume one electron per atom.) As regards the reconstructed structure, as illustrated in Fig. 1, two features are specially relevant for our discussion. One is that the lattice constant has now doubled to the value 2*a*, and the other is that the lattice is one with a basis, with two atoms, and thus two electrons, per primitive cell. The shortest vector of the reciprocal lattice has now a modulus equal to $2\pi/2a$, so that the first Brillouin zone ranges from $-\pi/2a$ to $\pi/2a$, as shown in Fig. 2*b*, where, of course, a band gap appears at the edge of the new Brillouin zone. Because there are two electrons per primitive cell, this Brillouin zone is full, thus leading to a new Fermi energy E_F' which is clearly *lower* than E_F.

We notice that, on reconstruction, the regular linear chain, which is metallic, changes into an insulator (because the Brillouin zone of the reconstructed structure is full). In order to decide which structure is more stable, we must consider the energy change on reconstruction. As we have seen, the Fermi energy is lowered by the quantity

$$\Delta E_F = E_F - E_F'. \tag{1}$$

The potential energy, on the other hand, is increased on reconstruction by the elastic energy of deformation, which must be proportional to δ^2, where δ is the modulus of the small displacement shown in Fig. 1. We shall see below, on the other hand, that ΔE_F is of the order of δ, and thus that it is the dominant term. It will therefore follow that the reconstructed, non-metallic structure, is the stable phase. It must be remembered, however, that the argument given here about the energetic balance is oversimplified and that more careful work is required in order to provide a realistic model of these systems (see Further Reading at the end of this chapter). The elastic energy of deformation, also, increases with the temperature so that it is possible that above a certain temperature the reconstructed structure ceases to be stable and thus that a *Peierls transition* takes place.

The electronic energy stabilization

In order to obtain the energy balance we should compare the total electronic energy E_T, corresponding to all the electrons occupying states up to E_F, with the elastic energy of deformation. It follows from (**1**-6.7) that E_T is proportional to the Fermi energy, so that it will be sufficient to consider this latter, as we have already done in (1), since we are concerned with orders of magnitude only.

It is clear from Fig. 2 that ΔE_F is one half of the energy gap, which is given by (**10**-1.33), in which we change the volume V of the crystal by the length L of the chain:

$$\Delta E_F = L^{-1/2} |\mathcal{V}_{\mathbf{g}}|. \tag{2}$$

In this equation, $\mathcal{V}_{\mathbf{g}}$ is the structure factor corresponding to the vector **g** that separates the two faces of the Brillouin zone at which the gap is computed, which vector, from Fig. 2*b*, is $\frac{1}{2}\mathbf{a}^{\#}$:

4-9.26
$$\mathcal{V}_{\mathbf{g}} = \sum_{\rho} V_{\mathbf{g}\rho} \exp(-\,\mathrm{i}\mathbf{g}\cdot\boldsymbol{\rho}). \tag{3}$$

The vector $\boldsymbol{\rho}$ ranges over the position vectors of each of the atoms of the basis, which, from Fig. 1 take the two values $\mathbf{0}$ and $\mathbf{a} - \boldsymbol{\delta}$. On expressing $\boldsymbol{\delta}$ in units of **a**, we shall write down the second $\boldsymbol{\rho}$ vector as $\mathbf{a} - \delta\mathbf{a}$. As regards the Fourier coefficients $V_{\mathbf{g}\rho}$ of the two atoms of the basis, they can both be taken as equal to V, say, since both atoms are identical. Therefore, on leaving out the index **g** (which is equal to $\frac{1}{2}\mathbf{a}^{\#}$), we have

3
$$\mathcal{V} = V + V\exp\{-\,\mathrm{i}\mathbf{g}\cdot(\mathbf{a} - \delta\mathbf{a})\}. \tag{4}$$

Here, $\mathbf{g}\cdot\mathbf{a}$ equals π, so that

4
$$\mathcal{V} = V + V(-1)\exp(\mathrm{i}\pi\delta). \tag{5}$$

On expanding the exponential in power series to first order in δ, we obtain

5
$$\mathcal{V} = V - V(1 + \mathrm{i}\pi\delta) = -\,V\pi\delta\mathrm{i}, \tag{6}$$

whence

$$|\mathcal{V}| = V\pi\delta. \tag{7}$$

On introducing this value into (2), we see that ΔE_F is of the order of δ, as asserted before. It shoud be remarked, however, that (7) is an expression of limited validity because a deformation of the chain must surely perturb the electron screening and thus the potential V itself. These electron-electron interactions affect (7) so substantially that, as pointed out by Littlewood and Heine (1981), the Peierls distortion cannot always be assumed to take place.

Electron–phonon interactions

The essence of the Peierls effect is this: a distortion of the linear chain is found to cause the wave function corresponding to the Fermi level $\psi_{\mathbf{k}_F}$, with $\mathbf{k}_F$ equal to $\frac{1}{2}\pi/a$, to change into another wave function with lower energy. We must now recognize that the chain distortion required cannot take place unless there is a normal mode which can drive the structure into the required configuration. It is pretty clear from Fig. 3 that the distortion from the equilibrium configuration shown in Fig. 3*a* to the one shown in *b* must take effect via the antisymmetric vibrational mode depicted in Fig. 3*a*.

We shall now sketch how this *vibronic interaction* between the vibrational and the electronic states takes place. The electronic Hamiltonian $\mathbb{H}$ must depend not only on the electronic coordinates $\mathbf{r}$ but also on the coordinates $\mathbf{q}$ of the atomic nuclei as parameters and should be written as $\mathbb{H}(\mathbf{r}, \mathbf{q})$. This *vibronic Hamiltonian* can be expanded in a Taylor series in the small displacements of the $\mathbf{q}$ parameters. It is convenient to take these small displacements to be given in terms of the normal coordinates $\mathbf{z}_\mathbf{q}$ of (**11**-2.19):

$$\mathbb{H}(\mathbf{r}, \mathbf{q}) = \mathbb{H}(\mathbf{r})_{\mathbf{q}=0} + \sum_{\mathbf{q}} U_\mathbf{q}(\mathbf{r})\mathbf{z}_\mathbf{q} + \sum_{\mathbf{q}\mathbf{q}'} W_{\mathbf{q}\mathbf{q}'}(\mathbf{r})\mathbf{z}_\mathbf{q}\mathbf{z}_{\mathbf{q}'} + \ldots . \tag{8}$$

The first term on the right here is the ordinary electronic Hamiltonian corresponding to the equilibrium configuration of the nuclei. The Hamiltonian $\mathbb{H}(\mathbf{r}, \mathbf{q})$, like it was the case for $\mathbb{H}(\mathbf{r})$, must be invariant under all the operations of the space group G, that is, it must belong to the totally symmetrical representation of this group. This must thus be true for each of the terms on (R8), as we know is the case for its first term. It is not difficult to satisfy oneself (see Problem 4.6) of a well-known result from group theory: for the second term in (R8) to belong to the totally symmetrical representation of G, it is necessary for $U_\mathbf{q}(\mathbf{r})$ to belong to the same representation as $\mathbf{z}_\mathbf{q}$.

We next need a result from quantum mechanics (see Landau and Lifshitz 1977, p. 407). An electronic state with wave function $\psi(\mathbf{r})$ will be able to interact with the vibrational state given by a normal mode $\mathbf{z}_\mathbf{q}$ if the *transition probability integral*

$$I = \int \psi_\mathbf{k}^*(\mathbf{r})\, U_\mathbf{q}(\mathbf{r})\, \psi_\mathbf{k}(\mathbf{r})\, \mathrm{d}\mathbf{r}, \tag{9}$$

where the integration is over all the electronic coordinates, does not vanish. In our case, the electronic state which we want to consider corresponds to the Fermi level, $\psi_{\mathbf{k}_F}(\mathbf{r})$, with $\mathbf{k}_F$ equal to $\frac{1}{2}\pi/a$, and the normal mode is the one of

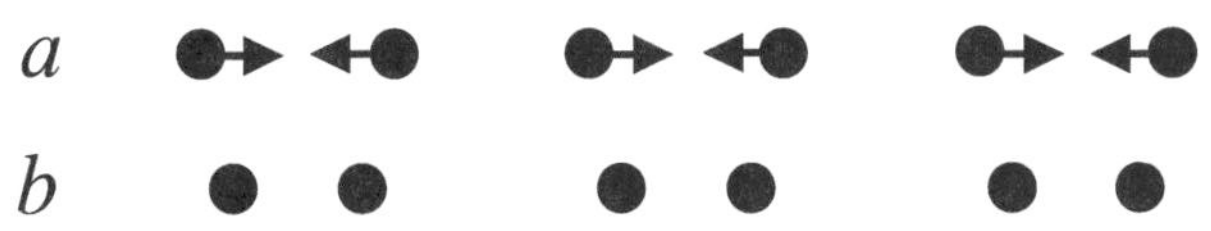

Fig. 12-2.3. Normal mode of vibration required to drive the Peierls reconstruction. The unreconstructed and reconstructed chains (*a* and *b* respectively) are identical with those shown in Fig. 1, except for the setting.

Fig. 3, which has been identified in Problem **11**-5.4 as corresponding to a phonon at the edge of the Brillouin zone, $\mathbf{q} = \pi/a$. The transition probability integral that we must prove does not vanish is thus

$$I = \int \psi^*_{\mathbf{k}_F}(\mathbf{r})\, U_{\mathbf{q}}(\mathbf{r})\, \psi_{\mathbf{k}_F}(\mathbf{r})\, d\mathbf{r}, \qquad \mathbf{k}_F = \tfrac{1}{2}\pi/a, \qquad \mathbf{q} = \pi/a = 2\mathbf{k}_F. \tag{10}$$

Here, from our discussion of eqn (8), $U_{\mathbf{q}}(\mathbf{r})$ belongs to the same irreducible representation as $\mathbf{z}_{\mathbf{q}}$ with the value of $\mathbf{q}$ in (10). If we regard this integral as the product of two factors $U_{\mathbf{q}}$ and $\psi^*_{\mathbf{k}_F}\psi_{\mathbf{k}_F}$, it is a well-known result of group theory that for the integral not to vanish the second factor, say, must belong to the same irreducible representation as the first or, if it belongs to a reducible representation, that this representation must reduce into a combination of irreducible representations which contains the irreducible representation to which the first factor belongs. The first thing we must do therefore is to find the symmetry of the product $\psi^*_{\mathbf{k}_F}\psi_{\mathbf{k}_F}$. The star of $\psi_{\mathbf{k}_F}$ contains $\psi_{\mathbf{k}_F}$ and $\psi_{-\mathbf{k}_F}$, the latter being identical with $\psi^*_{\mathbf{k}_F}$. We shall start by considering the irreducible representation spanned by the basis which contains $\psi^*_{\mathbf{k}_F}$ and $\psi_{\mathbf{k}_F}$:

8-8.2 basis: $\langle \psi_{\mathbf{k}_F} \psi^*_{\mathbf{k}_F} |$, matrix for $\{E|\mathbf{t}\}$:
$$\begin{bmatrix} e^{-i\mathbf{k}_F\cdot\mathbf{t}} & \\ & e^{i\mathbf{k}_F\cdot\mathbf{t}} \end{bmatrix}. \tag{11}$$

In order to be able to discuss the transformation properties of products of functions the standard trick in group theory is, given two bases $\langle f_1 \ldots f_m |$, (with matrix A), and $\langle h_1 \ldots h_m |$, (with matrix B), to form the *direct product basis*

$$\langle f_1 h_1, f_1 h_2, \ldots, f_1 h_m, f_2 h_1, \ldots, f_m h_1, \ldots, f_m h_m |. \tag{12}$$

The advantage of forming this basis is twofold: first, it contains, as we wanted, products of functions; secondly, the matrix by which this basis transforms, which is called the *direct product matrix* is given by the following simple rule: replace every matrix element A_{ij} in A by its product, $A_{ij}B$ with the *whole* of the second matrix. (See, for example, Altmann 1977, p. 56.) With these prescriptions, the direct product basis $\langle \psi_{\mathbf{k}_F} \psi^*_{\mathbf{k}_F} |$ times itself,

$$\langle \psi_{\mathbf{k}_F}\psi_{\mathbf{k}_F}, \psi_{\mathbf{k}_F}\psi^*_{\mathbf{k}_F}, \psi^*_{\mathbf{k}_F}\psi_{\mathbf{k}_F}, \psi^*_{\mathbf{k}_F}\psi^*_{\mathbf{k}_F} |, \tag{13}$$

transforms under the direct product of the matrix in (11) times itself:

$$\begin{bmatrix} e^{-i\mathbf{k}_F\cdot\mathbf{t}}e^{-i\mathbf{k}_F\cdot\mathbf{t}} & & & \\ & e^{-i\mathbf{k}_F\cdot\mathbf{t}}e^{i\mathbf{k}_F\cdot\mathbf{t}} & & \\ & & e^{i\mathbf{k}_F\cdot\mathbf{t}}e^{-i\mathbf{k}_F\cdot\mathbf{t}} & \\ & & & e^{i\mathbf{k}_F\cdot\mathbf{t}}e^{i\mathbf{k}_F\cdot\mathbf{t}} \end{bmatrix} =$$

$$= \begin{bmatrix} e^{-2\mathbf{k}_F\cdot\mathbf{t}} & & & \\ & 1 & & \\ & & 1 & \\ & & & e^{2\mathbf{k}_F\cdot\mathbf{t}} \end{bmatrix}. \tag{14}$$

Before we go any further here, a little remedial action is required, because the basis in (13) is defective, its second and third terms being identical. (Remember that bases, irreducible or not, must always be made up of linearly independent functions.) The medicine required can easily be administered: remove the third term in (13) and, since this term is responsible for the third diagonal elements in the matrices in (14), also these elements. The 3 by 3 purged matrix on (R14) must now reduce over the irreducible matrices of the group, and it is immediately seen that this reduction gives the representations in (**8**-8.1) and (**8**-8.2). Look at the second of these: it corresponds to the value of **k** (or **q**) equal to $2\mathbf{k}_F$, which agrees precisely with the value of **q** in (10). (Readers who might worry about details of the representations involved here should first consider the comment at the end of this section.)

We have thus proved that the electron-phonon interaction required for the Peierls reconstruction to take place is non vanishing. We can now summarize the principle involved in this work: if a system with a given symmetry has a top occupied electronic orbital which is degenerate, then this function will interact with a vibrational mode which has symmetry lower than that of the original system, causing it to distort from the original configuration and thus splitting the degeneracy (which results in a state of lower energy appearing). This principle is well known under the name of the *Jahn–Teller effect*. (See Problem 4.3.) Notice, in fact, that, after distortion, the doubly degenerate basis corresponding to $\psi_{\mathbf{k}_F}$ in (11), becomes singly degenerate, since $\mathbf{k}_F$ coincides then with the edge of the Brillouin zone.

One final comment about the previous work, in order to tie up a loose end. The representation which arises from the purged matrix from (R14), after the totally symmetric representation (**8**-8.1) is subtracted, is clearly spanned by the basis $\langle \psi_{\mathbf{q}}, \psi_{-\mathbf{q}} |$. It follows at once from (10) that these two functions differ by $2\pi/a$, that is, by a vector of the reciprocal lattice and thus that they are identical. This means that the alleged two-dimensional representation reduces into two copies of the same one-dimensional representation. This, of course, does not affect in the least our previous conclusion that the representation with **k** equal to $2\mathbf{k}_F$ is contained in the product $\psi^*_{\mathbf{k}_F}\psi_{\mathbf{k}_F}$, which is all that is necessary in order to prove that (10) does not vanish. In the case of normal vibrations, rather than just taking the representation to be that spanned by $\psi_{\mathbf{q}}$, with 'matrix' $\exp(2\mathbf{k}_F \cdot \mathbf{t})$, that is, $\exp(-i\pi t/a)$, it is customary to take as basis half the sum of the identical functions $\psi_{\mathbf{q}}$ and $\psi_{-\mathbf{q}}$, with 'matrix' $\cos(\pi t/a)$. Diligent readers who embark into Problem 4.3 will find this result useful, in order to tie up with the treatment of the Jahn–Teller effect done by means of standard recipes.

3 The Peierls instability for a quasi one-dimensional chain

Although the linear chain of § 2 is a very useful model, somewhat more complex structures are more common in real solids, specially for polymers.

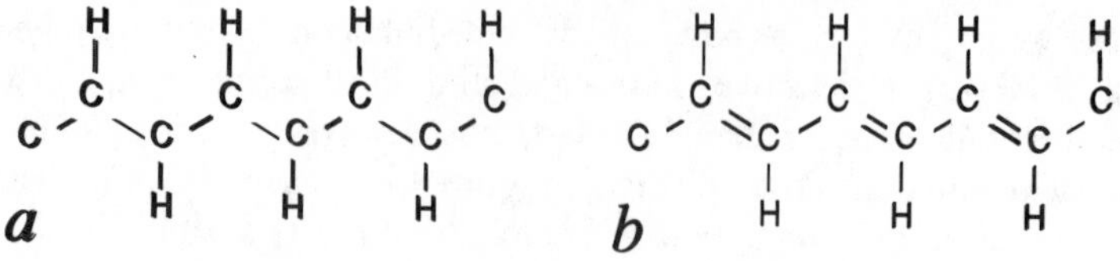

Fig. 12-3.1. Polyacetylene. A small section of the polymer chain is illustrated.

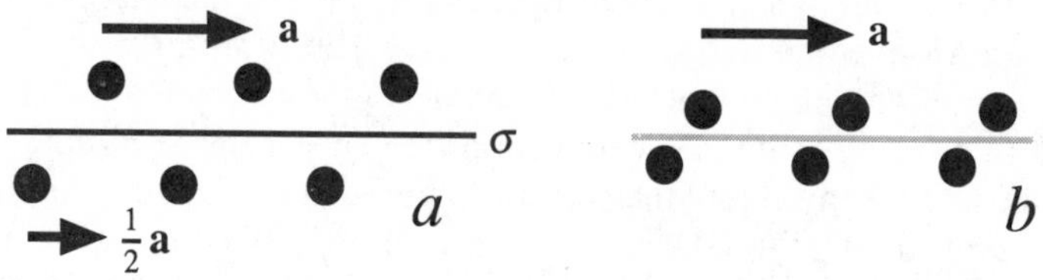

Fig. 12-3.2. A polyene-type chain, before reconstruction (a) with all bonds of equal length, and after reconstruction (b) with alternating bonds.

We illustrate in Fig. 1 one such polymer, polyacetylene, of much practical interest in the field of organic electrical conductors. Each carbon atom, in fact, has a π electron orbital (p_z), normal to the drawing, which is mobile. The classical structure of this molecule, however, presents alternating double and single bonds, as shown in Fig. 1b, which would suggest that Peierls reconstruction from the regular structure in a is likely. Long chains of the same form as shown in Fig. 1 appear frequently and they are called *polyene chains.* We illustrate these structures in Fig. 2a, in which all bonds are of equal length. The period here is $\mathbf{a}$ and we have a basis consisting of two atoms. The plane σ is not a reflection plane on its own, but the operation

$$\gamma = \{\sigma | \tfrac{1}{2}\mathbf{a}\} \tag{1}$$

is a symmetry operation of the infinite structure, in which a copy of the figure is first reflected under σ and then translated by $\frac{1}{2}\mathbf{a}$ in order to cover the original. This operation is a *glide plane.* (The space group of this chain, thus, is non-symmorphic.) It is important to notice that *there is no change of period* on reconstruction, as a difference with the linear chain of § 2. The relevant feature, instead, is that the glide plane is lost, as can be seen in Fig. 2b, where this operation no longer covers the system.

We shall first consider the unreconstructed structure. Its Brillouin zone, like the one of Fig. 2.2a, must extend from $-\pi/a$ to π/a, but now, because we have two electrons per primitive cell, it must be *full.* This is a surprising result which appears to contradict chemical experience: a full Brillouin zone indicates an insulator, that is a system in which the electrons are largely localized, whereas it is well known that the electron system of a polyene is mobile. The description of a full Brillouin zone as associated with a non mobile electron system, however, depends crucially on the existence of an energy gap at the Brillouin zone edge (§ **8**-4) and we shall now see that this feature disappears in the unreconstructed structure, owing to the presence of the glide plane γ: bands now stick together at the Brillouin zone edges, as in Fig. **8**-4.3*b*, a behaviour which was forbidden for the ordinary linear chain.

Sticking together of the bands at the Brillouin zone edge

Consider a Bloch function $\psi_{\mathbf{k}}$ for $\mathbf{k}$ at the Brillouin zone edge:

$$\mathbf{k} = \pi/a. \tag{2}$$

We know that there is no other Bloch function in the linear chain degenerate with this $\psi_{\mathbf{k}}$: whereas for general values of $\mathbf{k}$, $\psi_{\mathbf{k}}$ and $\psi_{-\mathbf{k}}$ are degenerate, this is not so for the value of $\mathbf{k}$ in (2) since these two functions differ by a vector of the reciprocal lattice and they are therefore identical. We shall now show, however, that in the zigzag chain this situation changes radically: *there is a function which belongs to the same* $\mathbf{k}$ *as* $\psi_{\mathbf{k}}$, *as given by* (2), *and which is degenerate with* $\psi_{\mathbf{k}}$.

Consider the function

$$\psi' = \gamma \mathbb{j} \psi_{\mathbf{k}} = \{\sigma|\tfrac{1}{2}\mathbf{a}\}\, \mathbb{j} \psi_{\mathbf{k}}, \tag{3}$$

where $\mathbb{j}$ is the conjugator operator defined in (**2**-3.11). Because ψ' is obtained from $\psi_{\mathbf{k}}$ by acting upon it with symmetry operators, it must belong to the same energy eigenvalue and thus it will be degenerate with $\psi_{\mathbf{k}}$ if it is linearly independent from it. (See Part 2 below.) In order to prove the assertion italicized above, we must therefore prove:

(1) that ψ' belongs to $\mathbf{k}$,
(2) that ψ' is linearly independent of $\psi_{\mathbf{k}}$.

Proof of Part 1. A Bloch function $\psi_{\mathbf{k}}$ belongs to the vector $\mathbf{k}$ if it satisfies the eigenvalue equation for the translations,

5-6.2
$$\{E|\mathbf{t}\}\psi_{\mathbf{k}} = \exp(-\,\mathrm{i}\mathbf{k}\cdot\mathbf{t})\psi_{\mathbf{k}}. \tag{4}$$

Let us act on ψ' with $\{E|\mathbf{t}\}$:

3
$$\{E|\mathbf{t}\}\psi' = \{E|\mathbf{t}\}\{\sigma|\tfrac{1}{2}\mathbf{a}\}\mathbb{j}\psi_{\mathbf{k}}. \tag{5}$$

We shall now prove that the product of the first two operators on (R5) commutes:

3-6.6
$$\{E|\mathbf{t}\}\{\sigma|\tfrac{1}{2}\mathbf{a}\} = \{\sigma|\tfrac{1}{2}\mathbf{a}\}\{E|\sigma\mathbf{t}\}. \tag{6}$$

(Remember in checking both sides of eqn 6 that σ^2 equals the identity.) It follows from Fig. 2*a*, however, that $\sigma\mathbf{t}$ must equal $\mathbf{t}$ for any translation vector

t of the system, which proves the asserted commutation. We also know that $\mathbb{j}$ commutes with all symmetry operators. Thus

2-3.17, 6|5 $$\{E|\mathbf{t}\}\psi' = \{\sigma|\tfrac{1}{2}\mathbf{a}\}\mathbb{j}\{E|\mathbf{t}\}\psi_{\mathbf{k}} \tag{7}$$

4|7 $$= \{\sigma|\tfrac{1}{2}\mathbf{a}\}\mathbb{j}\{\exp(-\,\mathrm{i}\mathbf{k}\cdot\mathbf{t})\psi_{\mathbf{k}}\} \tag{8}$$

2-3.11|8 $$= \exp(\mathrm{i}\mathbf{k}\cdot\mathbf{t})\{\sigma|\tfrac{1}{2}\mathbf{a}\}\mathbb{j}\psi_{\mathbf{k}} \tag{9}$$

3|9 $$= \exp(\mathrm{i}\mathbf{k}\cdot\mathbf{t})\psi'. \tag{10}$$

Comparison with (4) shows that ψ' belongs to $-\,\mathbf{k}$. We know, however, that for the value of **k** at the edge of the Brillouin zone, given by (2), $-\,\mathbf{k}$ and **k** are equivalent, so that ψ' belongs to **k** and shall henceforth be written as $\psi'_{\mathbf{k}}$.

Proof of Part 2. Suppose that $\psi'_{\mathbf{k}}$ were not linearly independent of $\psi_{\mathbf{k}}$, that is, that there is a constant α such that

$$\psi'_{\mathbf{k}} = \alpha\psi_{\mathbf{k}}. \tag{11}$$

3|11 $$\gamma\mathbb{j}\psi_{\mathbf{k}} = \alpha\psi_{\mathbf{k}}. \tag{12}$$

12 $$\gamma\mathbb{j}\gamma\mathbb{j}\psi_{\mathbf{k}} = \gamma\mathbb{j}\alpha\psi_{\mathbf{k}} \tag{13}$$

2-3.11|13 $$= \alpha^*\gamma\mathbb{j}\psi_{\mathbf{k}} \tag{14}$$

12|14 $$= \alpha^*\alpha\psi_{\mathbf{k}}. \tag{15}$$

On the other hand,

1, **3**-6.6 $$\gamma\mathbb{j}\gamma\mathbb{j} = \gamma^2\mathbb{j}\mathbb{j} = \gamma^2 = \{\sigma|\tfrac{1}{2}\mathbf{a}\}\,\{\sigma|\tfrac{1}{2}\mathbf{a}\} = \{E|\mathbf{a}\}. \tag{16}$$

16|L15 $$\gamma\mathbb{j}\gamma\mathbb{j}\psi_{\mathbf{k}} = \{E|\mathbf{a}\}\psi_{\mathbf{k}} \tag{17}$$

4|17 $$= \exp(-\,\mathrm{i}\mathbf{k}\cdot\mathbf{a})\psi_{\mathbf{k}} \tag{18}$$

2|18 $$= \exp(-\,\mathrm{i}\pi)\psi_{\mathbf{k}} = -\psi_{\mathbf{k}}. \tag{19}$$

Therefore,

19|15 $$\alpha^*\alpha = -1, \tag{20}$$

which is impossible, since (L20) is always positive. It follows that (11) is also impossible and the proof is completed.

The double Brillouin zone

What we have proved so far is that at the edge **k** equal to π/a of the Brillouin zone there are two degenerate eigenfunctions, which means that the first and second bands must stick together at this point of the Brillouin zone. In other words, there is no energy gap at this edge, as shown in Fig. 3*a*. Because of this,

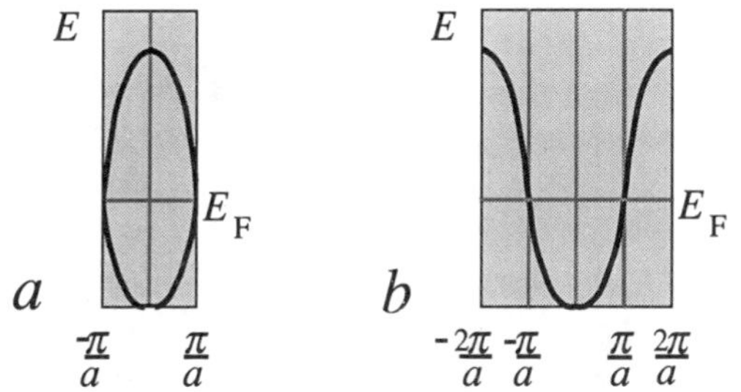

Fig. 12-3.3. Brillouin zone and bands for the unreconstructed polyene chain of Fig. 2*a*. The curve shown in *b* from π/a to $2\pi/a$ is the branch of the second band given in *a* from $-\pi/a$ to 0. Similarly for the other branch of the second band. The shaded areas indicate Brillouin zones, whether single (*a*) or double (*b*).

it is best to display the bands in the extended zone scheme of Fig. 3*b*, where continuity at the Brillouin zone edge is shown. (Notice that, therefore, $\partial E/\partial \mathbf{k}$ does not vanish here: if the bands of Fig. 3*a* are repeated in the periodically repeated scheme, a cusp at the Brillouin zone edge would appear, as in Fig. **8**-2.3.) Because in many problems, and the Peierls reconstruction is one of them, the determining feature of the Brillouin zone face is the fact that it entails energy discontinuities, the effective Brillouin zone (sometimes called a *Jones zone*) is one where all the faces determine energy gaps. This is in our case the *double zone* of Fig. 3*b*. This double zone contains four electrons (including spin) per primitive cell and, since we have only two electrons in it, it is half full, from $-\pi/a$ to π/a, thus giving the Fermi energy E_F shown. The material is a *conductor*, as we had expected, thus solving the difficulty posed within the elementary Brillouin zone theory.

Reconstruction

We shall discuss the mechanism whereby reconstruction of the polyene chain takes place by means of Fig. 4. In part *a* of this figure we reproduce again the band for the unreconstructed case in the double Brillouin zone, as shown in Fig. 3*b*, and in Fig. 4*b* we deal with the reconstructed case. As we have seen in Fig. 1, reconstruction does not change the period *a* of the polyene-type chain. Thus, the *first* Brillouin zone extends from $-\pi/a$ to π/a, as for the unreconstructed case. On the other hand, as also follows from Fig. 1, the significant feature of the reconstruction is that the glide plane disappears. Since it was the presence of this glide that caused the sticking together of the bands on the edge of the first Brillouin zone (Fig. 4*a*), it follows at once that, on reconstruction, this edge becomes a point of discontinuity, requiring an energy gap in the usual way, as shown in Fig. 4*b*. Since we still have two electrons per primitive cell in the reconstructed structure, the whole of the first Brillouin

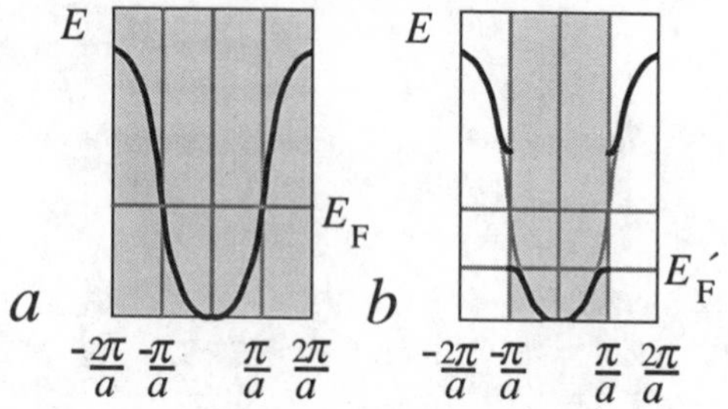

Fig. 12-3.4. Bands and Fermi energies for the polyene-type regular chain (a) and its reconstructed, alternating, form (b). The shaded areas indicate Brillouin zones, whether double (a) or single (b).

zone must be full, thus lowering down the Fermi energy from E_F in the unreconstructed case to E'_F in the alternating, reconstructed, chain. The latter is thus the stable, non-metallic, phase.

4 Problems

1. Prove that the free-electron sphere that touches the hexagonal face of the face-centred cubic Brillouin zone contains 1.36 electrons per atom.
2. State why, in order to obtain the character of an operation g in an irreducible basis $\langle u|, \chi(g|u)$, it is always possible to assume that the matrix for g is diagonal. Thus, if the basis runs from u_1 to u_m, prove that

$$\chi(g|u) = \sum_i U_i, \tag{1}$$

where the U_i are the diagonal elements of the matrix representative. Given the direct product basis

$$\langle u_1, \ldots, u_m| \otimes \langle v_1, \ldots, v_n| = \langle u_1 v_1, \ldots, u_1 v_n, u_m v_1, \ldots, u_m v_n|, \tag{2}$$

prove that, with $i = 1, \ldots, m$, and $j = 1, \ldots, n$,

$$\chi(g|u \otimes v) = \sum_{ij} U_i V_j = \chi(g|u)\,\chi(g|v). \tag{3}$$

In order to eliminate redundancies when u equals v, define the *symmetrized direct product* basis

$$\langle u_1, \ldots, u_m| \,\bar{\otimes}\, \langle u_1, \ldots, u_m| =_{\text{def}} \langle u_i v_j|,$$
$$i = 1, \ldots, m; \quad j = i, \ldots, m. \tag{4}$$

Prove that

$$\chi(g|u \,\bar{\otimes}\, u) = \tfrac{1}{2}[\{\chi(g|u)\}^2 + \chi(g^2|u)]. \tag{5}$$

3. Call $\langle\psi|$ the basis $\langle\psi_{\mathbf{k}_F}\psi^*_{\mathbf{k}_F}|$ of (2.11), and $\langle\psi|\bar{\otimes}\langle\psi|$ the symmetrized direct product basis

4|2.13 $$\langle\psi_{\mathbf{k}_F}\psi_{\mathbf{k}_F}, \psi_{\mathbf{k}_F}\psi^*_{\mathbf{k}_F}, \psi^*_{\mathbf{k}_F}\psi^*_{\mathbf{k}_F}|, \tag{6}$$

and prove that for an operation t,

2.11 $$\chi(t|\psi) = 2\cos(\pi t/2a), \chi(t^2|\psi) = \chi(2t|\psi) = 2\cos(\pi t/a). \tag{7}$$

On using (5) prove that

$$\chi(t|\psi \bar{\otimes} \psi) = 2\cos^2(\pi t/2a) + \cos(\pi t/a) = 1 + 2\cos(\pi t/a). \tag{8}$$

Prove thus that the transition probability integral (2.10) does not vanish. What happens when the inversion is included?

4. Find the normal vibration that will cause the distortion required for the quasi one-dimensional chain of Fig. 3.2, and establish its symmetry type. Show that, with the correct value of $\mathbf{k}_F$, the integral (2.10) does not vanish and thus that electron–phonon coupling takes place. Discuss the effect on this result of inversion symmetry.
5. Notice in Fig. **4**-9.1 that if the intermediate layer of a hexagonal-closed packed (hcp) structure slides until it registers exactly above and below the other two, then the planes in question all become the (111) planes of a face-centred cubic (fcc) structure. Notice also that in the hcp structure there is a screw axis parallel to the **c** axis. Prove that the existence of this screw axis determines sticking together of the bands at the top face of the hcp Brillouin zone. Consider a hcp crystal with four electrons per atom, and discuss whether its transformation into a fcc structure would be energetically favourable. How would the possibility of relaxation of the crystal structure affect this transformation?
6. Call ψ_i and ψ_j two functions that belong respectively to the representations ${}^i\hat{G}$ and ${}^j\hat{G}$ of a group G. Prove that the condition for the product $\psi_i\psi_j$ to transform like the totally symmetrical representation of G is that the representations ${}^i\hat{G}$ and ${}^j\hat{G}$ are one the complex conjugate of the other. Notice that for real Hamiltonians, in which case the conjugator operator belongs to the Schrödinger group, the bases of complex conjugate representations are degenerate and must be considered to span the same, degenerate, representation. Therefore the condition required in this problem can be more simply given when the Hamiltonian is real: ψ_i and ψ_j must belong to the same representation.

Further reading

The Peierls instability was first discussed by R. E. Peierls in pp. 108–112 of his book (Peierls, 1955). Whangbo (1982) and Jerome and Schulz (1982) deal with the Peierls instability in quasi one-dimensional models. The paper by Littlewood and Heine (1981) is very important in stressing the limitations of the simple-minded theory when electron–electron interactions are included. The relation between the Peierls and the Jahn–Teller effects has been pointed out by Salem (1966) and Božović (1985). An elementary introduction to the Jahn–Teller effect can be found in Chapter 17 of Knox and Gold (1964) or in p. 407 of Landau and Lifshitz (1977). A good early review of the Jahn–Teller effect in solids, although containing no discussion of its relation to the Peierls instability, is given by Sturge (1967). The effects of screw rotations and glide planes have been studied by Tobin (1960), Parry and Thomas (1975), and Altmann (1982). Full group-theoretical treatments of the line groups are provided by Božović *et al.* (1978, 1981). Electron–phonon coupling is treated by Apostol and Baldea (1982). An elementary but very good treatment of phase stability is given in Cottrell (1988). Hoffmann (1988) contains a very useful treatment of the Peierls instability from the point of view of chemical bonding.

The question of the alternation of bond lengths in polyacetylenes has had a chequered history. An early paper by Longuet-Higgins and Salem (1959) showed by means of a tight-binding (linear combination of molecular orbitals) calculation that alternation was energetically favourable. This work, however, like all methods discussed in this book, was based on the one-electron approximation in which the motion of the electrons is uncorrelated, each electron moving in the averaged field of all others. When electron–electron correlations were introduced by Harris and Falicov (1969) alternation was shown to be negligible. A classical paper on the alternation of polyacetylenes in relation to the Peierls instability is by Whangbo, Hoffmann, and Woodward (1979). Ovchinnikov, Ukrainskiĭ, and Kventsel' (1973) discuss in detail the various effects that determine the stability of one-dimensional models.

13

Wannier functions and Löwdin orbitals

Plane waves and Bloch functions, which have so far been our tools for describing electron states in solids, are delocalized functions which extend throughout the crystal. When dealing with localized phenomena, such as impurities or surface states, it is useful to replace the Bloch functions or the plane waves by functions which are, as far as possible, localized. This means that whereas the Bloch functions are labelled by the different **k** vectors of the Brillouin zone, the functions we seek, which are called the *Wannier functions*, must be labelled in terms of the different primitive cells of the crystal. This does not mean that they are entirely localized within each cell but, because each function is associated with one primitive cell, a degree of localization might be achieved, as we shall see later on (§ 2) is in fact the case.

In order to transform Bloch functions into functions which are associated with the different crystal cells, we shall define and construct sets of functions which shall be called *equivalent functions*, of which the Wannier functions will be special case. When precisely the same construction is applied to Bloch sums, or the linear combinations of atomic orbitals such as are used in the tight-binding method, new localized functions will be obtained, now written in terms of atomic orbitals rather than Bloch functions, which have some useful properties as functions to be fed into a tight-binding calculation. These functions are called Löwdin orbitals. First of all, we shall define equivalent functions and give a method for their construction.

1 Equivalent functions

It will be useful to illustrate the concepts which we want to develop by discussing the molecule of ammonia, NH_3, as an example. We show in Fig. 1*b* three hybrids p_1, p_2, p_3, around the nitrogen atom and directed respectively to the three hydrogen atoms. As shown in the figure, the symmetry group G of ammonia is $\mathbf{C}_{3v}$:

$$G = \mathbf{C}_{3v}\text{: } E,\ C_3^+,\ C_3^-,\ \sigma_{v1},\ \sigma_{v2},\ \sigma_{v3}. \tag{1}$$

This group has a subgroup H which is the cyclic group $\mathbf{C}_3$ (see Fig. **2**-7.1), which will play a fundamental role in the work that follows:

$$H = \mathbf{C}_3\text{: } E,\ C_3^+,\ C_3^-. \tag{2}$$

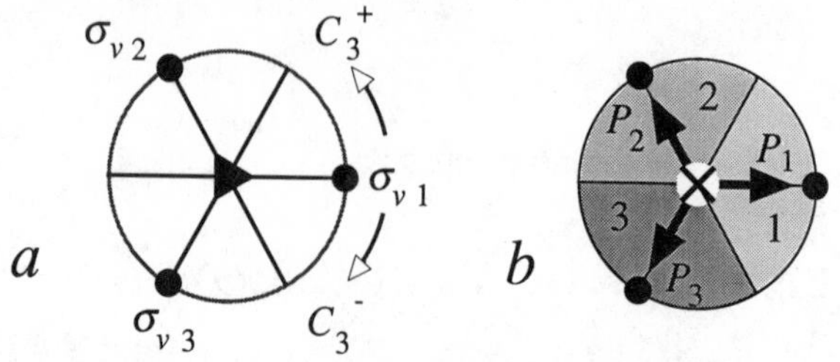

Fig. 13-1.1. The symmetry operations of ammonia (*a*), and equivalent functions and cells (*b*). The symmetry planes $\sigma_{v1}, \sigma_{v2}, \sigma_{v3}$, are normal to the plane of the figure. The hydrogen atoms are represented in *b* by the black circles and they are on the plane of the drawing. The nitrogen atom, represented with a cross, is above that plane. The hybrids p_1, p_2, p_3, are symbolically represented with thick arrows. The three differently shaded areas represent the cells 1, 2, and 3.

We notice at once, in fact, two things. First, if we apply the operations of $\mathbf{C}_3$ on any of the functions of the set

$$\{p_i\} = p_1, p_2, p_3, \tag{3}$$

then precisely the same functions are obtained, none repeated more than once:

$$Ep_1 = p_1, \qquad C_3^+ p_1 = p_2, \qquad C_3^- p_1 = p_3, \tag{4}$$

and similarly for p_2 and p_3. (This would not be the case if $\mathbf{C}_{3v}$ were used: both σ_{v1} and E produce p_1 when they act on p_1.) The second thing that we notice is that $\mathbf{C}_3$ partitions the configuration space of the molecule into three disjoint cells, displayed in Fig. 1*b*. This is so because the operations listed in (2) increase the angular coordinates by 0, $2\pi/3$, and $4\pi/3$, respectively, and if we call two points of space rotationally equivalent when they are separated by such angles, then cells can be defined as each being a maximal set of rotationally inequivalent points. Any two cells, on the other hand, are rotationally equivalent. (This partitioning of the space, which is characteristic of cyclic groups, is entirely similar to the construction of primitive cells through the properties of the translation group, as done in §3-1.)

Definition of equivalent functions

We shall say that a set of N functions,

$$\{\varphi\} = \varphi_0, \varphi_1, \ldots, \varphi_{N-1}, \tag{5}$$

is *equivalent* under a group H, also of order N,

$$H: h_0 \equiv E, h_1, \ldots, h_{N-1}, \tag{6}$$

if, for any operation h in H, and any function φ_i in $\{\varphi\}$,

$$h\varphi_i = \varphi_j, \tag{7}$$

where φ_j also belongs to $\{\varphi\}$ and it equals φ_i only when h is the identity. This means, of course, that any operation of the group not the identity effects a permutation of the set $\{\varphi\}$ in which no function is left invariant, the identity of course leaving all the functions of the set invariant. We assert that the above definition is equivalent to saying that each function of $\{\varphi\}$ can be generated from the single function φ_0 by acting on it with all the N operations of H, in such a way that the label i of the function φ_i is identified with the label i of the operation h_i which generates it from φ_0:

$$h_i\varphi_0 = \varphi_i. \tag{8}$$

We must also require that the following condition obtains:

$$h_j\varphi_0 = \varphi_0 \qquad \Rightarrow \qquad h_j = E. \tag{9}$$

It is easy to verify that (8) and (9) entail (7) and the conditions attached to it. On acting, in fact, with h on both sides of (8), and remembering that $h\,h_i$ must be some operation h_j of H, we obtain

$$h\varphi_i = hh_i\varphi_0 = h_j\varphi_0 = \varphi_j, \tag{10}$$

where, in the last step, we use (8) again. Equation (10) agrees with (7), so that all that remains is to prove, from (8) and (9) that φ_j in (10) equals φ_i only when h is the identity. We leave this to the reader. (Problem 4.1.)

It must be clearly understood that we do not expect the above definitions to work for any group H. We shall see, however, that cyclic groups, and their direct products, are acceptable, as we have seen in our example for the cyclic group $\mathbf{C}_3$.

Properties of cyclic groups and of their direct products

We recall (see Problem **2**-7.3) that a *cyclic group*, say H, is one in which any element h_n of it can be taken to be the n-th power of some generating element h. That is, h_n is h^n, the product of n elements equal to h. Thus:

$$H: h_0 = E, \qquad h_1 = h, \qquad h_2 = h^2, \ldots, \qquad h_{N-1} = h^{N-1}, \tag{11}$$

with the condition

$$h_N = h^N = h_0 = E. \tag{12}$$

In $\mathbf{C}_3$, for example,

$$h_0 = E, \quad h_1 = h = C_3^+, \quad h_2 = h^2 = (C_3^+)^2 = C_3^-, \quad h_3 = h^3 = (C_3^+)^3 = E. \tag{13}$$

Another example of a cyclic group is the translation group in one dimension with periodic boundary conditions. (See §**5**-1.) The multiplication rules for cyclic groups are very easy:

11 $$h_i h_j = h^i h^j = h^{i+j} = h_{i+j}. \qquad (14)$$

It is quite clear that $h_j h_i$ will also lead to (R14), which means that all cyclic groups are abelian. It follows that a cyclic group of order N will have N one-dimensional irreducible representations in total. (See Problem **2**-7.2.) The form of these representations is easy to obtain, since they must satisfy the multiplication rule (14), which indicates that the representatives must be exponentials in h_i or, more properly, in some variable w_i which must be associated with the operations h_i in such a way that $w_i + w_j$ should give w_{i+j}. (This, for example, is precisely the way in which rotation angles and translation vectors are associated with their corresponding operations.) We thus write

$$\hat{H}(h_i) = \exp(\alpha w_i), \qquad (15)$$

where the constant α must vary for the different irreducible representations. (Compare eqn 15 with the irreducible representations of the one-dimensional translation group, given by eqn **5**-2.3.) The importance of (15) will soon be understood.

Cyclic groups have three properties which are important from our point of view. First, it is easy, for a point group, to choose a function φ_0 which, as (9) requires, is left invariant only by the identity. (In a cyclic rotation group, for example, where all operations are around the same axis, it suffices to take φ_0 away from this axis of symmetry. In a one-dimensional translation group along the x axis, any function not periodic in x will do.) As a consequence of this property, it is easy to prove (see Problem 4.3), that the N functions generated by eqn (8) from the N operations of H in (11) are all distinct, a result which is intuitively true in a cyclic rotation group as in Fig. 1.

The second important property which we need is that, as already discussed, a cyclic group partitions the configuration space into disjoint cells, as well known by now for the translation group, and as illustrated for $\mathbf{C}_3$ in Fig. 1. The point we want to stress now is that each of the N elements of the cyclic group can uniquely be associated with a primitive cell. Thus, the identity corresponds to the cell 0, say, and the elements $h_1, h_2, \ldots, h_{N-1}$, generate from cell 0 the successive cells number $1, 2, \ldots, N-1$.

The third and final property with which we shall be concerned is that the irreducible representations of a cyclic group are (with a single exception) *faithful* representations of the group, a point which we shall now discuss. If you look at any table of irreducible representations of a group, you will see that different operations are often represented by the same representative within the same representation. There is an extreme example of this situation which appears in all groups, and this is the *trivial representation*, in which all operations of the group are represented by the unity. At the other extreme, there are representations in which all representatives are distinct, and these representations are called *faithful*. If we look at (15) we see that either α equals zero, in which case $\hat{H}(h_i)$ equals unity for all i (trivial representation), or α does not vanish, in which case no two representatives can be equal. That is, all irreducible representations of all cyclic groups, except for the trivial representation, are faithful, as we had already anticipated.

The translation group in three dimensions is not a cyclic group, but it has three cyclic subgroups, T_x, T_y, T_z, of which it is a direct product. (See eqn **5**-1.6.) It is not difficult to prove that the three properties of cyclic groups which we have just discussed are also valid for direct products of cyclic groups. The first two properties are, in any case, intuitively true for the translation group, and that its irreducible representations are faithful follows as once from (**5**-2.9).

Construction of the equivalent functions

Consider a group G with a subgroup H which is either cyclic or a direct product of cyclic groups. Let N be the order of H. We know that this group must possess N one-dimensional irreducible representations ${}^s\hat{H}$, $(s = 0, 1, \ldots, N-1)$, with bases ${}^s\psi$, $(s = 0, 1, \ldots, N-1)$, so that

$$h_i\, {}^s\psi = {}^s\psi\, {}^s\hat{H}(h_i). \tag{16}$$

Clearly, because the representations are one-dimensional, the functions ${}^s\psi$ are eigenfunctions of the operations h_i. From (16), we can construct a $N \times N$ reducible representation:

$$h_i \langle {}^0\psi, {}^1\psi, \ldots, {}^{N-1}\psi| =$$

$$\langle {}^0\psi, {}^1\psi, \ldots, {}^{N-1}\psi| \begin{bmatrix} {}^0\hat{H}(h_i) & & & & \\ & {}^1\hat{H}(h_i) & & & \\ & & \cdot & & \\ & & & \cdot & \\ & & & & \cdot \\ & & & & {}^{N-1}\hat{H}(h_i) \end{bmatrix}, \tag{17}$$

which we rewrite in a compact notation as follows:

$$h_i \langle \psi | = \langle \psi | \hat{H}(h_i), \tag{18}$$

where the diagonal matrix $\hat{H}(h_i)$ has diagonal elements

$$\hat{H}_{ss} = {}^s\hat{H}(h_i). \tag{19}$$

Consider now an *arbitrary* N-dimensional column vector $|v\rangle$. We assert that the functions

$$\varphi_i = \langle \psi | \hat{H}(h_i) | v \rangle, \qquad i = 0, 1, \ldots, N-1, \tag{20}$$

satisfy the generating relations (8) for the equivalent functions,

$$h_i \varphi_0 = \varphi_i, \tag{21}$$

that is, that they form an equivalent set under H. The proof of this result is very simple. From (20),

$$\varphi_0 = \langle \psi | \hat{H}(h_0) | v \rangle = \langle \psi | \hat{H}(E) | v \rangle = \langle \psi | 1 | v \rangle = \langle \psi | v \rangle. \tag{22}$$

22,18,20 $$h_i \varphi_0 = h_i \langle \psi | v \rangle = \langle \psi | \hat{H}(h_i) | v \rangle = \varphi_i. \tag{23}$$

We have required that all functions of an equivalent set be distinct and for this to be true it is clear from (20) that the representation $\hat{H}(h_i)$ must be faithful. Since the matrix elements in (17) form always faithful representations, $\hat{H}(h_i)$ is also faithful.

Orthogonality of the equivalent functions

What we have so far done is this. We have started with the bases of the irreducible representations of the group H, ${}^s\psi$, which are orthogonal for different values of s, since functions that belong to different irreducible representations are orthogonal. From the ${}^s\psi$, we have constructed the equivalent functions φ_i of eqn (20). It is clear from this equation that φ_i is uniquely determined by the operation h_i which, as we know, is uniquely associated with the i-th primitive cell of the configuration space of the system. Thus the function φ_i, for all i, is uniquely associated with the i-th primitive cell of the system. Since these functions are obtained by transformation of the ${}^s\psi$, which are orthogonal, (because they are bases of different irreducible representations), it should not be difficult to ensure that the φ_i be orthogonal for different values of i. That is, we shall have equivalent functions corresponding to each of the different primitive cells of the system and such that two functions pertaining to two different cells are orthogonal. We shall now carry out the orthogonalization of these functions.

In order to show up more clearly the relation between the functions φ_i and ${}^s\psi$, we shall carry out the matrix multiplication on (R20) on using the matrix $\hat{H}(h_i)$ defined in (18):

$$\varphi_i = \langle \psi | {}^0\hat{H}(h_i)v_0, {}^1\hat{H}(h_i)v_1, \ldots, {}^s\hat{H}(h_i)v_s, \ldots \rangle \tag{24}$$

$$= \sum_s {}^s\psi\, {}^s\hat{H}(h_i)v_s =_{\text{def}} \sum_s {}^s\psi\, \Lambda_{si}, \tag{25}$$

where we have written the matrix

$$\Lambda_{si} =_{\text{def}} {}^s\hat{H}(h_i)v_s. \tag{26}$$

It is a well known result in matrix theory that a transformation such as (25) of the orthogonal functions ${}^s\psi$ onto the functions φ_i by a matrix Λ will preserve orthogonality if Λ is unitary, that is if

$$\Lambda\Lambda^{\dagger} = \mathbf{1}, \tag{27}$$

which means that

$$\sum_i \Lambda_{si}(\Lambda^{\dagger})_{it} = \sum_i \Lambda_{si}\Lambda_{ti}^{*} = \delta_{st}. \tag{28}$$

In order to satisfy this condition we introduce (26) on (L28):

$$\sum_i \Lambda_{si}\Lambda_{ti}^{*} = \sum_i {}^s\hat{H}(h_i)v_s\, {}^t\hat{H}(h_i)^{*}v_t^{*} = v_s v_t^{*} \sum_i {}^s\hat{H}(h_i)\, {}^t\hat{H}(h_i)^{*}. \tag{29}$$

We can now apply the famous orthogonality theorem for irreducible representations (eqn **2**-5.2; remember that the ${}^s\hat{H}(h_i)$, being one-dimensional, coincide with the characters), in the form

$$\sum_i {}^s\hat{H}(h_i)\, {}^t\hat{H}(h_i)^{*} = N\,\delta_{st}. \tag{30}$$

30|29

$$\sum_i \Lambda_{si}\Lambda_{ti}^{*} = v_s v_t^{*} N\,\delta_{st}, \tag{31}$$

so that it is sufficient to take

$$v_s = N^{-1/2}, \qquad s = 0, 1, \ldots, N-1, \tag{32}$$

for Λ to satisfy (28) and thus be unitary. On introducing (32) into (25), the equivalent functions

$$\varphi_i = N^{-1/2} \sum_s {}^s\psi\, {}^s\hat{H}(h_i) \tag{33}$$

are orthogonal for different values of i, that is, for different primitive cells. It is this property that makes these functions largely localized, as we shall soon verify by means of an example.

2 The Wannier functions

The Wannier functions are equivalent functions under the translation group T constructed, as equivalent functions must be, from the eigenfunctions of T, that is from the Bloch functions. The operation h_i in (1.33) is now $\{E|\mathbf{t}\}$, with $\mathbf{t}$ ranging over the Bravais lattice, and, accordingly, the primitive cell i is now denoted with the translation vector $\mathbf{t}$ (which takes the cell at the origin into the cell labelled by $\mathbf{t}$). The equivalent function φ_i will thus be written as $\varphi_{\mathbf{t}}(\mathbf{r})$. As we know, the irreducible representations of T are labelled by the $\mathbf{k}$ vector, so that the ${}^s\psi$ in (1.33) must be replaced by the Bloch functions $\psi_{\mathbf{k}}(\mathbf{r})$. Thus

(1.33) takes the following form:

$$\varphi_{\mathbf{t}}(\mathbf{r}) = N^{-1/2} \sum_{\mathbf{k}} \psi_{\mathbf{k}}(\mathbf{r})\, {}_{\mathbf{k}}\hat{T}\{E|\mathbf{t}\} \tag{1}$$

5-2.9

$$= N^{-1/2} \sum_{\mathbf{k}} \psi_{\mathbf{k}}(\mathbf{r}) \exp(-\,\mathrm{i}\mathbf{k}\cdot\mathbf{t}), \tag{2}$$

which defines the Wannier functions. The summation over **k** here ranges over all the N values of **k** in the Brillouin zone. From the orthogonality property of equivalent functions, Wannier functions corresponding to different cells must be orthogonal:

$$\int_V \varphi_{\mathbf{t}}(\mathbf{r})^* \varphi_{\mathbf{t}'}(\mathbf{r})\, \mathrm{d}v = \delta_{\mathbf{t}\mathbf{t}'}, \tag{3}$$

where dv is the volume element and V is the volume of the crystal.

Whereas the Bloch functions $\psi_{\mathbf{k}}(\mathbf{r})$ are largely delocalized, it is easy to see from (2) that the Wannier functions $\varphi_{\mathbf{t}}(\mathbf{r})$ must be to some extent localized in the cell labelled by **t**, that is that their largest value will be found in that cell. In fact,

5-6.3|2

$$\varphi_{\mathbf{t}}(\mathbf{r}) = N^{-1/2} \sum_{\mathbf{k}} u_{\mathbf{k}}(\mathbf{r}) \exp\{\mathrm{i}\mathbf{k}\cdot(\mathbf{r}-\mathbf{t})\}, \tag{4}$$

and it is clear that the largest *real* value of (R4) will be obtained when **r** equals **t**.

Once the Bloch functions are transformed into the Wannier functions (2), if an impurity is introduced into the cell **t**′ of the crystal, it is comparatively easy to replace the Wannier function of the crystal matrix in the cell **t**′ by that of the impurity. In order to show how the Wannier functions are localized, we shall work out the Wannier functions for plane waves, in which the $\psi_{\mathbf{k}}(\mathbf{r})$ in (2) are taken to be the plane waves $\exp(\mathrm{i}\mathbf{k}\cdot\mathbf{r})$. (This will be a very good test of the localization of the Wannier functions, since the plane waves from which we shall construct them are extremely delocalized.)

The Wannier functions for free electrons

We shall calculate the Wannier function $\varphi_0(\mathbf{r})$ for the first unit cell (**t** equal to **0**) of a linear chain of lattice constant a and of N cells. The volume Ω in the normalized plane waves $\Omega^{-1/2}\exp(\mathrm{i}\mathbf{k}\cdot\mathbf{r})$ of (**4**-10.2) has to be replaced by the crystal length Na. Therefore, for **t** equal to **0**,

2

$$\varphi_0(\mathbf{r}) = N^{-1/2} \sum_{\mathbf{k}} (Na)^{-1/2} \exp(\mathrm{i}\mathbf{k}\cdot\mathbf{r}) \tag{5}$$

$$= N^{-1} a^{-1/2} \sum_{\mathbf{k}} \exp(\mathrm{i}\mathbf{k}\cdot\mathbf{r}). \tag{6}$$

Here,

5-4.3 $\mathbf{k} = k_x \mathbf{a}^{\#}; \qquad k_x = 0, N^{-1}, 2N^{-1}, \ldots, (N-1)N^{-1}; \qquad \mathbf{r} = x\mathbf{a}.$ (7)

7|6 $$\varphi_0(\mathbf{r}) = N^{-1} a^{-1/2} \sum_{k_x} \exp(2\pi \mathrm{i} k_x x). \tag{8}$$

The range in k_x can be taken for large N to go from 0 to 1 or, more conveniently, on using equivalence in the $\mathbf{k}$ space, from $-\frac{1}{2}$ to $\frac{1}{2}$ (first Brillouin zone). Also, k_x varies in steps Δk_x equal to N^{-1}. Thus, on multiplying and dividing (8) by Δk_x we get

7 $$\varphi_0(\mathbf{r}) = a^{-1/2} \sum_{k_x} \exp(2\pi \mathrm{i} k_x x)\, \Delta k_x. \tag{9}$$

The increment Δk_x is an infinitesimal for large N and the summation in (8) may be replaced by an integral:

9 $$\varphi_0(\mathbf{r}) = a^{-1/2} \int_{-1/2}^{1/2} \exp(2\pi \mathrm{i} k_x x)\, \mathrm{d}k_x \tag{10}$$

10 $$= (2\pi \mathrm{i} x)^{-1} a^{-1/2} (\mathrm{e}^{\mathrm{i}\pi x} - \mathrm{e}^{-\mathrm{i}\pi x}) = a^{-1/2} \frac{\sin \pi x}{\pi x}. \tag{11}$$

This function is very easy to compute and we display it in Fig. 1, where we have taken a as unity. It is clear from the figure that the Wannier function $\varphi_0(\mathbf{r})$ is largely localized on the cell at the origin. On starting from ordinary Bloch functions rather than plane waves one should expect an even better degree of localization.

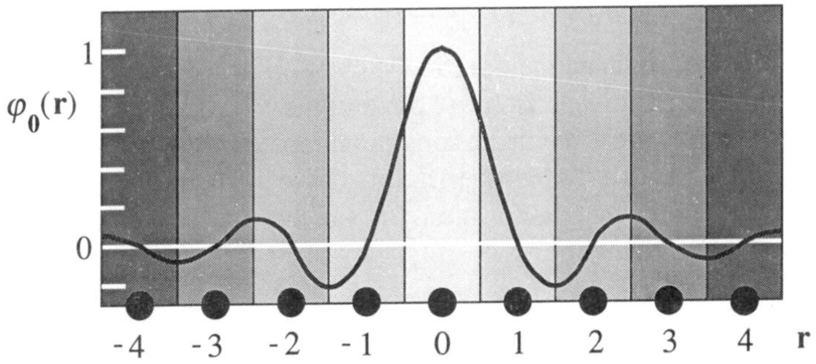

Fig. 13-2.1. The Wannier function $\varphi_0(\mathbf{r})$ for free electrons for the successive unit cells of a linear chain, of lattice constant equal to unity.

3 Löwdin orbitals

We have taken in the tight-binding method linear combinations of atomic orbitals $|m\rangle$ placed at the various primitive cells labelled by m. Since we are now labelling our cells by means of the translation vectors $\mathbf{t}$, it is convenient to change our notation a little, the orbital at the cell $\mathbf{t}$ being now denoted by the symbol $\phi_{\mathbf{t}}(\mathbf{r})$. (As in our discussion of the tight-binding method, we consider the case when there is only one orbital per primitive cell.) The overlap integral between the orbitals at different cells is defined by

$$S_{\mathbf{tt}'} = \int_V \phi_{\mathbf{t}}(\mathbf{r})^* \, \phi_{\mathbf{t}'}(\mathbf{r}) \, \mathrm{d}v, \tag{1}$$

where the integration is over the crystal volume. We had to introduce in the tight-binding method the very drastic approximation of overlap neglect, in which $S_{\mathbf{tt}'}$ was assumed to vanish whenever $\mathbf{t}$ and $\mathbf{t}'$ are different cells. What we now want to do is to construct linear combinations of the atomic orbitals $\phi_{\mathbf{t}}(\mathbf{r})$ which are actually orthogonal for different cells and which are called *Löwdin orbitals*. We shall denote such combinations with the symbol $\varphi_{\mathbf{t}}(\mathbf{r})$. The motivation for doing this job should be clear: once the $\varphi_{\mathbf{t}}(\mathbf{r})$ become available, they can be used instead of the $\phi_{\mathbf{t}}(\mathbf{r})$ in a tight-binding calculation and the neglect of overlap approximation becomes now exact.

It is very easy to see how we must proceed. Because the $\varphi_{\mathbf{t}}(\mathbf{r})$ must be orthogonal for different cells, it is sufficient to construct them as equivalent functions under the translation group. We have written them in (2.2):

$$\varphi_{\mathbf{t}}(\mathbf{r}) = N^{-1/2} \sum_{\mathbf{k}} \psi_{\mathbf{k}}(\mathbf{r}) \exp(-\,\mathrm{i}\mathbf{k}\cdot\mathbf{t}). \tag{2}$$

The functions $\psi_{\mathbf{k}}(\mathbf{r})$ here are translation eigenfunctions which we took in (2.2) to be Bloch functions. We can equally well take them to be the Bloch sums

10-2.6

$$\psi_{\mathbf{k}}(\mathbf{r}) = \sum_{\mathbf{t}} \exp(\mathrm{i}\mathbf{k}\cdot\mathbf{t})\, \phi_{\mathbf{t}}(\mathbf{r}). \tag{3}$$

(We have here changed the notation slightly, in order to make it agree with our present conventions.) It is pretty obvious now that, when (3) is introduced into (2) we shall express the orthogonal functions $\varphi_{\mathbf{t}}(\mathbf{r})$ in terms of the atomic orbitals $\phi_{\mathbf{t}}(\mathbf{r})$. All that remains, therefore, is to turn the handle. It will be helpful to introduce matrix notation, for which purpose we define in (2) the following matrix element:

$$U_{\mathbf{kt}} = N^{-1/2} \exp(-\,\mathrm{i}\mathbf{k}\cdot\mathbf{t}). \tag{4}$$

Before we introduce it into eqn (2), we shall verify that this matrix is unitary, that is that $UU^{\dagger}$ equals the unit matrix. We must remember for this purpose that $\exp(-\,\mathrm{i}\mathbf{k}\cdot\mathbf{t})$ is nothing else that the term ${}_{\mathbf{k}}\hat{T}\{E|\mathbf{t}\}$ which we had in (2.1) and that, because of the orthogonality theorem for irreducible representations,

2-5.2 $$\sum_{\mathbf{t}} {}_{\mathbf{k}}\hat{T}\{E|\mathbf{t}\}^* \, {}_{\mathbf{k}'}\hat{T}\{E|\mathbf{t}\} = N\delta_{\mathbf{kk}'}. \tag{5}$$

Thus, on identifying ${}_{\mathbf{k}}\hat{T}\{E|\mathbf{t}\}$ in (4), we obtain

4|5 $$\sum_{\mathbf{t}} N^{1/2}\, U^*_{\mathbf{kt}}\, N^{1/2}\, U_{\mathbf{k}'\mathbf{t}} = N\,\delta_{\mathbf{kk}'}. \tag{6}$$

6 $$\sum_{\mathbf{t}} U_{\mathbf{k}'\mathbf{t}}\, U^\dagger_{\mathbf{tk}} = \delta_{\mathbf{kk}'} \qquad \Rightarrow \qquad UU^\dagger = \mathbf{1}. \tag{7}$$

We can now safely introduce (4) into (2):

4|2 $$\varphi_{\mathbf{t}}(\mathbf{r}) = \sum_{\mathbf{k}} \psi_{\mathbf{k}}(\mathbf{r})\, U_{\mathbf{kt}}. \tag{8}$$

On denoting with $\langle\varphi|$ the row vector formed by all the equivalent functions for all **t**, and with $\langle\psi|$ that formed by all the translation eigenfunctions for all **k**, we have,

8 $$\langle\varphi| = \langle\psi|\, U. \tag{9}$$

We can also introduce (4) into (3):

4|3 $$\psi_{\mathbf{k}}(\mathbf{r}) = \mathcal{N}_{\mathbf{k}} \sum_{\mathbf{t}} U^*_{\mathbf{kt}}\, \phi_{\mathbf{t}}(\mathbf{r}), \tag{10}$$

where $\mathcal{N}_{\mathbf{k}}$ is a normalization constant which will be chosen so as to normalize the functions $\psi_{\mathbf{k}}(\mathbf{r})$, and which incorporates the factor $N^{-1/2}$ in (4). In order to get the suffices in (10) to agree with the matrix multiplication rules, we rewrite this equation as

10 $$\psi_{\mathbf{k}}(\mathbf{r}) = \sum_{\mathbf{t}} \phi_{\mathbf{t}}(\mathbf{r})\, U^\dagger_{\mathbf{tk}}\, \mathcal{N}_{\mathbf{k}}, \tag{11}$$

so that, if we define a *diagonal* matrix $\mathcal{N}$ of diagonal elements $\mathcal{N}_{\mathbf{k}}$, then, with the notation defined underneath the eqn (8), we have

11 $$\langle\psi| = \langle\phi|\, U^\dagger \mathcal{N}. \tag{12}$$

12|9 $$\langle\varphi| = \langle\phi|\, U^\dagger \mathcal{N}\, U. \tag{13}$$

We have solved with this equation the major part of our problem, since (13) expresses the orthogonal orbitals $\varphi_{\mathbf{t}}(\mathbf{r})$ in terms of the non-orthogonal atomic orbitals $\phi_{\mathbf{t}}(\mathbf{r})$. All that remains is to determine $\mathcal{N}$ so that

$$I = \int_V \psi_{\mathbf{k}}(\mathbf{r})^* \psi_{\mathbf{k}}(\mathbf{r})\, \mathrm{d}v = 1. \tag{14}$$

The functions in the integrand here are obtained from (10) in which we shall write $|\mathcal{N}_{\mathbf{k}}|$ as $\mathcal{N}_{\mathbf{k}}$, ignoring its phase factor since it is irrelevant for the normalization. We are thus taking $\boldsymbol{N}_\mathrm{k}$ as real in (10), so that, with small changes of notation, we have:

10 $$\psi_\mathrm{k}(\mathbf{r})^* = \boldsymbol{N}_\mathrm{k} \sum_\mathrm{t} U_\mathrm{kt}\, \phi_\mathrm{t}(\mathbf{r})^*, \qquad \psi_\mathrm{k}(\mathbf{r}) = \boldsymbol{N}_\mathrm{k} \sum_\mathrm{t'} \phi_\mathrm{t'}(\mathbf{r})\, U^\dagger_\mathrm{t'k}. \tag{15}$$

15|14 $$I = (\boldsymbol{N}_\mathrm{k})^2 \sum_\mathrm{tt'} U_\mathrm{kt}\, U^\dagger_\mathrm{t'k} \int_V \phi_\mathrm{t}(\mathbf{r})^* \phi_\mathrm{t'}(\mathbf{r})\, \mathrm{d}v \tag{16}$$

1|16 $$= (\boldsymbol{N}_\mathrm{k})^2 \sum_\mathrm{tt'} U_\mathrm{kt}\, S_\mathrm{tt'}\, U^\dagger_\mathrm{t'k} = (\boldsymbol{N}_\mathrm{k})^2\, (U S U^\dagger)_\mathrm{kk} = 1. \tag{17}$$

We must remember here that $\boldsymbol{N}$ is diagonal and that $\boldsymbol{N}_\mathrm{k}$ is its diagonal element $\boldsymbol{N}_\mathrm{kk}$:

17 $$\boldsymbol{N}^2\, U S U^\dagger = \mathbf{1} \qquad \Rightarrow \qquad \boldsymbol{N}^{-2} = U S U^\dagger. \tag{18}$$

The matrix $\boldsymbol{N}$ itself can now easily be obtained (see Appendix) as

$$\boldsymbol{N} = U S^{-1/2} U^\dagger. \tag{19}$$

On substituting (19) into (13), and making use of the unitary property of U from eqn (7), we obtain our final expansion of the orthogonal orbitals $\varphi_\mathrm{t}(\mathbf{r})$ in terms of the atomic orbitals $\phi_\mathrm{t}(\mathbf{r})$:

19, 7 |13 $$\langle \varphi | = \langle \phi | S^{-1/2}. \tag{20}$$

This is Löwdin's formula, in a slightly changed notation.

Appendix: The matrix $\boldsymbol{N}$

The matrix $\boldsymbol{N}$ appears in (18) as the *matrix function* $\boldsymbol{N}^{-2}$. The procedure to deal with a matrix function $f(A)$ of a matrix A (say $\sin A$) is very simple. You expand the function $f(x)$ (say $\sin x$) in power series

$$f(x) = \sum_n a_n x^n, \tag{21}$$

and then write $f(A)$ by substituting A for x in (21):

$$f(A) = \sum_n a_n A^n, \tag{22}$$

an expression which can easily be worked out since powers of matrices are simple to obtain. With this definition we shall prove that, if

$$M = U S U^\dagger, \qquad U U^\dagger = \mathbf{1}, \tag{23}$$

then, for any function $f(M)$,

$$f(M) = U f(S)\, U^\dagger. \tag{24}$$

Proof

22,23
$$f(M) = \sum_n a_n M^n = \sum_n a_n (USU^\dagger)^n. \tag{25}$$

Here:

$$(USU^\dagger)^n = (USU^\dagger)(USU^\dagger)\ldots(USU^\dagger),\ (n \text{ factors}) \tag{26}$$

23|26
$$= US^nU^\dagger. \tag{27}$$

27|25
$$f(M) = \sum_n a_n\, US^nU^\dagger = U\left(\sum_n a_n S^n\right)U^\dagger. \tag{28}$$

From the first equality in (25), the bracket in (R28) is $f(S)$, thus verifying (24).

We can now apply (24) for $f(M)$ equal to $M^{-1/2}$:

23,24
$$M = USU^\dagger \qquad \Rightarrow \qquad M^{-1/2} = U\,S^{-1/2}\,U^\dagger. \tag{29}$$

If we take M as equal to $\boldsymbol{N}^{-2}$, as in (18), $M^{-1/2}$ is $(\boldsymbol{N}^{-2})^{-1/2}$, that is $\boldsymbol{N}$, and (29) gives (19). In this equation, like in the final expression (20), $S^{-1/2}$ must be written as a power series. We recognize for this purpose that $S_{\mathbf{tt}}$ in (1) can always be taken as unity on choosing normalized orbitals. Thus S can be written as $\mathbf{1} + S'$, where S' is the matrix in (1) for $\mathbf{t}$ different from $\mathbf{t}'$, the matrix elements of which are all smaller than unity. Thus $(\mathbf{1} + S')^{-1/2}$ can be expanded in the well-known Taylor series.

4 Problems

1. Prove in eqn (1.10), that if $h\varphi_i$ equals φ_i, then h must be the identity.
2. Prove that the inverse h_i^{-1} of an element h_i of (1.11) is h^{N-i}.
3. Prove that, if H is a cyclic group of order N and φ_0 is a function which is left invariant only by the identity of H, then all the N functions defined by (1.8) are distinct.

   Hint. Apply (1.8) for h_j to obtain φ_j and assume it is equal to φ_i from (1.8). Deduce therefore that

   $$\varphi_0 = h_i^{-1}\,h_j\,\varphi_0. \tag{1}$$

   Write h_i^{-1} from Problem 2 and h_j as h^j. You will then find from Problem 1 that h^{N-i+j} must equal the identity, which requires that i and j be equal.

Further reading

Wannier functions were defined by G. Wannier (1937) and Löwdin orbitals by Per-Olov Löwdin (1950). Applications of the Wannier functions in the study of impurity levels are discussed by Ziman (1964), pp. 151–155. Applications of the Löwdin orbitals are discussed in Löwdin (1956). The definition and method of construction given here for the equivalent functions is due to Altmann (1958). For the application of Wannier functions in tight-binding calculations, see Bullett (1980).

14

Surface and impurity states

We have so far seen that the electron states of a periodic structure are described by Bloch functions which form energy bands as a function of **k**, and that these bands are such that, in general, forbidden energy gaps exist between them. It is reasonable to expect, on the other hand, that electrons which arise from impurities or from surface states in the crystal will not give rise to periodic states and will not, therefore, be described by Bloch functions. We must expect, accordingly, that the corresponding energies may appear in the forbidden energy gaps, as will be confirmed in this chapter. The first thing which we shall do is to look for the possible existence of states in the gap, and this will lead in § 1 to the *Tamm states.* We shall then consider a simple model of surface states (*Shockley states*) in § 2. A more detailed method of calculation of surface and impurity states, proposed by Koster and Slater, will be discussed in § 3 and will serve as a simple introduction to the use of Green functions. A practical example of this treatment will be given in § 4.

1 States in the energy gap. Tamm states

Let us first review the work we did in order to obtain the band gap. We used nearly free-electron theory, and wrote the wave functions in terms of plane waves:

10-1.10 $$\psi_{\mathbf{k}}(\mathbf{r}) = \sum_{\mathbf{g}} c_{\mathbf{g}} \,|\, \mathbf{k} + \mathbf{g} \rangle. \qquad (1)$$

We then simplified the problem by keeping only the first two values of **g** in (1), whereby we got the following secular determinant:

10-1.28 $$\begin{vmatrix} E_{\mathbf{k}} - \mathscr{E} & \mathscr{V}_{\mathbf{g}}^{*} \\ \mathscr{V}_{\mathbf{g}} & E_{\mathbf{k}+\mathbf{g}} - \mathscr{E} \end{vmatrix} = 0. \qquad (2)$$

For convenience, we have now taken V as unity in (2), and it must be remembered that the $E_{\mathbf{k}}$ are free-electron eigenvalues which depend only on $|\mathbf{k}|$. We now take **k** to be a vector ending at a Brillouin zone face, in which case we know that $\mathbf{k} + \mathbf{g}$ must be a vector of the same modulus:

8-2.1 $$|\mathbf{k} + \mathbf{g}| = |\mathbf{k}|, \qquad (3)$$

where **g** is the vector of the reciprocal lattice that separates the two parallel faces of the Brillouin zone. It follows from (3) and our previous remark about the free-electron energies that

$$E_{\mathbf{k}+\mathbf{g}} = E_{\mathbf{k}}. \tag{4}$$

In order to stress that (4) is valid only when **k** ends at a Brillouin zone face, we shall denote such vectors with the letter **b**. On introducing this notation in (4) and the resulting equation into (2), we have

$$(E_{\mathbf{b}} - \mathscr{E})^2 = \mathscr{V}_{\mathbf{g}}^* \mathscr{V}_{\mathbf{g}} = |\mathscr{V}_{\mathbf{g}}|^2. \tag{5}$$

5 $$\mathscr{E} = E_{\mathbf{b}} \pm |\mathscr{V}_{\mathbf{g}}|. \tag{6}$$

We know that the free-electron energy $E_{\mathbf{k}}$ is proportional to $\mathbf{k}^2$ and we can always adjust the units so that the proportionality constant be unity, so that, in (6),

6 $$\mathscr{E} = \mathbf{b}^2 \pm |\mathscr{V}_{\mathbf{g}}|. \tag{7}$$

We are looking for states in the band gap, that is, with energy in the range from $\mathbf{b}^2 - |\mathscr{V}_{\mathbf{g}}|$ to $\mathbf{b}^2 + |\mathscr{V}_{\mathbf{g}}|$, the energy of which must therefore have the form

7 $$\mathscr{E} = \mathbf{b}^2 + \varepsilon, \qquad |\varepsilon| < |\mathscr{V}_{\mathbf{g}}|. \tag{8}$$

We want now to find out what form the value of **k** must take for such states. Since they are related to the **k** vector **b** at the Brillouin zone face, we shall write

$$\mathbf{k} = \mathbf{b} + \boldsymbol{\kappa}, \tag{9}$$

with $\boldsymbol{\kappa}$ a small vector normal to the Brillouin zone face. Accordingly,

9, F1 $$\mathbf{k} + \mathbf{g} = \mathbf{b} + \mathbf{g} + \boldsymbol{\kappa} = \mathbf{b}' + \boldsymbol{\kappa}. \tag{10}$$

On introducing (9) and (10) into the secular equation (2), this takes the form

2 $$\begin{vmatrix} E_{\mathbf{b}+\boldsymbol{\kappa}} - \mathscr{E} & \mathscr{V}_{\mathbf{g}}^* \\ \mathscr{V}_{\mathbf{g}} & E_{\mathbf{b}'+\boldsymbol{\kappa}} - \mathscr{E} \end{vmatrix} = 0. \tag{11}$$

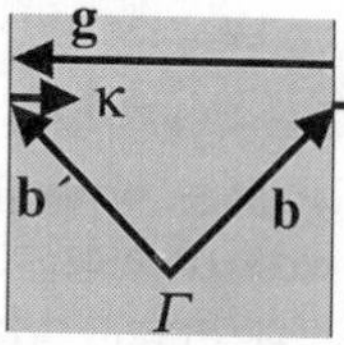

Fig. 14-1.1. The **k** vector $\mathbf{b} + \boldsymbol{\kappa}$ for Tamm states. The vertical lines are Brillouin zone faces.

With the units in present use in the free-electron approximation, we have, neglecting $\boldsymbol{\kappa}^2$,

$$E_{\mathbf{b}+\boldsymbol{\kappa}} = (\mathbf{b} + \boldsymbol{\kappa})^2 = \mathbf{b}^2 + 2\mathbf{b}\cdot\boldsymbol{\kappa}, \qquad (12)$$

$$E_{\mathbf{b}'+\boldsymbol{\kappa}} = (\mathbf{b}' + \boldsymbol{\kappa})^2 = \mathbf{b}^2 + 2\mathbf{b}'\cdot\boldsymbol{\kappa} = \mathbf{b}^2 - 2\mathbf{b}\cdot\boldsymbol{\kappa}. \qquad (13)$$

We use in eqn (13) the fact that $\mathbf{b}'$ and $\mathbf{b}$ have the same modulus, and the scalar product $\mathbf{b}'\cdot\boldsymbol{\kappa}$ is obtained from Fig. 1. We can now get the secular determinant:

12, 13, 8 | 11

$$\begin{vmatrix} 2\mathbf{b}\cdot\boldsymbol{\kappa} - \varepsilon & \mathscr{V}_{\mathbf{g}}^{*} \\ \mathscr{V}_{\mathbf{g}} & -2\mathbf{b}\cdot\boldsymbol{\kappa} - \varepsilon \end{vmatrix} = 0. \qquad (14)$$

The solution of this equation is

$$(\mathbf{b}\cdot\boldsymbol{\kappa})^2 = \tfrac{1}{4}(\varepsilon^2 - |\mathscr{V}_{\mathbf{g}}|^2). \qquad (15)$$

It follows from (8) that (R15) is negative, whence $\boldsymbol{\kappa}$ must be an imaginary vector, which we shall write as i $\boldsymbol{\mu}$. The meaning of a propagation vector with an imaginary part is easier to grasp than it might appear at first sight: the plane wave corresponding to (9) is, for V unity,

9 | **4**-10.2

$$|\mathbf{b} + \mathrm{i}\,\boldsymbol{\mu}\rangle = \exp(\mathrm{i}\mathbf{b}\cdot\mathbf{r})\exp(-\boldsymbol{\mu}\cdot\mathbf{r}). \qquad (16)$$

This means that the states in the gap are represented by a travelling plane wave (with **k** corresponding to a vector **b** ending at a Brillouin zone face) modulated by a rapidly decaying factor. Solutions such as (16) must, like all wave functions, satisfy continuity conditions. Imagine there is a perturbation potential V', that is a small departure from the periodic potential V, at the crystal surface. Equation (16) applied to a region near the crystal surface, will then have to be fitted to the wave functions which satisfy V' outside the crystal and to the periodic solutions (Bloch functions) in the bulk of it. Because the functions (16) are rapidly decaying, the fitting to the periodic solutions can only be done in a very small region, of the same order of magnitude as that in which the perturbing potential V' is significant. The corresponding states will thus be localized states at the surface which decay rapidly away from the region where the surface perturbation potential is significant and which have energies in the forbidden gap. Such states are often called *Tamm states.*

2 Shockley states

We have seen in § 1 how surface states (Tamm states) arise when we have a perturbation potential V localized on the crystal surface. We shall now consider surface states which arise from a different point of view, first formulated by Shockley. Consider a covalent crystal, such as silicon, in which we have strong covalent bands, and in which each atom of the bulk is in

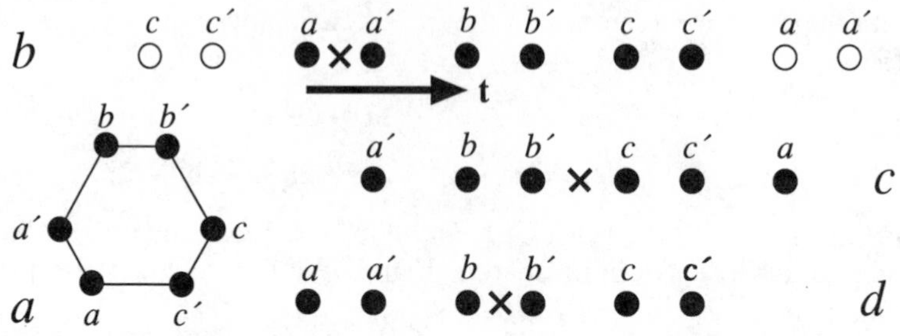

Fig. 14-2.1. A model for bond breaking and surface states. The crosses indicate centres of inversion.

a four-coordinated state with sp^3 hybrids. The atoms at the surface, however, are linked only to three ligands, so that, in principle, one sp^3 hybrid is left sticking out of the surface at each crystal site, forming what is called a *dangling bond*. Such dangling bonds will correspond to states localized on the surface with energies in the forbidden gap, that is to surface states. Some necessary conditions are required for this to happen, which we shall illustrate by means of an example. We show in Fig. 1*a* a molecule with six identical atoms with one orbital each but with alternating bonds, three short and three long. (This molecule could be thought of as a Kekulé structure in benzene, say.) This molecule is entirely equivalent to the infinite chain with periodic boundary conditions shown in Fig. 1*b*. (Compare with Figs **7**-1.1 and **7**-1.2.) This infinite chain is of course a lattice with basis (the atoms *a* and *a*′, say) and the translation vector is the one from *a* to *b*, which in order to avoid confusion we shall call **t** instead of the usual **a**. The states in the molecule of Fig. 1*a* can therefore be considered as the 'bulk' states, that is the crystal states which are represented by Bloch functions. In order to consider surface states of the Shockley type we shall break a bond, which can either be a strong (short) bond like *aa*′, (Fig. 1*c*), or a weak (long) bond like *ac*′, (Fig. 1*d*). We shall see that it is only when a strong bond is broken that a surface state arises.

The bulk states

The molecule of Fig. 1*a* is of symmetry $\mathbf{C}_{3v}(E, C_3^{\pm}, 3\sigma_v$, the latter being the reflection planes through the mid-points of opposite bonds). Its irreducible representations are A_1 (totally symmetric), A_2 (antisymmetric with respect to σ_v) and E (doubly degenerate). In the molecular orbital LCAO approximation (linear combination of atomic orbitals, which is identical with the tight-binding method) it is easy to obtain from group theory the symmetrized molecular orbitals, and their energies. We need for the latter the Coulomb, exchange and overlap integrals, which we write in terms of the atomic orbitals, designated with the symbols a, a', etc:

Coulomb: $$\langle a|\mathbb{H}|a\rangle = \langle a'|\mathbb{H}|a'\rangle = \alpha. \quad (1)$$

Exchange: $$\langle a|\mathbb{H}|a'\rangle = \beta, \quad (2)$$

$$\langle a|\mathbb{H}|c'\rangle = \gamma. \quad (3)$$

Overlap: $$\langle a|a\rangle = \langle a'|a'\rangle = 1, \quad (4)$$

$$\langle a|a'\rangle = 0. \quad (5)$$

In all these expressions a and a' can be replaced cyclically by b, c, and b', c', respectively. Also, all the interaction terms between the six orbitals a, a', b, b', c, c', that are not listed in (1) to (5) are neglected, as it is explicitly done for the overlap in eqn (5). The results are displayed in Table 1. (See Problems 5.1 and 5.6 for their derivation.) In order to represent these results graphically we must remember, as is well known in molecular orbital theory, that the exchange integrals are negative. Also, because β and γ correspond to the short and long bonds respectively, β must be numerically larger than γ: $|\beta| > |\gamma|$. Bearing this is in mind, the energy levels are represented in the usual manner on the left of Fig. 2. In order to identify them in terms of bands, that is as the energy functions given over the Brillouin zone, we notice from Fig. 1 that there are only three primitive cells in the 'crystal', so that the Brillouin zone will contain only three $\mathbf{k}$ values, identified in Fig. 2 as 1, 0, − 1, in convenient units. (Notice from eqn **5**-7.4 that the occupied states do not reach the edges of the Brillouin zone. It is convenient, however, to represent them as we do in the figure, over the whole of the Brillouin zone, as it is indeed possible when the number of atoms in the chain is large.) It is now clear that the energy levels on the left of Fig. 2 form two bands, the nature of which can be identified as follows. Consider in Table 1 the two lowest levels, that is, A_1 and the first of the E states listed. Their nature must be investigated by considering what happens to the wave function of the primitive cell when the lattice constant (in this case the vector $\mathbf{t}$) goes to infinity. This must be

Table 14-2.1. Wave functions and energies for the bulk states.

Rep.	Molecular Orbital	Energy
A_1	$a+b+c+a'+b'+c'$	$\alpha+\beta+\gamma$
A_2	$a+b+c-a'-b'-c'$	$\alpha-\beta-\gamma$
E	$2a-b-c+2a'-b'-c'$ $b-c+b'-c'$	$\alpha+\beta-\frac{1}{2}\gamma$
E	$2a-b-c-2a'+b'+c'$ $b-c-b'+c'$	$\alpha-\beta+\frac{1}{2}\gamma$

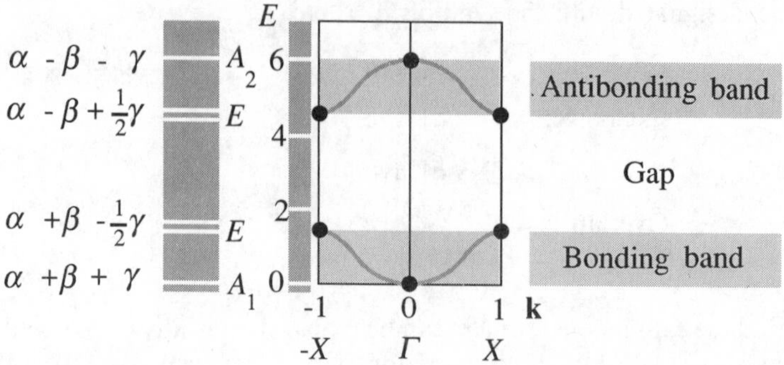

Fig. 14-2.2. The energy levels of the bulk states from Table 1. Both parts of the figure are on the same energy scale. The energy levels of Table 1 are given on the left in the conventional representation for molecules, whereas they are depicted on the right over the Brillouin zone. (Since this zone contains only three **k** values the grey lines, depicting bands, are purely symbolic.) The values 3, − 2, and − 1, have been taken for α, β, and γ, respectively, in order to draw the figure, all of them in the conventional energy units used on the vertical scale.

done without altering the primitive cell, that is, without altering the bond aa'. This means that the limiting process must be carried out by blowing up the long bond, in which case A_1 in Table 1 will separate out into three orbitals $a + a', b + b', c + c'$, (because interactions such as between a' and b cancel when the long bonds expand, so that these orbitals must not be kept together). When we look at both components of the first (lowest) E level in Table 1, we see that precisely these combinations remain when the long bond is stretched out to infinity. If we do the same for A_2 and the second (highest) E level, the orbitals that remain in the primitive cell are $a - a', b - b', c - c'$. For lattices without basis, the limiting cell function that denotes a band is an atomic orbital but, in the present case, because there is a basis, this limiting cell function is a molecular orbital. Since $a + a'$ and $a - a'$ are, as well known, *bonding* and *antibonding* molecular orbitals, respectively, for the short bond of the basis in the primitive cell, it is natural to call the lower band in Fig. 2 the *bonding band* and the upper one the *antibonding band*. Between them there is an energy gap. (More details about these bands and of the derivation of the wave functions in Table 1 can be obtained from Problems 5.4, 5.5, and 5.6. Some important cautionary notes about too enthusiastic an interpretation of the discussion here are treated in Problems 5.7 and 5.8.)

Breaking a strong bond

When we break the bond $a\,a'$ in the periodic linear chain of Fig. 1*b* (or, what is the same, in Fig. 1*a*), we are left with the molecule shown in Fig. 1*c*, which is of symmetry $\mathbf{C}_i$, the inversion being the point marked with a cross in the figure. The group $\mathbf{C}_i$ has only two irreducible representations, A_g (gerade or symmetric with respect to the inversion) and A_u (ungerade or antisymmetric). It is very easy to construct symmetry-adapted orbitals from the six orbitals of Fig. 1*c*, which are:

$$A_g: \qquad a + a', b + c', c + b', \tag{6}$$

$$A_u: \qquad a - a', b - c', c - b'. \tag{7}$$

Since the three orbitals in each of the equations (6) and (7) are of the same symmetry, they can be combined in linear combinations in order to obtain the final molecular orbitals ψ_g and ψ_u, from (6) and (7) respectively:

$$\psi_g = c_1(a + a') + c_2(b + c') + c_3(c + b'), \tag{8}$$

$$\psi_u = c_1(a - a') + c_2(b - c') + c_3(c - b'), \tag{9}$$

in which the coefficients are obtained from the usual secular determinant, on neglecting all interaction terms between non neighbouring orbitals (see Problem 5.9):

$$A_g: \begin{bmatrix} \alpha - E & \gamma & 0 \\ \gamma & \alpha - E & \beta \\ 0 & \beta & \alpha + \gamma - E \end{bmatrix} = 0;$$

$$A_u: \begin{bmatrix} \alpha - E & \gamma & 0 \\ \gamma & \alpha - E & \beta \\ 0 & \beta & \alpha - \gamma - E \end{bmatrix} = 0. \tag{10}$$

Because γ is small we can neglect γ^2, in which case both determinants admit the approximate root

$$E = \alpha, \tag{11}$$

which is precisely at the centre of the energy gap of Fig. 2. In order to see whether this is a surface state or not we must obtain the values of the coefficients c_1, c_2, c_3 in (8). This is done by solving the homogeneous equations for which the determinant in the left of (10) is the determinant of the coefficients of the unknowns c_1, c_2, c_3. This is a straightforward piece of work, but a little care is required because (11) is not an exact root so that the determinant does not vanish precisely, as necessary to avoid the homogeneous equations having only the trivial solution in which all unknowns vanish. It is possible, nevertheless, by eliminating c_2 from the first two equations, to obtain the rough solution

$$c_1 = -\beta/\gamma, \qquad c_2 = 0, \qquad c_3 = 1, \tag{12}$$

which can be seen at once to satisfy also the third equation, as long as γ is sufficiently small to be neglected. If we accept this solution as sufficient for our present purposes, we can see at once that c_1 must be large, which means from eqn (8) that the wave function is largely localized at the end of the chain. The physical interpretation of this result is easy from Fig. 1*c*: because the interaction across the long bonds from a and a' is very weak, the orbitals a and a' are largely non-bonded and thus become effectively dangling bonds, with energies in the gap. These are Shockley states. It is sufficient to look at Fig. 1*d* to realize that such states cannot arise in this case, in which a weak bond has been broken, as we shall now verify.

Breaking a weak bond

The molecule of Fig. 1*d* has again $\mathbf{C}_i$ symmetry, and the symmetry-adapted orbitals can easily be formed:

$$A_g: \qquad a + c', a' + c, b + b', \tag{13}$$

$$A_u: \qquad a - c', a' - c, b - b'. \tag{14}$$

The molecular orbitals, therefore, are:

$$\psi_g = c_1(a + c') + c_2(a' + c) + c_3(b + b'), \tag{15}$$

and similarly for ψ_u. The two secular determinants are, because of our approximations, accidentally identical:

$$\begin{bmatrix} \alpha - E & \beta & 0 \\ \beta & \alpha - E & \gamma \\ 0 & \gamma & \alpha - E \end{bmatrix} = 0. \tag{16}$$

The roots of this determinant are

$$E = \alpha, \qquad E = \alpha \pm (\beta^2 + \gamma^2)^{1/2}. \tag{17}$$

The first of these roots is in the gap, but its corresponding coefficients are:

$$c_1 = -\gamma/\beta, \qquad c_2 = 0, \qquad c_3 = 1. \tag{18}$$

We see that the coefficient c_1, which corresponds to the end atoms in Fig. 1*d*, is now small, so that we do not have a surface state at all.

Comments. Shockley and Tamm states

We can see from our example that Shockley states arise when we have strong bonding in the crystal, in which case the energy gaps are small, of the same order as the bond energy. (Compare with the gap in Fig. 2, which is $-2\beta + \gamma$.)

Breaking a strong bond in this case will result in a surface state. Because we are in a situation of strong bonding (short bond lengths), we are dealing here with gaps of the type I as defined in Fig. **10**-1.2. When we have, instead gaps of type II, that is long bonds and weak bonding, the Shockley picture loses meaning and surface states must be considered which are created by perturbations of the potential at the surface, that is Tamm states. The distinction between Tamm and Shockley states, however, is not clear cut and there is a continuous gradation between them. This distinction, as well as the classification of gaps into types I and II, must thus be considered only as a rough guidance as to the mechanism whereby surface states are created.

It should be appreciated that the dangling bonds associated with surface states often tend to couple in pairs. This leads, of course, to a relaxation of the geometry of the surface atoms and thus to a *reconstruction* of the surface, an effect which can now be observed experimentally (by means of the tunnelling scanning microscope) in silicon crystals.

3 The Koster and Slater method for impurities and surface states

We shall discuss in this section a powerful method due to Koster and Slater for dealing with a perturbation potential in a crystal or a molecule. This method will serve as an introduction to the use of Green functions and it will be explicitly applied in detail in § 4 in order to obtain surface states in a linear chain.

Definitions and basic formulae

Consider a regular linear chain with sites labelled $1, 2, \ldots, m, \ldots, N$, at each of which there is only one orbital, which shall be labelled with the ket $|m\rangle$, $(m = 1, 2, \ldots, N)$. The bra $\langle m|$ will also be used when necessary to denote $|m\rangle^*$. The idea is to start with the unperturbed Hamiltonian $\mathbb{H}$ and wave function ψ^i for the whole chain:

$$\mathbb{H}\,\psi^i = E^i\,\psi^i. \tag{1}$$

(Notice that we now denote the states and their energies with superscripts, in order to avoid confusion with labels, such as m, denoting orbitals or sites. This will permit a clearer reading of running indices of these two types.) We shall then introduce a perturbed Hamiltonian $\mathbb{H}$ through a *perturbation* potential V:

$$\mathbb{H} = \mathbb{H} + V. \tag{2}$$

The unperturbed wave function ψ^i is written in the tight-binding (LCAO) approximation:

$$\psi^i = \sum_m c^i(m)\,|m\rangle, \tag{3}$$

with the usual notation and approximations, (see eqns **10**-2.3 and **10**-2.4),

$$H_{mn} = \langle m|\mathbb{H}|n\rangle, \qquad V_{mn} = \langle m|V|n\rangle, \qquad S_{mn} = \langle m|n\rangle = \delta_{mn}. \tag{4}$$

We know that any two eigenfunctions ψ^i, ψ^j of (1) must be orthogonal, and we shall take them to be normalized, so that,

3 $$\int (\psi^i)^* \psi^j \, \mathrm{d}x = \sum_{mn} c^i(m)^* c^j(n) \langle m|n\rangle \tag{5}$$

4|5 $$= \sum_{mn} c^i(m)^* c^j(n)\, \delta_{mn} \tag{6}$$

$$= \sum_{m} c^i(m)^* c^j(m) = \delta^{ij}. \tag{7}$$

It can also be proved in a similar way that

$$\sum_i c^i(m)^* c^i(n) = \delta_{mn}. \tag{8}$$

The secular equations for the unperturbed Hamiltonian are

$$\sum_{n=1}^{N} (H_{mn} - \varepsilon S_{mn})\, c^i(n) = 0, \qquad m = 1, 2, \ldots, N, \tag{9}$$

where the roots ε will lead to the energy levels $E^1, E^2, \ldots, E^i, \ldots, E^N$. (Notice that the position of the coefficients in eqn 9 is highly significant.) The superscript i on the coefficients $c^i(n)$ indicates that E^i is to be substituted for ε in (9). On using as well the orthogonality condition from (4) we obtain

$$\sum_n (H_{mn} - E^i \delta_{mn})\, c^i(n) = 0. \tag{10}$$

10 $$\sum_n H_{mn}\, c^i(n) = E^i c^i(m). \tag{11}$$

The fundamental idea of the method is to write the perturbed wave function Ψ corresponding to the perturbed Hamiltonian $\mathbb{H}$ in (2) as a linear combination of the orbitals $|m\rangle$, as ψ^i was written in (3), but now with perturbed coefficients $u(m)$ replacing the $c^i(m)$:

$$\Psi = \sum_m u(m)\,|m\rangle. \tag{12}$$

The secular equation for this perturbed wave function is exactly as in (10), with $\mathbb{H}$ replaced by $\mathbb{H} + V$ and with the c's replaced by the u's:

$$\sum_n (H_{mn} - E\,\delta_{mn})\, u(n) = -\sum_n V_{mn}\, u(n). \tag{13}$$

The energy E here, naturally, is the perturbed energy. Equation (13) is the fundamental equation which we must solve.

The Green function

The method for obtaining the coefficients $c^i(n)$, $(n = 1, 2, \ldots, N)$, in (9) or (10) is very well known, since these are systems of *homogeneous* linear equations. We have written (13) instead, as a inhomogeneous system because this allows us to treat the right-hand side as a perturbation term with respect to (10). We have in mind, in doing this, that there is a classic method for dealing with such inhomogeneous problems, which is based in the construction of the so-called *Green function*. We shall first explain what the general idea of this method is but, the problem at hand being quite simple, we shall at a second stage be able to construct its solution (and thus the Green function) entirely from first principles without any special mathematical machinery.

The inhomogeneous term in (R13) is clearly some function $F(m)$ of the site index m:

$$\sum_n (H_{mn}\, E\, \delta_{mn})\, u(n) = F(m), \qquad m = 1, 2, \ldots, N. \tag{14}$$

We want to find one of the unknown coefficients, say $u(p)$, at the site p. Let us call G_{pm} the *response* at p (that is, the solution at p) *due to a unit perturbation at the site m*. Since the perturbation at m is $F(m)$, it follows at once that the response at p due to the site at m will be $G_{pm}\,F(m)$. (This is so because the equations are *linear*.) In the system of equations (14), on the other hand, each site m contains a perturbation $F(m)$. In order to get the total response at p, which is the solution sought $u(p)$, we must therefore add up at p all the responses at this site originating from the perturbations at each site m:

$$u(p) = \sum_m G_{pm}\, F(m). \tag{15}$$

(Again, the addition of the individual responses is effected because the equations are linear.) The function G_{pm} in (15), which is the response at p due to a unit perturbation at m, is called the Green function of the problem in hand. On replacing in (15) the value of $F(m)$ which results from comparison of (R14) and (R13), we get

15

$$u(p) = -\sum_{mn} G_{pm}\, V_{mn}\, u(n). \tag{16}$$

The notation here must be made a little more explicit. From (14), which defines the Green function G_{pm}, this function must depend on the value of the perturbed energy E in these equations and should thus be written as $G(E)_{pm}$.

Therefore,

$$u(p) = -\sum_{mn} G(E)_{pm}\, V_{mn}\, u(n). \tag{17}$$

Explicit form of the Green function

We shall now recover from first principles the form (17) of our solutions, and this work will provide us at the same time with an explicit form for the Green function. What we are now going to do is entirely independent from the discussion following eqn (14) but, having that discussion in mind at the appropriate time (eqn 30 below) will give us an insight into the general validity of what might otherwise appear to be an *ad hoc* definition. In comparing (13) with (10) we regard the inhomogeneous term on (R13) as a perturbation with respect to (10). We therefore write the perturbed coefficient $u(n)$ as a linear combination of the unperturbed coefficients $c^i(n)$,

$$u(n) = \sum_i x^i\, c^i(n), \tag{18}$$

the x^i being some coefficients to be found. Before introducing (18) into (13), we rewrite the latter in the following form:

13 $$\sum_n H_{mn}\, u(n) - E\, u(m) = -\sum_n V_{mn} u(n). \tag{19}$$

18|19 $$\sum_{ni} H_{mn}\, x^i\, c^i(n) - E \sum_i x^i\, c^i(m) = -\sum_n V_{mn} u(n). \tag{20}$$

11|20 $$\sum_i x^i\, E^i\, c^i(m) - E \sum_i x^i\, c^i(m) = -\sum_n V_{mn} u(n). \tag{21}$$

21 $$\sum_i x^i (E^i - E)\, c^i(m) = -\sum_n V_{mn} u(n). \tag{22}$$

22 $$\sum_{im} x^i (E^i - E)\, c^i(m)\, c^j(m)^* = -\sum_{mn} c^j(m)^*\, V_{mn} u(n). \tag{23}$$

7|23 $$\sum_i x^i (E^i - E)\, \delta^{ij} = -\sum_{mn} c^j(m)^*\, V_{mn} u(n). \tag{24}$$

24 $$x^j (E^j - E) = -\sum_{mn} c^j(m)^*\, V_{mn} u(n). \tag{25}$$

We are almost there! (And the reason for the trick used in deriving eqn 24 should now be apparent.) From (25),

$$x^j = -(E^j - E)^{-1} \sum_{mn} c^j(m)^*\, V_{mn} u(n). \tag{26}$$

We can now introduce (26) into (18), but it will first be convenient to rewrite this latter equation:

18 $$u(p) = \sum_j x^j c^j(p). \tag{27}$$

26|27 $$= -\sum_{mnj} (E^j - E)^{-1} c^j(m)^* V_{mn} u(n) c^j(p) \tag{28}$$

28 $$= -\sum_{mnj} c^j(p) c^j(m)^* (E^j - E)^{-1} V_{mn} u(n). \tag{29}$$

Define the following function in eqn (29):

$$G(E)_{pm} = \sum_j c^j(p) c^j(m)^* (E^j - E)^{-1}. \tag{30}$$

(We use here the Green function notation because the next step will show that this summation behaves exactly as predicted for the Green function.)

30|29 $$u(p) = -\sum_{mn} G(E)_{pm} V_{mn} u(n). \tag{31}$$

Equation (31) is indeed identical with (17), but the advantage is that eqn (30) gives the explicit form for the until then unspecified Green function.

Our next task is to obtain a system of N equations in the N unknowns $u(n)$ in (13), for which purpose we rewrite (31):

31 $$\sum_n \delta_{pn} u(n) + \sum_n \left\{ \sum_m G(E)_{pm} V_{mn} \right\} u(n) = 0. \tag{32}$$

32 $$\sum_n \left\{ \delta_{pn} + \sum_m G(E)_{pm} V_{mn} \right\} u(n) = 0. \tag{33}$$

33 $$n = 1, 2, \ldots, N; \qquad p = 1, 2, \ldots, N. \tag{34}$$

This is the required system of equations, the compatibility condition for which is

$$\det \left\{ \delta_{pn} + \sum_m G(E)_{pm} V_{mn} \right\} = 0. \tag{35}$$

The problem is now fully solved and all that remains is to apply the results obtained to the study of a localized perturbation, such as an impurity or an edge perturbation leading to a surface state.

Localized perturbation

A localized perturbation given by a perturbation potential V equal to $\Delta\alpha$ at the site q may be written as follows

4 $$V_{mn} = \langle m \mid V \mid n \rangle = \langle m \mid \Delta\alpha\, \delta_{nq} \mid n \rangle = \Delta\alpha\, \delta_{nq} \langle m \mid q \rangle \tag{36}$$

4|36 $$= \Delta\alpha\, \delta_{nq}\, \delta_{mq}\,. \tag{37}$$

On introducing (37) into (33) we obtain,

37|33 $$u(p) + G(E)_{pq}\,\Delta\alpha\, u(n) = 0, \tag{38}$$

whereas the compatibility condition (35) takes the form

37|35 $$\det\{\delta_{pq} + G(E)_{pq}\,\Delta\alpha\} = 0; \quad n = q; \quad p = 1, 2, \ldots, N. \tag{39}$$

The matrix in the curly brackets here has only one column, the q-th one. Determinants, of course, are only defined for square arrays, so we take the 1 by 1 determinant for the row with p equal to q. This gives the compatibility condition

39 $$1 + G(E)_{qq}\,\Delta\alpha = 0, \tag{40}$$

at the perturbation site q. This relation will allow us to work out the energy level corresponding to such perturbation.

4 Surface states in a linear chain

As an application of the Koster and Slater method discussed in § 3 we shall consider a finite linear chain (that is a chain with non-periodic boundary conditions) of N equally spaced atoms with one electron per atom and we shall show that a localized perturbation of the potential at one end of the chain creates a surface state. This work follows closely a treatment given by G. F. Kventsel (1976).

The wave functions and eigenvalues of the described linear chain in the LCAO approximation were determined in the 1930's by J. E. Lennard-Jones and C. A. Coulson, and their results are quoted in McWeeny (1979), pp. 240–241. The molecular orbital ψ^k, written as a linear combination of atomic orbitals $|m\rangle$, as in (3.3), is given by

$$\psi^k = \sum_m \left(\frac{2}{N+1}\right)^{1/2} \sin km\,|m\rangle, \tag{1}$$

and its energy E^k is

$$E^k = E_0 + 2\beta \cos k, \tag{2}$$

with

$$k = \pi s(N+1)^{-1}, \qquad s = 1, 2, \ldots, N\,. \tag{3}$$

The parameter β in (2) is the resonance integral between the orbitals on two contiguous sites (see eqn 2.2), all other interactions being neglected.

The molecular-orbital coefficients in (1) allow us to write down the Green function (3.30) which we first quote in a form adapted to the present work:

$$G(E)_{mn} = \sum_k c^k(m)c^k(n)^*\,(E^k - E)^{-1}, \quad m > n. \tag{4}$$

The condition $m > n$ imposed is merely convenient in order to keep track of the atomic sites. Notice that, if we ignore the factor $(E^k - E)^{-1}$, (R4) is a sort of bond order or combined charge density, *except* that we are adding up over *all* the states in the band rather than over the occupied states only. The factor $(E^k - E)^{-1}$ in this expression will allow us to extract from it the perturbed eigenvalue, and it is called a *resolvent*. On introducing in (4) the $c^k(m)$ coefficients implicit in (1) and the E^k from (2), we get:

$$G(E)_{mn} = \frac{2}{N+1}\sum_k \frac{\sin km \sin kn}{E_0 - E + 2\beta \cos k}, \; k = \frac{\pi s}{N+1}, \; s = 1, 2, \ldots, N. \tag{5}$$

The summation index k goes in steps of $\pi(N+1)^{-1}$, which we shall call Δk. For large N, k is in the range 0 to π and Δk is small so that we shall introduce it in the summation in (5) with a view to converting the latter into an integral:

$$G(E)_{mn} = \frac{2}{\pi}\sum_k \frac{\sin km \sin kn}{E_0 - E + 2\beta \cos k}\Delta k, \tag{6}$$

$$G(E)_{mn} \approx \frac{2}{\pi}\int_0^{\pi} \frac{\sin km \sin kn}{E_0 - E + 2\beta \cos k}\,\mathrm{d}k. \tag{7}$$

We now express the sine product in (7) as the difference between two cosines and obtain

$$G(E)_{mn} = \frac{1}{\pi}\left\{\int_0^{\pi} \frac{\cos(m-n)k}{E_0 - E + 2\beta\cos k}\,\mathrm{d}k - \int_0^{\pi} \frac{\cos(m+n)k}{E_0 - E + 2\beta\cos k}\,\mathrm{d}k\right\}. \tag{8}$$

These integrals must be treated with great care. Their denominator arises from the resolvent $(E^k - E)^{-1}$, which goes to infinity every time E hits one of the unperturbed eigenvalues E^k. The range in these eigenvalues can easily be obtained from (2), on recognizing from (3) that for large N, $0 \leq k \leq \pi$:

$$E_0 + 2\beta \leq E^k \leq E_0 - 2\beta. \tag{9}$$

(Remember that β is negative!) The range in (9) of 4β around E_0 is the band width, and it is when E is in this range that the integrals in (8) can go to infinity. Fortunately, this is not a possibility that should worry us: we are interested in surface states, for which E is in a band gap, so that we need not consider values of E in the range in (9). All that we need to do in our work is to ensure that this range is safely excluded from it. We do this with the help of Fig. 1, where we can see at once that the energy regions for which (8) must be calculated are those areas shaded in the figure, whereas the clear areas must be excluded. We would be interested in practice in only one of the band gaps, namely the one above the band discussed, that is for E larger than $E_0 - 2\beta$.

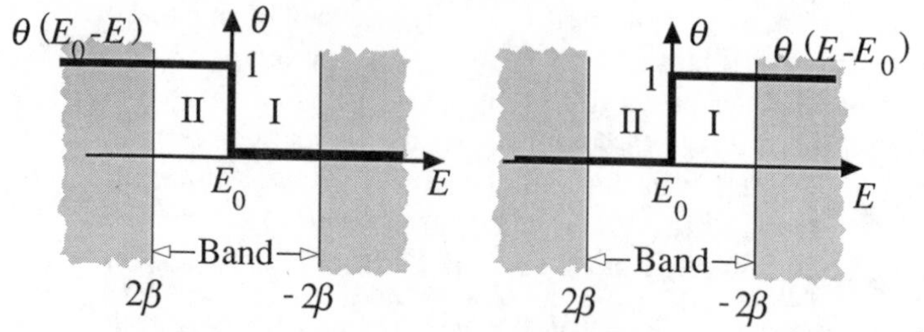

Fig. 14-4.1. The step functions $\Theta(x)$ and the energy regions over which the Green functions must be computed.

(This is the gap to the right of the pictures shown in Fig. 1.) For generality of calculation of the Green function we shall however determine the latter for the two gaps shown in grey in the figure. We shall achieve the necessary selection of the range in E by means of two successive tricks. In the first one, we shall use the well known *step function* $\Theta(x)$, which is unity for positive x and it is zero otherwise:

$$\Theta(x) = 1, \qquad x \geq 0, \tag{10}$$

$$= 0, \qquad x < 0. \tag{11}$$

By means of this trick we can divide the computation of $G(E)_{mn}$ into the two regions for which $E - E_0$ is respectively positive and negative:

$$G(E)_{mn} = G(E)_{mn}\Theta(E - E_0) + G(E)_{mn}\Theta(E_0 - E). \tag{12}$$

We can see, in fact, that the first term on (R12) gives the Green function in region I of Fig. 1, whereas the second term gives it in region II. This is not all, however, because in each region we must take only the range of E outside the band, that is in the shaded area. That is, the following conditions must be satisfied by $E - E_0$:

$$\text{Region I.} \quad (E - E_0 > 0). \qquad E - E_0 > -2\beta, \tag{13}$$

$$\text{Region II.} \quad (E - E_0 < 0). \qquad E - E_0 < 2\beta. \tag{14}$$

In order to establish these restrictions we need our second trick, which is based on the fact that, for all values of ω, $\cosh\omega$ is always larger than or equal to unity. Conditions (13) and (14), therefore, are satisfied by writing:

$$\text{Region I.} \quad (E > E_0). \qquad E - E_0 = -2\beta \cosh \omega, \tag{15}$$

$$\text{Region II.} \quad (E < E_0). \qquad E - E_0 = 2\beta \cosh \omega. \tag{16}$$

Notice that these equations mean that we are now replacing E by the new variable ω. In order to compute the first and second terms in (12) the transformations (15) and (16) must be respectively introduced into (8). When this is done, and the necessary integrals are obtained from tables (see Problem 5.10), the result is

$$G(E)_{mn} = \frac{\exp(-m\omega)\sinh n\omega}{\beta \sinh \omega}\{(-1)^{m-n}\Theta(E - E_0) - \Theta(E_0 - E)\}. \quad (17)$$

The surface state

We are now ready to use the results of § 3 for a localized perturbation $\Delta\alpha$ at the site q:

3.38 $$u(p) + G(E)_{pq}\Delta\alpha\, u(q) = 0, \quad (18)$$

3.40 $$1 + G(E)_{qq}\Delta\alpha = 0. \quad (19)$$

(Remember that the u's here are the coefficients that define the perturbed wave function at the various sites.) Since we want a surface state we consider a perturbation at one end of the chain, for which q equals 1.

18 $$u(p)/u(1) = -G(E)_{p1}\Delta\alpha = 0, \quad (20)$$

19 $$1 + G(E)_{11}\Delta\alpha = 0. \quad (21)$$

The Green function $G(E)_{11}$ here is obtained from (17):

$$G(E)_{11} = \beta^{-1}\exp(-\omega)\{\Theta(E - E_0) - \Theta(E_0 - E)\}. \quad (22)$$

We must work this out in two parts, for each step function. Consider the first term, for $E > E_0$,

15 $$E_0 - E = 2\beta\cosh\omega. \quad (23)$$

We must relate ω to E, for which purpose we use the relation

$$\exp(-\omega) = \cosh\omega - (\cosh^2\omega - 1)^{1/2} = x - (x^2 - 1)^{1/2}, \quad (24)$$

where, for convenience, we write $\cosh\omega$ as x. Thus,

24|22, 21 $$G(E)_{11} = \beta^{-1}\{x - (x^2 - 1)^{1/2}\} = -(\Delta\alpha)^{-1}. \quad (E > E_0). \quad (25)$$

The unknown x is easily eliminated from this equation and therefore, from (23), E is obtained:

$$E = E_0 + \left(\Delta\alpha + \frac{\beta^2}{\Delta\alpha}\right), \qquad (E > E_0). \quad (26)$$

On repeating the same work for $E < E_0$, that is, on using the second term in (22), the same result as in (26) is obtained, except for the opposite sign for the bracket. Thus, the final result is

$$E = E_0 \pm \left(\Delta\alpha + \frac{\beta^2}{\Delta\alpha}\right). \quad (27)$$

Notice that, $\Delta\alpha$ must be sufficiently large with respect to $|\beta|$ in order to ensure that the energy eigenvalues be outside the range of the band (see eqn 9), as we wanted, and thus in the band gaps. In order to see whether they correspond to surface states we must look at the perturbed wave functions. We must find for this purpose the ratio $u(p)/u(1)$, given by (20) in terms of $G(E)_{p1}$, which it is easy to obtain

17 $$G(E)_{p1} = \beta^{-1}\exp(-p\omega)\{(-1)^{p-1}\Theta(E-E_0)-\Theta(E_0-E)\}. \quad (28)$$

This expression, clearly, goes rapidly to zero for increasing p. It therefore follows immediately from (20) that

$$u(p)/u(1)\to 0, \qquad p\to\infty, \quad (29)$$

which shows that this is a state which decays rapidly from the surface into the bulk: it is thus a surface state.

5 Problems

1. Given that the characters for the representations A_1 and A_2 of the $\mathbf{C}_{3v}$ group are:

$$A_1\colon\ \chi(E)=1, \qquad \chi(C_3^{\pm})=1, \qquad \chi(\sigma_v)=1, \quad (1)$$

$$A_2\colon\ \chi(E)=1, \qquad \chi(C_3^{\pm})=1, \qquad \chi(\sigma_v)=-1, \quad (2)$$

apply the projection operator (**2**-5.4) on the orbital a of Fig. 2.1a and verify the first two molecular orbitals in Table 2.1.
(Remember to use *all* the group operations, not just one of each class!)

2. Write the energy E of the degenerate orbital $b-c+b'-c'$ in Table 2.1 as

$$E=\langle b-c+b'-c'|\mathbb{H}|b-c+b'-c'\rangle/\langle b-c+b'-c'|b-c+b'-c'\rangle, \quad (3)$$

and, on using the relations (2.1) to (2.5), derive its energy as listed in Table 2.1.

3. Show, by considering the transforms of the orbitals of the system, that the inversion operation at the centre of inversion shown in Fig. 2.1b is the same as the reflection σ_{v1}, in Fig. 2.1a, on a plane through the centres of the bonds aa' and cb'.
4. Discuss the stars and groups of the $\mathbf{k}$ vectors of the Brillouin zone of the chain in Fig. 2.1b. Show that the values of $\mathbf{k}$ are given in terms of the indices 0 (Γ), and ± 1 ($\pm X$). Show also that Γ is singly degenerate with group $\mathbf{C}_i$ whereas X is doubly degenerate (belonging to a star of two elements) with group $\mathbf{C}_1$ (the identity only).
5. On starting from the orbitals a and a' in Fig. 2.1b, form all their corresponding Bloch sums and show that, with the notation of Problem

4, they are:

$$\Gamma: \quad a + \quad b + \quad c, \tag{4}$$

$$a' + \quad b' + \quad c', \tag{5}$$

$$X: \quad a + \omega b + \omega^* c, \tag{6}$$

$$a' + \omega b' + \omega^* c', \tag{7}$$

$$-X: \quad a + \omega^* b + \omega c, \tag{8}$$

$$a' + \omega^* b' + \omega c', \tag{9}$$

where

$$\omega = \exp(2\pi\mathrm{i}/3). \tag{10}$$

Hint. You will need the coefficients $\exp(\mathrm{i}\mathbf{k} \cdot \mathbf{t}_m)$ in eqn (**10**-2.6). Show that they are of the form $\exp(2\pi\mathrm{i}\kappa m/3)$, with $\kappa = 0, \pm 1$, write them in terms of ω and form a table which should be identical to Table **2**-7.1. (Think why.) From this table, (4) to (9) follow at once.

6. Construct, starting from the Bloch sums (4) to (9), the six molecular orbitals in Table 2.1. Prove that the two second functions of the two E bases are exactly orthogonal. (That is, orthogonal irrespective of the approximation of eqn 2.5.) Why is it possible to achieve this orthogonalization?

Hints. (i) Because the group of Γ is $\mathbf{C}_i$ (Problem 4), construct combinations of (4) and (5) which are g and u with respect to the inversion shown in Fig. 2.1*b*. (ii) From Problem 4, X and $-X$ must be degenerate: notice that the pairs (6), (8) and (7), (9) are indeed conjugate. Replace those pairs by their sums and differences, ignoring constant factors, to form the new degenerate pairs

$$2a - b - c, \qquad b - c, \tag{11}$$

$$2a' - b' - c', \qquad b' - c'. \tag{12}$$

(iii) You must form two *orthogonal* degenerate bases, and within each basis the two functions must also be orthogonal. Exploit for this purpose the fact that g and u functions are orthogonal. (It is the same and simpler, to make them symmetrical and antisymmetrical, respectively, with respect to the σ_{v1} plane identified with the inversion in Problem 3.) It is sufficient for this purpose to form sums and differences of the first and second components in (11) and (12) and to form two pairs, one in which the first component is g and the second is u, and vice versa for the second pair. This provides the final result. Notice that the functions of the two E bases need not have been orthogonal to each other, since they belong to the same irreducible representations but that having made them orthogonal considerably simplifies the work, since it means that they need not be mixed variationally.

7. (i) Give an argument to show that the g and u symmetries of the degenerate levels E for X and $-X$ of Problem 5 are not significant. (ii) Why is it nevertheless legitimate to call the lower and upper bands in Fig. 2.2 bonding and antibonding respectively?
8. In order to form Bloch sums for a lattice with basis, such as Fig. 2.1*b*, it is tempting to argue that, rather than starting with single orbitals, the Bloch sums should be constructed from the bonding and antibonding orbitals, respectively, in the primitive cell. Verify that the results thus obtained disagree with those of Problems 5 and 6. Why is this idea wrong?

   Hint. Consider whether the inversion symmetry, which defines the bonding and antibonding orbitals, is present in all stars.
9. Write the 11 matrix element of the A_g determinant in (2.10) as

$$\langle a + a'|\mathbb{H}|a + a'\rangle - E\langle a + a'|a + a'\rangle, \tag{13}$$

   and verify the value given in that determinant. (Use Fig. 2.1*c*, *not* Fig. 2.1*a*.)
10. Verify eqn (4.17).

   Hints. In region I, from (4.15), $E_0 - E = 2\beta \cosh \omega$. This value, on (4.8), gives,

$$G(E)_{mn} = \frac{1}{2\beta\pi}\left\{\int_0^\pi \frac{\cos(m-n)k}{\cosh\omega + \cos k}\,\mathrm{d}k - \int_0^\pi \frac{\cos(m+n)k}{\cosh\omega + \cos k}\,\mathrm{d}k\right\}. \tag{14}$$

   Use the integral (Gröbner and Hofreiter, 1966, p. 112),

$$\int_0^\pi \frac{\cos px}{a + b\cos x}\,\mathrm{d}x = \frac{\pi}{b^p}\frac{\{(a^2 - b^2)^{1/2} - a\}^p}{(a^2 - b^2)^{1/2}}, \quad a \geqslant |b| > 0, \tag{15}$$

   to find

$$G(E)_{mn} = \frac{1}{2\beta}\left\{\frac{(\sinh\omega - \cosh\omega)^{m-n}}{\sinh\omega} - \frac{(\sinh\omega - \cosh\omega)^{m+n}}{\sinh\omega}\right\}. \tag{16}$$

   On introducing $(\sinh\omega - \cosh\omega)$ equal to $-\exp(-\omega)$, you will find

$$G(E)_{mn} = (\beta\sinh\omega)^{-1}(-1)^{m-n}\exp(-m\omega)\sinh n\omega. \tag{17}$$

   In the same manner, for region II,

$$G(E)_{mn} = -(\beta\sinh\omega)^{-1}\exp(-m\omega)\sinh n\omega. \tag{18}$$

   Eqns (17) and (18) into (4.12) give the final result.

Further reading

A good discussion of Tamm states may be found in Ziman (1964), pp. 166–168. The method of §3 was given in Koster and Slater (1954). See Davison and Levine (1970) for a review of surface states. García Moliner and Flores (1979) provide a detailed introduction to the theory of solid surfaces, with emphasis on their mathematical treatment. A very useful and more chemical approach is given by Hoffmann (1988). A formal but nevertheless readable introduction to Green functions may be found in Lukes (1969).

15

Solutions to the problems

1 Problems 1-9

Problem 1

$$\mathbb{A}\psi = a\psi,\ \mathbb{A} = \sum_i \mathbb{A}_i,\ \mathbb{A}_i\psi_i = a_i\psi_i, \qquad i = 1, 2, \ldots, n. \tag{1}$$

$$\mathbb{A}(\psi_1 \ldots \psi_i \ldots \psi_n) = \sum_i \mathbb{A}_i(\psi_1 \ldots \psi_i \ldots \psi_n) = \sum_i (\psi_1 \ldots \mathbb{A}_i\psi_i \ldots \psi_n) \tag{2}$$

1|2
$$= \sum_i a_i(\psi_1 \ldots \psi_i \ldots \psi_n). \tag{3}$$

Notice in (2) that when $\mathbb{A}_i$ acts on the product $\psi_1 \ldots \psi_i \ldots \psi_n$, all the functions except ψ_i act as constants, since $\mathbb{A}_i$ does not contain any of their coordinates.

Problem 2

Call $\varphi(\mathbf{r}) = N \exp(\mathrm{i}\mathbf{k}\cdot\mathbf{r})$, where N will be chosen so as to normalize $\varphi(\mathbf{r})$ over the volume V. Then:

$$N^2 \int_V \exp(-\,\mathrm{i}\mathbf{k}\cdot\mathbf{r}) \exp(\mathrm{i}\mathbf{k}\cdot\mathbf{r})\, \mathrm{d}v = 1 \;\Rightarrow\; N^2 V = 1 \;\Rightarrow\; N = V^{-1/2}. \tag{4}$$

Problem 3

The first two eigenstates correspond to the values shown in Table 1. Follow on in the same manner.

Table 15-1.1. Degeneracies

κ_x	κ_y	κ_z	$\mathbf{k}^2$	degeneracy	deg. with spin
0	0	0	0	1	2
0	0	± 1	1		
0	± 1	0	1	6	12
0	0	± 1	1		

Problem 4

Use the values of $\hbar$ and of the energy equivalent of the electron mass shown in the table on the inside back cover.

Problem 5

As in Problem 4, prove from (**1**-6.4) that

$$E_F = 0.365\,\mathcal{N}^{2/3}\,\mathrm{eV}\,(\mathrm{nm})^{-2}. \tag{5}$$

Sodium: $\mathcal{N} = 2(0.428)^{-3}\,(\mathrm{nm})^{-3}$; copper: $\mathcal{N} = 4(0.361)^{-3}\,(\mathrm{nm})^{-3}$.

Problem 6

The solutions of (**1**-3.9), with a normalization factor $\boldsymbol{N}$, are

$$\varphi(x) = \boldsymbol{N} \sin(\kappa\pi x/L), \qquad \kappa = 1, 2, 3, \ldots. \tag{6}$$

6

$$\int_0^L \varphi(x)^* \varphi(x)\,\mathrm{d}x = \boldsymbol{N}^2 \int_0^L \sin^2(\kappa\pi x/L)\,\mathrm{d}x = 1. \tag{7}$$

On integration, $\boldsymbol{N}$ is $(2/L)^{1/2}$.

Problem 7

Write $\mathbb{p}$ as $\mathbb{p}_x + \mathbb{p}_y + \mathbb{p}_z$. Use Problem 1 with the wave functions (**1**-9.5).

Problem 8

From (**1**-9.7) and (**1**-9.8) the elementary cube corresponding to each state has side π/L, whence v' equals $(\pi/L)^3$. Divide by 2 to obtain v.

Explanation: From note (i) of Problem **1**-9.7 the volume of the new **k** space is 1/8 of the old one. We must have the same number of states whence each state must occupy 1/8 of the previous volume.

Problem 9

The volume of the shell required in the new **k** space is 1/8 of the volume of the shell in (**1**-5.2). From Problem 8, the volume v occupied by each state is 1/8 of the volume used in (**1**-5.2). Thus (**1**-5.3) is unaltered and the rest of the proof in § **1**-5 goes through unchanged.

2 Problems 2-7

Problem 1

,2-2.7 $$\hat{t}\phi_1 = \hat{t}\phi(\mathbf{r} - \mathbf{r}_1) = \phi(t^{-1}(\mathbf{r} - \mathbf{r}_1))$$
$$= \phi(\mathbf{r} - \mathbf{r}_1 - t) = \phi(\mathbf{r} - (\mathbf{r}_1 + t)) \tag{1}$$
F2-2.1 $$= \phi(\mathbf{r} - \mathbf{r}_6). \tag{2}$$

Problem 2

Call g_i a single operation of G (i not ranging) and use in the second step below the commutation property of abelian groups:

$$C(g_i) = \{g g_i g^{-1}\} = \{g g^{-1} g_i\} = \{g_i\} = g_i. \tag{3}$$

This proves the first part of the problem. For the second part, identify the notation from Chapter **0**.

3 $$|C(G)| = |G|. \tag{4}$$

§ **2**-5, 4 $$|I| = |C(G)| \qquad \Rightarrow \qquad |I| = |G|. \tag{5}$$

§ **2**-5 $$\sum_j |{}^j\hat{G}|^2 = |G|, \qquad j = 1, 2, \ldots, |I| = |G|. \tag{6}$$

For the sum of $|G|$ integers, all larger than zero, to be equal to $|G|$, each integer must be the unity. This proves the second part.

Problem 3

If C_n is a rotation by $2\pi/n$, then $(C_n)^2$ is a rotation by $2 \times (2\pi/n)$ and so on. Therefore, in order to prove closure:

$$(C_n)^r(C_n)^s = (C_n)^{r+s}. \tag{7}$$

If $r + s < n$, (R7) is in the set **2**-7.1. If $r + s > n$, then

$$(C_n)^{r+s} = (C_n)^n(C_n)^{r+s-n} = (C_n)^{r+s-n}. \tag{8}$$

If $r + s - n$ is still larger than n, repeat the trick in (8) a sufficient number of times until this property is satisfied. Thus closure is verified. Similarly the other group properties. That the group is abelian follows because (R7) is also obtained when the elements in (L7) are commuted.

Problem 4

Irreducibility: Use (**2**-5.1). For ${}^1E$, for example,

$$1^*1 + \omega^*\omega + \omega\omega^* = 1 + 1 + 1 = 3, \tag{9}$$

which is the order of the group. *Multiplication rules*: Prove geometrically, for example, that

$$C_3^+ C_3^+ = C_3^- . \tag{10}$$

The product of the corresponding matrices from ^1E is:

$$\omega\omega = \omega^2 = \exp(\tfrac{4}{3}\pi \mathrm{i}) = \exp(\tfrac{4}{3}\pi \mathrm{i} - 2\pi \mathrm{i}) = \exp(-\tfrac{2}{3}\pi \mathrm{i}) = \omega^*, \tag{11}$$

which is the matrix for C_3^-.

Problem 5

Equation (**2**-7.2) is immediate from (**2**-5.4). The factor ω on (R**2**-7.3) is the correct matrix from Table **2**-7.1.

Problem 6

If φ is an eigenfunction of a real Hamiltonian $\mathbb{H}$ then, from (**2**-3.13), φ^* and φ are degenerate. From (**2**-7.2), with real orbitals ϕ,

$$(\psi^{^1E})^* = \phi_1 + \omega\phi_2 + \omega^*\phi_3 = \psi^{^2E}, \tag{12}$$

whence $\psi^{^1E}$ and $\psi^{^2E}$ are degenerate under the conditions stated. Notice that the representatives of one representation are the complex conjugate of the other: representations related in this way are called *conjugate*.

Problem 7

Equation (**2**-7.4) follows from Fig. **2**-6.1 and (**2**-7.5) follows from the matrix multiplication rule. The character is the sum of the diagonal elements of the matrix.

Problem 8

We treat as an example the basis (**2**-7.7), the two functions of which will be called ψ_1 and ψ_2 respectively.

$$\psi_1 = 2\phi_1 - \phi_2 - \phi_3, \qquad \psi_2 = \phi_2 - \phi_3. \tag{13}$$

F**2**-7.2 $$C_3^+\psi_1 = C_3^+(2\phi_1 - \phi_2 - \phi_3) = 2\phi_2 - \phi_3 - \phi_1 \tag{14}$$

14 $$= -\tfrac{1}{2}(2\phi_1 - \phi_2 - \phi_3) + \tfrac{3}{2}(\phi_2 - \phi_3)$$

$$= -\tfrac{1}{2}\psi_1 + \tfrac{3}{2}\psi_2. \tag{15}$$

F**2**-7.2 $$C_3^+\psi_2 = C_3^+(\phi_2 - \phi_3) = \phi_3 - \phi_1 = -\tfrac{1}{2}\psi_1 - \tfrac{1}{2}\psi_2. \tag{16}$$

15, 16 $$C_3^+ \langle\psi_1\psi_2| = \langle\psi_1\psi_2|\begin{bmatrix} -1/2 & -1/2 \\ 3/2 & -1/2 \end{bmatrix} \Rightarrow \chi(C_3^+) = -1. \tag{17}$$

F2-7.2 $$\sigma_{v1}\psi_1 = \sigma_{v1}(2\phi_1 - \phi_2 - \phi_3) = 2\phi_1 - \phi_3 - \phi_2 = \psi_1. \tag{18}$$

F2-7.2 $$\sigma_{v1}\psi_2 = \sigma_{v1}(\phi_2 - \phi_3) = \phi_3 - \phi_2 = -\psi_2. \tag{19}$$

18, 19 $$\sigma_{v1}\langle\psi_1\psi_2| = \langle\psi_1\psi_2|\begin{bmatrix} 1 & \\ & -1 \end{bmatrix} \Rightarrow \chi(\sigma_{v1}) = 0. \tag{20}$$

Note that the representation matrices should be unitary (that is orthogonal in our case, since they are real). To achieve this the functions of the basis should be normalized but this would make the arithmetic harder and it entails an orthogonal transformation which in any case leaves the characters invariant.

Problem 9

In the notation of (**4**-9.3) and (**4**-9.4) we must prove that $\langle\psi_1|\psi_2\rangle$ vanishes:

$$\langle\phi_1 + \phi_2 + \phi_3|\phi_2 - \phi_3\rangle = \langle\phi_1|\phi_2\rangle + \langle\phi_2|\phi_2\rangle + \langle\phi_3|\phi_2\rangle - \langle\phi_1|\phi_3\rangle - \langle\phi_2|\phi_3\rangle - \langle\phi_3|\phi_3\rangle. \tag{21}$$

By symmetry (see Fig. **2**-7.2),

$$\langle\phi_1|\phi_2\rangle = \langle\phi_1|\phi_3\rangle, \quad \langle\phi_2|\phi_2\rangle = \langle\phi_3|\phi_3\rangle, \quad \langle\phi_3|\phi_2\rangle = \langle\phi_2|\phi_3\rangle. \tag{22}$$

3 Problems 3-10

Problem 1

Make a copy of the pattern points 1 to 6 on a transparent sheet. Displace it as required for the translations and relabel the sites for the point operations. Different colour sticks help.

Problem 2

The space group operators required are:

$$\{g_1|\mathbf{v}\}, \{g_2|\mathbf{v}\}, \{g_3|\mathbf{0}\}, \{g_4|\mathbf{v}\}, \{\sigma_1|\mathbf{v}\}, \{\sigma_2|\mathbf{0}\}, \{\sigma_3|\mathbf{v}\}, \{\sigma_4|\mathbf{0}\}. \tag{1}$$

3-10.1 $$g_1 = (-v)C_2' = \{C_2'|-\mathbf{v}\}. \tag{2}$$

3-6.6, 2 $$\{g_1|\mathbf{v}\} = \{E|\mathbf{v}\}\{g_1|\mathbf{0}\} = \{E|\mathbf{v}\}\{C_2'|-\mathbf{v}\} = \{C_2'|\mathbf{0}\}. \tag{3}$$

3-6.6, **3**-10.1 $$\{g_2|\mathbf{v}\} = \{E|\mathbf{v}\}\{g_2|\mathbf{0}\} = \{E|\mathbf{v}\}\{C_2'|\mathbf{v}\} = \{C_2'|\mathbf{a}'\}. \tag{4}$$

In the same manner:

$$\{g_3|\mathbf{0}\} = \{C_2'|\mathbf{a}'\}, \{g_4|\mathbf{v}\} = \{C_2'|2\mathbf{a}'\}. \qquad (5)$$

$$\{\sigma_1|\mathbf{v}\} = \{\sigma_y'|\mathbf{a}'\}, \quad \{\sigma_2|\mathbf{0}\} = \{\sigma_y'|\mathbf{a}'\}, \quad \{\sigma_3|\mathbf{v}\} = \{\sigma_y'|2\mathbf{a}'\},$$
$$\{\sigma_4|\mathbf{0}\} = \{\sigma_y'|2\mathbf{a}'\}. \qquad (6)$$

Problem 3

3-6.6 $$\{p|\mathbf{w}\}(\mathbf{r} + \mathbf{r}') = p(\mathbf{r} + \mathbf{r}') + \mathbf{w} = p\mathbf{r} + p\mathbf{r}' + \mathbf{w}. \qquad (7)$$

3-6.6 $$\{p|\mathbf{w}\}\mathbf{r} + \{p|\mathbf{w}\}\mathbf{r}' = p\mathbf{r} + \mathbf{w} + p\mathbf{r}' + \mathbf{w}. \qquad (8)$$

Problem 4

$$E = \{E|\mathbf{0}\}, \quad C_2 = \{C_2|\mathbf{0}\}, \quad v\sigma_x = \{\sigma_x|\mathbf{v}\}, \quad v\sigma_y = \{\sigma_y|\mathbf{v}\}. \qquad (9)$$

3-6.6 $$\{\sigma_x|\mathbf{v}\}\{\sigma_y|\mathbf{v}\} = \{\sigma_x\sigma_y|\sigma_x\mathbf{v} + \mathbf{v}\} = \{C_2|\mathbf{a}\} = \{E|\mathbf{a}\}\{C_2|\mathbf{0}\}. \qquad (10)$$

Similarly for the other products. In multiplying point group operations graphically by transforming points, use crosses and circles, respectively, for points above and below the plane of the drawing.

Problem 5

3-6.6 $$\{p|\mathbf{w}\}\{E|\mathbf{t}\} = \{p|p\mathbf{t} + \mathbf{w}\}. \qquad (11)$$

3-6.6 $$\{E|\mathbf{t}\}\{p|\mathbf{w}\} = \{p|\mathbf{w} + \mathbf{t}\}. \qquad (12)$$

(R11) will be equal to (R12) if: (i) p equals E in which case $\{p|\mathbf{w}\}$ is a translation; (ii) if $p\mathbf{t}$ equals $\mathbf{t}$, in which case p must be a rotation around $\mathbf{t}$ or a reflection through a plane which contains $\mathbf{t}$.

4 Problems 4-11

Problem 1

4-3.3 $$\mathbf{a}^{\#} = 2\pi(\mathbf{b} \times \mathbf{c})(\mathbf{a}\cdot\mathbf{b} \times \mathbf{c})^{-1}. \qquad (1)$$

1 $$\mathbf{a}^{\#}\cdot\mathbf{a} = 2\pi(\mathbf{b} \times \mathbf{c}\cdot\mathbf{a})(\mathbf{a}\cdot\mathbf{b} \times \mathbf{c})^{-1} = 2\pi. \qquad (2)$$

1 $$\mathbf{a}^{\#}\cdot\mathbf{b} = 2\pi(\mathbf{b} \times \mathbf{c}\cdot\mathbf{b})(\mathbf{a}\cdot\mathbf{b} \times \mathbf{c})^{-1} = 0. \qquad (3)$$

See the inside back cover for the vector properties used.

Problem 2

$$\Omega = \mathbf{a}\cdot\mathbf{b} \times \mathbf{c}, \qquad \Omega^{\#} = \mathbf{a}^{\#}\cdot\mathbf{b}^{\#} \times \mathbf{c}^{\#}, \qquad (4)$$

4-3.3 $$\Omega^{\#} = (2\pi)^3\{(\mathbf{b} \times \mathbf{c})\cdot(\mathbf{c} \times \mathbf{a}) \times (\mathbf{a} \times \mathbf{b})\}(\mathbf{a}\cdot\mathbf{b} \times \mathbf{c})^{-3} \qquad (5)$$

4|5 $$\Omega^{\#} = (2\pi)^3\,(\mathbf{b} \times \mathbf{c})\cdot\{(\mathbf{c} \times \mathbf{a}\cdot\mathbf{b})\mathbf{a} - (\mathbf{c} \times \mathbf{a}\cdot\mathbf{a})\mathbf{b}\}\,\Omega^{-3} \qquad (6)$$

6 $$= (2\pi)^3\,(\mathbf{b} \times \mathbf{c}\cdot\mathbf{a})(\mathbf{c} \times \mathbf{a}\cdot\mathbf{b})\Omega^{-3} \qquad (7)$$

7 $$= (2\pi)^3\,(\mathbf{a}\cdot\mathbf{b} \times \mathbf{c})^2\,\Omega^{-3} = (2\pi)^3\,\Omega^2\Omega^{-3} = (2\pi)^3\Omega^{-1}. \qquad (8)$$

Problem 3

From Fig. **4**-7.1*b* the lattice constant is $2|\mathbf{a}^{\#}|$, that is $4\pi a^{-1}$.

Problem 4

Draw a figure like Fig. **4**-7.1*b* with a cube of side *a*. Take

$$\mathbf{u} = \tfrac{1}{2}a[1\bar{1}1], \qquad \mathbf{v} = \tfrac{1}{2}a[11\bar{1}], \qquad \mathbf{w} = \tfrac{1}{2}a[\bar{1}11]. \qquad (9)$$

You will find

$$\mathbf{u}^{\#} = 4\pi a^{-1}[\tfrac{1}{2}0\tfrac{1}{2}], \quad \mathbf{v}^{\#} = 4\pi a^{-1}[\tfrac{1}{2}\tfrac{1}{2}0], \quad \mathbf{w}^{\#} = 4\pi a^{-1}[0\tfrac{1}{2}\tfrac{1}{2}]. \qquad (10)$$

On comparison with (**4**-7.1) this is a face-centred cubic lattice of lattice constant $4\pi a^{-1}$. Alternatively, draw a figure.

Problem 5

4-11.1 $$|a|^2 = |b|^2 = |c|^2 = m^2 + n^2 + p^2 \quad \Rightarrow \quad \text{equal lengths.} \qquad (11)$$

4-11.1 $$\mathbf{a}\cdot\mathbf{b} = \mathbf{b}\cdot\mathbf{c} = \mathbf{c}\cdot\mathbf{a} = pm + mn + np \quad \Rightarrow \quad \text{equal angles.} \qquad (12)$$

You will find

$$\mathbf{a}^{\#} = \lambda[\mu\nu\rho], \qquad \mathbf{b}^{\#} = \lambda[\rho\mu\nu], \qquad \mathbf{c}^{\#} = \lambda[\nu\rho\mu], \qquad (13)$$

$$\lambda = 2\pi(m^3 + n^3 + p^3 - 3mnp),$$
$$\mu = m^2 - np,\ \nu = n^2 - pm,\ \rho = p^2 - mn. \qquad (14)$$

Equation (13) is of the same form as (**4**-11.1) so that these vectors form a trigonal lattice.

5 Problems 5-7

Problem 1

3-6.6 $$\{E|\mathbf{t}\}\,\{E|\mathbf{t}'\} = \{E|\mathbf{t} + \mathbf{t}'\}. \qquad (1)$$

5-2.9 $${}_{\mathbf{k}}\hat{T}\{E|\mathbf{t}\}\,{}_{\mathbf{k}}\hat{T}\{E|\mathbf{t}'\} = \exp(-\,i\mathbf{k}\cdot\mathbf{t})\exp(-\,i\mathbf{k}\cdot\mathbf{t}') \qquad (2)$$

$$= \exp\{-\,i\mathbf{k}\cdot(\mathbf{t} + \mathbf{t}')\} = {}_{\mathbf{k}}\hat{T}\{E|\mathbf{t} + \mathbf{t}'\}. \qquad (3)$$

Problem 2

New form of **5**-2.9:

$$_{\mathbf{k}}\hat{T}\{E|\mathbf{t}\} = \exp(\mathrm{i}\mathbf{k}\cdot\mathbf{t}). \tag{4}$$

Assert now that the bases of these representations (eqn **5**-6.3) are

$$\psi_{\mathbf{k}}(\mathbf{r}) = \exp(-\,\mathrm{i}\mathbf{k}\cdot\mathbf{r})\,u_{\mathbf{k}}(\mathbf{r}). \tag{5}$$

5-6.6 $$\{E|\mathbf{t}\}\,\psi_{\mathbf{k}}(\mathbf{r}) = \psi_{\mathbf{k}}(\mathbf{r}-\mathbf{t}) = \exp(-\,\mathrm{i}\mathbf{k}\cdot\mathbf{r})\exp(\mathrm{i}\mathbf{k}\cdot\mathbf{t})u_{\mathbf{k}}(\mathbf{r}-\mathbf{t}) \tag{6}$$

5-6.4 $$= \exp(-\,\mathrm{i}\mathbf{k}\cdot\mathbf{r})\exp(\mathrm{i}\mathbf{k}\cdot\mathbf{t})u_{\mathbf{k}}(\mathbf{r}) \tag{7}$$

5 $$= \exp(\mathrm{i}\mathbf{k}\cdot\mathbf{t})\psi_{\mathbf{k}}(\mathbf{r}). \tag{8}$$

Compare with (**5**-6.2).

Problem 3

Introduce the functions $|\mathbf{k}\rangle$ in the form $\exp(\mathrm{i}\mathbf{k}\cdot\mathbf{r})$ in eqns (**5**-6.6) to (**5**-6.8) and the result follows at once.

Problem 4

To prove:

$$I = \langle\mathbf{k}|\mathbf{k}'\rangle = \int_V \exp(-\,\mathrm{i}\mathbf{k}\cdot\mathbf{r})\exp(\mathrm{i}\mathbf{k}'\cdot\mathbf{r})\,\mathrm{d}v \tag{9}$$

$$= \int_V \exp\{\mathrm{i}(\mathbf{k}'-\mathbf{k})\cdot\mathbf{r}\}\,\mathrm{d}v = 0, \quad \mathbf{k}\neq\mathbf{k}', \tag{10}$$

5-4.3 $$\mathbf{k} = \frac{\kappa_x}{N_x}\mathbf{a}^{\#} + \frac{\kappa_y}{N_y}\mathbf{b}^{\#} + \frac{\kappa_z}{N_z}\mathbf{c}^{\#}, \quad \kappa = 1, 2, \ldots, N, \tag{11}$$

where κ and N must be given the subscripts x, y, or z.

The vectors $\mathbf{k}'$ and $\mathbf{k}'-\mathbf{k}$ will have the same form (11). We can thus simplify the notation and write

$$I = \int_V \exp(\mathrm{i}\mathbf{k}\cdot\mathbf{r})\,\mathrm{d}v,\ \mathbf{k} = \frac{\kappa_x}{N_x}\mathbf{a}^{\#} + \frac{\kappa_y}{N_y}\mathbf{b}^{\#} + \frac{\kappa_z}{N_z}\mathbf{c}^{\#}, \kappa = 1, \ldots, N;\ \mathbf{k}\neq\mathbf{0}. \tag{12}$$

At least one of the values of κ in (12) must not vanish and the integration over V means integrating over each direction x, y, z from 0 to the appropriate value of N. Consider therefore the corresponding integral, say along x:

$$I'' = \int_0^{N_x} \exp\left(2\pi i \frac{\kappa_x}{N_x} x\right) dx = \left(2\pi i \frac{\kappa_x}{N_x}\right)^{-1} \exp\left(2\pi i \frac{\kappa_x}{N_x} x\right)\Bigg|_0^{N_x} = 0. \quad (13)$$

Notice that the plane waves $|\mathbf{g}\rangle$ of (**4**-9.6) are orthogonal over the unit cell and it is because of this that they lead to functions that are periodic over the crystal. The plane waves discussed here, instead, being orthogonal over the crystal, satisfy automatically periodicity with respect to the boundary conditions but permit the expansion of functions which are not periodic from one cell to another, that is which are not periodic over the crystal. This is as it should be, because the general wave functions (as opposite to the cell functions) are not necessarily periodic over the crystal.

See Problem **10**-5.1 for an alternative proof of the orthogonality property.

Problem 5

Use (**5**-4.3) with κ up to $\frac{1}{2}N$ or $\frac{1}{2}(N-1)$ and subtracting N from the rest. The range for $\mathbf{k}$ in (**5**-7.1) and (**5**-7.3) follows from p. 94 and conditions (**5**-7.2) and (**5**-7.4).

6 Problems 6-6

Problem 1

Take x and y along and σ_{v1} and σ_{v2} respectively in Fig. **6**-3.1a. Show graphically that

$$C_4^+ \langle p_x\, p_y| = \langle p_y, -p_x| = \langle p_x\, p_y| \begin{bmatrix} & -1 \\ 1 & \end{bmatrix}, \quad (1)$$

$$C_2 \langle p_x\, p_y| = \langle -p_x, -p_y| = \langle p_x\, p_y| \begin{bmatrix} -1 & \\ & -1 \end{bmatrix}, \quad (2)$$

and so on. From (1) and (2) the characters for C_4^+ and C_2 are, respectively, 0 and -2, and in the same way you will find that the other characters agree with Table **6**-3.2.

Problem 2

Read the operations from Fig. **6**-3.1a. Only E and σ_{v1} leave X invariant, whence its group is C_s.

Problem 3

From Fig. **6**-3.2c, on taking M as M_1, $M = [\frac{1}{2}\,\frac{1}{2}]^{\#}$, whence we can compute the value, $\exp(-i\mathbf{k}\cdot\mathbf{t})$, by which the Bloch functions get multiplied on translation (eqn **5**-6.2), for different translations $\{E|\mathbf{t}\}$:

$\{E\|\mathbf{t}\}$	$\exp(-i\mathbf{k}\cdot\mathbf{t})$,
[11]	1
[10]	-1
[01]	-1

Draw a figure like Fig. **6**-4.2*a* as follows. Copy the bottom left-hand square. Then repeat it upwards and sideways (translations [10] and [01]) on changing all the signs, as required by the last two rows of the above list. From the remaining row, the top right-hand square must repeat the pattern of the first square drawn. You must also obtain the group of M, which is $\mathbf{C}_{4v}$ (eqn **6**-3.2). You can now argue as follows.

(i) The operation C_4^+, which is acceptable since it belongs to the group of the **k** vector, exchanges p_x and p_y, whence these functions must be degenerate. (Notice, moreover, that the number of nodes along p_x and p_y are equal.) (ii) In X (see Fig. **6**-4.2*b*) the p_x function do not have any nodes between the atoms either along x or y. In M, instead, there are nodes between the atoms along the y direction. Therefore the energy for the p_x band must go up from X to M. In X, p_y has a node in the middle of the bonds along y, but this node disappears for M, whence the energy should go down from X to M for the p_y band. Remember that these guesses provide only plausible trends.

7 Problems 7-4

Problem 1

Show that $\{\bar{\imath}|3\mathbf{a}\}$ and $\{\bar{\imath}|5\mathbf{a}\}$, on acting on the point labelled **r** in Fig. **7**-1.1, take it to the black sites labelled 1 and 3 respectively, whereas $\{\bar{\imath}|4\mathbf{a}\}$ takes it to the black site 2. The results stated follow at once.

Problem 2

Conjugate $\{E|\mathbf{a}\}$ under $\{\bar{\imath}|\mathbf{0}\}$:

3-6.7 $$\{\bar{\imath}|\mathbf{0}\}^{-1} = \{\bar{\imath}|-\mathbf{0}\} = \{\bar{\imath}|\mathbf{0}\}. \tag{1}$$

3-6.6 $$\{\bar{\imath}|\mathbf{0}\}\{E|\mathbf{a}\}\{\bar{\imath}|\mathbf{0}\} = \{\bar{\imath}|\mathbf{0}\}\{\bar{\imath}|\mathbf{a}\} = \{E|-\mathbf{a}\} = \{E|6\mathbf{a}\}\{E|-\mathbf{a}\} = \{\bar{\imath}|5\mathbf{a}\}. \tag{2}$$

We use here the fact that $\{E|6\mathbf{a}\}$ coincides with the identity because of the cyclic boundary conditions. (See eqn **7**-1.1.)

Problem 3

Self-explanatory.

Problem 4

From (**7**-3.6) the matrices for the inversion will be $\begin{bmatrix} & 1 \\ 1 & \end{bmatrix}$ and $\begin{bmatrix} & -1 \\ -1 & \end{bmatrix}$ respectively, leading to characters both equal to zero.

Problem 5

The bands should be of the form shown in Fig. **8**-4.2, the lower band being the s band. They cannot touch or cross at any **k** in the Brillouin zone because no small representation is two-dimensional. Hence, no two Bloch functions can belong to the same **k**.

Problem 6

Draw a picture similar to Fig. **10**-2.1.

8 Problems 8-8

Problem 1

Follow the hint.

Problem 2

Follow the hint.

Problem 3

From (**6**-5.6), $E(\delta)$ equals $E(-\delta)$. In order to obtain the result, use this relation in (**8**-2.4) for **k** equal to **0** and δ in any direction around Γ. A cusp cannot exist here because in the limiting case of the free-electron approximation, which can safely be assumed to be valid in a small region around the origin, the energy must be parabolic. It is because this limiting case is not valid for phonons that a cusp can exist here (see § **11**-4).

Problem 4

Follow the hint.

Problem 5

Cusps appear because in the strictly free-electron case periodicity has no physical meaning and, correspondingly, the Brillouin zone faces are not physically significant. What is significant, instead, is the continuity of the energy in **k** space, of which the cusps are a consequence.

Problem 6

Always translate the appropriate segment of the Brillouin zone by a vector of the reciprocal lattice.

Problem 7

Start underneath the bottom point labelled '1' in Fig. **8**-6.1. The first triangle below it belongs to the fifth zone, as do the two triangles which appear below those of the fourth zone. Move out of these triangles to find two more belonging to the sixth zone, and proceed in the same manner over the rest of the figure.

Problem 8

$$E = \hbar^2 \mathbf{k}^2/2m \qquad \Rightarrow \qquad \frac{\mathrm{d}^2 E}{\mathrm{d}k^2} = \hbar^2/m. \tag{1}$$

On introducing (1) into (**8**-7.4) the result follows.

Problem 9

The stars of Γ and X contain only one element, Γ and X respectively and the star for Δ contains Δ and $-\Delta$, whence the bases shown in (**8**-8.1) to (**8**-8.3) follow. The corresponding matrices are obtained from (**5**-6.2). The inversion belongs to the groups of Γ and X, so that the corresponding bases and representations split into two, one gerade and one ungerade. In both cases the matrices for the translations are those shown and those for the inversion are $+1$ and -1 respectively for the gerade and the ungerade representations. The inversion does not belong to the group of Δ and there is thus only one representation for it. The matrices for the translations are those shown and the matrix for the inversion is the two by two matrix with unity in the secondary diagonal. Both gerade and ungerade bases lead to equivalent representations in this case.

Problem 10

From Problem **8**-8.3, $\partial E/\partial \mathbf{k}$ vanishes at Γ. It does not vanish at Z because it is only the normal component that vanishes on the face to which Z belongs. It vanishes at X by considering the symmetry of the energy at points slightly above and below X in Fig. **6**-4.1a. It vanishes at M because the normal component of the derivative on the top face in Fig. **6**-4.1a vanishes, since there is a reflection plane parallel to this face.

9 Problems 9-4

Problem 1

The first bond shown in Fig. **9**-1.1 goes from $[000]$ to $[\frac{1}{4}\frac{1}{4}\frac{1}{4}]a$, whence its length d is $\frac{\sqrt{3}}{4}a$:

$$d = 0.433a \;\Rightarrow\; \mathrm{C}{:}\,d = 0.155\ \mathrm{nm};\quad \mathrm{Si}{:}\,d = 0.235\ \mathrm{nm};\quad \mathrm{Ge}{:}\,d = 0.245\ \mathrm{nm}. \tag{1}$$

Problem 2

Draw the diagonal cross section of the Brillouin zone of Fig. **8**-1.1, as shown in Fig. 1. The lattice constant of the reciprocal lattice, i.e., the side of the cube in Fig. **8**-1.1, is $2\mathbf{a}^\#$:

$$2\mathbf{a}^\# = 4\pi a^{-1} = (4\pi/0.361)\ \mathrm{nm}^{-1} = 34.8\ \mathrm{nm}^{-1};\quad 2\sqrt{2}\mathbf{a}^\# = 49.2\ \mathrm{nm}^{-1}. \tag{2}$$

The point marked '1' in the figure is the first neighbour in the first octant of Fig. **8**-1.1. Draw the perpendicular bisector to $\Gamma 1$ and do the same for the other lines that join Γ to the first neighbours. Notice that X, U, L, and K are precisely the points so labelled in Fig. **9**-1.3. X' is a point belonging to the star of X in an adjoining Brillouin zone. In order to obtain the radius r of the sphere required it is important to remember that the volume $\Omega^\#$ of the Brillouin zone must be obtained from the volume of the *primitive cell*, which is one quarter of the unit cell:

4.8 $$\tfrac{4}{3}\pi r^3 = \tfrac{1}{2}\Omega^\# = \tfrac{1}{2}(2\pi)^3(\tfrac{1}{4}\Omega)^{-1} = 16\pi^3 a^{-3}\ \mathrm{nm}^{-3}. \tag{3}$$

3 $$r = (12\pi^2)^{1/3} a^{-1} = 4.91\,a^{-1}\ \mathrm{nm}^{-1} = 13.6\ \mathrm{nm}^{-1}. \tag{4}$$

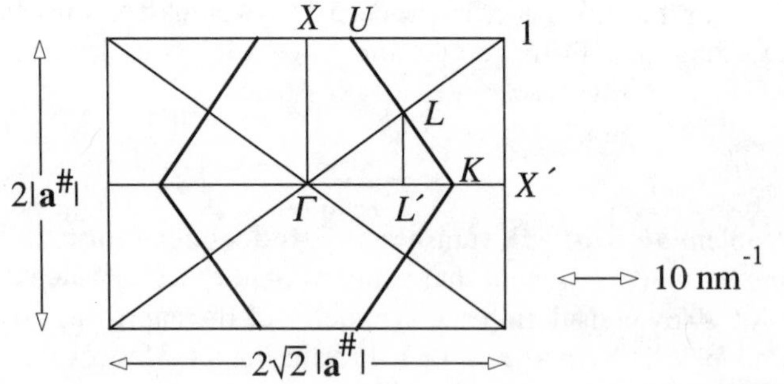

Fig. 15-9.1. Cross-section of the face-centred cubic Brillouin zone of Fig. **8**-1.1.

Problem 3

In units of $2|\mathbf{a}^{\#}|$,

F1 $$\Gamma 1 = \frac{\sqrt{3}}{2}, \Gamma L = \frac{\sqrt{3}}{4}, \Gamma K = \frac{3\sqrt{2}}{8}, \Gamma L' = \frac{\sqrt{2}}{4},$$

$$XU = \Gamma X' - \Gamma K = \frac{\sqrt{2}}{8}. \tag{5}$$

In order to get components of the **k** vectors along the *orthogonal* directions $\mathbf{a}^{\#}, \mathbf{b}^{\#}, \mathbf{c}^{\#}$ of Fig. **8**-1.1, XU, $\Gamma L'$, and ΓK must be multiplied by cos 45° to get the $\mathbf{b}^{\#}$, $\mathbf{c}^{\#}$ components of U, L, and K, respectively. Like U in Fig. 1, W is taken on the top face of Fig. **8**-1.1 and it differs from X only in its $\mathbf{b}^{\#}$ component. In units of $2|\mathbf{a}^{\#}|$:

$$\Gamma = [000], \quad X = [00\tfrac{1}{2}], \quad L = [\tfrac{1}{4}\tfrac{1}{4}\tfrac{1}{4}], \quad U = [\tfrac{1}{8}\tfrac{1}{8}\tfrac{1}{2}],$$
$$K = [\tfrac{3}{8}\tfrac{3}{8}0], \; W = [0\tfrac{1}{4}\tfrac{1}{2}]. \tag{6}$$

6 $$\Gamma X = [00\tfrac{1}{2}], \quad XW = [0\tfrac{1}{4}0], \quad WL = [\tfrac{1}{4}0\bar{\tfrac{1}{4}}],$$
$$\Gamma L = [\tfrac{1}{4}\tfrac{1}{4}\tfrac{1}{4}], \quad \Gamma K = [\tfrac{3}{8}\tfrac{3}{8}0]. \tag{7}$$

7 $$\Gamma X = \tfrac{1}{2}2|\mathbf{a}^{\#}| = 17.4\ \mathrm{nm}^{-1}, \qquad XW = \tfrac{1}{4}2|\mathbf{a}^{\#}| = 8.7\ \mathrm{nm}^{-1}, \tag{8}$$

7 $$WL = \frac{1}{\sqrt{8}}2|\mathbf{a}^{\#}| = 12.3\ \mathrm{nm}^{-1}, \qquad \Gamma L = \frac{\sqrt{3}}{4}2|\mathbf{a}^{\#}| = 15.1\ \mathrm{nm}^{-1}, \tag{9}$$

7 $$\Gamma K = \frac{3\sqrt{2}}{8}2|\mathbf{a}^{\#}| = 18.5\ \mathrm{nm}^{-1}. \tag{10}$$

4,9 $$r/\Gamma L = 0.90. \tag{11}$$

Problem 4

8, 9, 10|**1**-9.4 $$E_X = 11.5\ \mathrm{eV}, \quad E_L = 8.7\ \mathrm{eV}, \quad E_K = 13.0\ \mathrm{eV}. \tag{12}$$

On comparing these values with those in Fig. **9**-2.1, remember that they refer to the bottom of the band, that is to the lowest eigenvalue at Γ. About 1.6 eV should thus be added to the values in (12) for this purpose.

Problem 5

Use Problem **8**-8.1. Notice that the line UL is parallel to a binary axis which joins two vertices of the cube in Fig. **8**-1.1.

10 Problems 10-5

Problem 1

On acting with $\{E|\mathbf{t}\}$ on $H_{\mathbf{kk}'}$ in (**10**-5.1), this operator transforms all three factors in it. The operator $\mathbb{H}$ transforms as required by (**2**-2.13) but, because $\{E\mathbf{t}\}$ is a symmetry operator, $\mathbb{H}$ is left invariant (see eqn **2**-3.1). As regards the first factor in (**10**-5.1), it must be remembered that it is implicitly conjugated, so that any constant factor arising from this term which is taken out of the integral must be conjugated:

10-5.1 $$\{E|\mathbf{t}\}\langle\psi_{\mathbf{k}}(\mathbf{r})|\mathbb{H}|\psi_{\mathbf{k}'}(\mathbf{r})\rangle = \langle\{E|\mathbf{t}\}\psi_{\mathbf{k}}(\mathbf{r})|\mathbb{H}|\{E|\mathbf{t}\}\psi_{\mathbf{k}'}(\mathbf{r})\rangle \quad (1)$$

5-6.2 $$= \exp(\mathrm{i}\mathbf{k}\cdot\mathbf{t})\exp(-\mathrm{i}\mathbf{k}'\cdot\mathbf{t})\langle\psi_{\mathbf{k}}(\mathbf{r})|\mathbb{H}|\psi_{\mathbf{k}'}(\mathbf{r})\rangle. \quad (2)$$

No integral, however, can be changed by a symmetry operation (otherwise a volume, for example, might be altered by such an operation, which is absurd). Thus (R2) must equal the original matrix element, which is only possible when $\mathbf{k}$ and $\mathbf{k}'$ are equal. In all other cases the equality of (R2) to the original matrix element requires the vanishing of the latter. Clearly, this proof applies in precisely the same manner to the overlap integral in (**10**-5.1). Since the plane waves $\exp(\mathrm{i}\mathbf{k}\cdot\mathbf{r})$ satisfy exactly the same eigenvalue equation of the translations as the Bloch functions do (see Problem **5**-7.3), the same result is valid for them. A direct proof of the orthogonality of the plane waves was already given in Problem **5**-7.4.

This result is a particular case of a general theorem of group theory which states that functions which belong to different irreducible representations of a group are orthogonal (that is that the overlap integral between them vanishes), and that the matrix element between them is zero.

Problem 2

It follows from Problem 1 that all matrix elements and overlaps vanish for $\mathbf{k}$ different from $\mathbf{k}'$. Moreover, for $\mathbf{k}$ equal to $\mathbf{k}'$, $S_{\mathbf{kk}'}$ becomes unity, owing to the normalization of eqn (**10**-1.3). Thus the secular determinant (**10**-1.7) contains only diagonal elements $H_{\mathbf{kk}} - \varepsilon$, n in number, whence the product of such terms for varying $\mathbf{k}$ must vanish. The solution of this equation gives n eigenvalues, each equal to $H_{\mathbf{kk}}$ (remember that eqn **10**-1.5 contains n plane waves, i.e. n values of $\mathbf{k}$):

10-1.8, **10**-1.1 $$\varepsilon_{\mathbf{k}} = H_{\mathbf{kk}} = \langle\mathbf{k}|\mathbb{H}|\mathbf{k}\rangle = \langle\mathbf{k}|\mathbb{T}|\mathbf{k}\rangle + \langle\mathbf{k}|\mathbb{V}|\mathbf{k}\rangle. \quad (3)$$

10-1.4 $$\mathbb{T}|\mathbf{k}\rangle = E_{\mathbf{k}}|\mathbf{k}\rangle. \quad (4)$$

Write $\mathbb{V}$ as $V(\mathbf{r})$ as in (**10**-1.18):

$$\langle \mathbf{k} | \mathbb{V} | \mathbf{k} \rangle = V^{-1} \int_V \exp(-\mathrm{i}\mathbf{k}\cdot\mathbf{r})\, V(\mathbf{r}) \exp(\mathrm{i}\mathbf{k}\cdot\mathbf{r})\, \mathrm{d}v \tag{5}$$

$$= V^{-1} \int_V V(\mathbf{r})\, \mathrm{d}v =_{\mathrm{def}} \bar{V}. \tag{6}$$

Since $\bar{V}$ does not depend on $\mathbf{k}$, we can shift the energy origin and get rid of this term:

4|3
$$\varepsilon_{\mathbf{k}} = E_{\mathbf{k}} \langle \mathbf{k} | \mathbf{k} \rangle = E_{\mathbf{k}}. \tag{7}$$

Problem 3

As in Fig. **10**-1.1, take a value of **k** in the second Brillouin zone equal to, say, $\mathbf{a}^{\#} - \boldsymbol{\delta}$. This will give you a **k** at $-\boldsymbol{\delta}$ with an energy in the second band near the top of the figure. This state will interact with **k** equal to $-\boldsymbol{\delta} - \mathbf{a}^{\#}$, which is in the third zone (third band: draw this addition to the figure and you will see that the two states for **k** equal to $-\boldsymbol{\delta}$ are near in energy).

Problem 4

10-2.6
$$\psi_{\mathbf{k}}(\mathbf{r}) = \sum_m \exp(\mathrm{i}\mathbf{k}\cdot\mathbf{t}_m) | m \rangle \tag{8}$$

$$\{E|\mathbf{t}_n\}\, \psi_{\mathbf{k}}(\mathbf{r}) = \sum_m \exp(\mathrm{i}\mathbf{k}\cdot\mathbf{t}_m) \{E|\mathbf{t}_n\}\, | m \rangle \tag{9}$$

$$= \sum_m \exp(\mathrm{i}\mathbf{k}\cdot\mathbf{t}_m)\, | m + n \rangle \tag{10}$$

$$= \exp(-\mathrm{i}\mathbf{k}\cdot\mathbf{t}_n) \sum_{m+n} \exp\{\mathrm{i}\mathbf{k}\cdot(\mathbf{t}_m + \mathbf{t}_n)\}\, | m + n \rangle \tag{11}$$

$$= \exp(-\mathrm{i}\mathbf{k}\cdot\mathbf{t}_n) \psi_{\mathbf{k}}(\mathbf{r}). \tag{12}$$

Compare (12) with (**5**-6.2).

Problem 5

It follows at once from the fact that these Bloch sums are eigenfunctions of the translations belonging to different eigenvalues for different values of **k**.

11 Problems 11-5

Problem 1

11-2.19
$$\mathbf{z_q} = N^{-1/2} \sum_{\mathbf{t}'} \exp(\mathrm{i}\mathbf{q}\cdot\mathbf{t}')\mathbf{s}_{\mathbf{t}'}. \tag{1}$$

2-2.7
$$t\mathbf{z_q} = N^{-1/2} \sum_{\mathbf{t}'} \exp\{\mathrm{i}\mathbf{q}\cdot(\mathbf{t}' - \mathbf{t})\}\mathbf{s}_{\mathbf{t}'} \tag{2}$$

$$= \exp(-\mathrm{i}\mathbf{q}\cdot\mathbf{t})N^{-1/2} \sum_{\mathbf{t}'} \exp(\mathrm{i}\mathbf{q}\cdot\mathbf{t}')\mathbf{s}_{\mathbf{t}'} = \exp(-\mathrm{i}\mathbf{q}\cdot\mathbf{t})\mathbf{z_q}. \tag{3}$$

Problem 2

11-2.19
$$\mathbf{z_q} = N^{-1/2} \sum_{\mathbf{t}'} \chi(\mathbf{t}'|_{\mathbf{q}}\hat{T})^* \, \mathbf{s}_{\mathbf{t}'}. \tag{4}$$

$$\sum_{\mathbf{q}} \chi(\mathbf{t}|_{\mathbf{q}}\hat{T})^* \, \mathbf{z_q} = N^{-1/2} \sum_{\mathbf{q}} \chi(\mathbf{t}|_{\mathbf{q}}\hat{T})^* \sum_{\mathbf{t}'} \chi(\mathbf{t}'|_{\mathbf{q}}\hat{T})^* \mathbf{s}_{\mathbf{t}'} \tag{5}$$

$$= N^{-1/2} \sum_{\mathbf{t}'} \left\{ \sum_{\mathbf{q}} \chi(\mathbf{t}|_{\mathbf{q}}\hat{T})^* \, \chi(\mathbf{t}'|_{\mathbf{q}}\hat{T})^* \right\} \mathbf{s}_{\mathbf{t}'} \tag{6}$$

2-5.3, **5**-1.8
$$= N^{-1/2} \sum_{\mathbf{t}'} N \delta_{\mathbf{t}\mathbf{t}'} \mathbf{s}_{\mathbf{t}'} = N^{1/2} \mathbf{s}_{\mathbf{t}}. \tag{7}$$

7
$$\mathbf{s_t} = N^{-1/2} \sum_{\mathbf{q}} \chi(\mathbf{t}|_{\mathbf{q}}\hat{T})^* \mathbf{z_q}. \tag{8}$$

Problem 3

11-2.15
$$\boldsymbol{T} = \tfrac{1}{2} \sum_{\mathbf{t}i} \dot{s}^*_{\mathbf{t}i} \, \dot{s}_{\mathbf{t}i} \tag{9}$$

$$= \tfrac{1}{2} N^{-1} \sum_{\mathbf{q}\mathbf{q}'i} \chi(\mathbf{t}|_{\mathbf{q}}\hat{T}) \, \chi(\mathbf{t}|_{\mathbf{q}'}\hat{T})^* \, \dot{z}^*_{\mathbf{q}i} \, \dot{z}_{\mathbf{q}'i} \tag{10}$$

2-5.2
$$= \tfrac{1}{2} N^{-1} \sum_{\mathbf{q}\mathbf{q}'i} N \delta_{\mathbf{q}\mathbf{q}'} \dot{z}^*_{\mathbf{q}i} \dot{z}_{\mathbf{q}'i} = \tfrac{1}{2} \sum_{\mathbf{q}i} \dot{z}^*_{\mathbf{q}i} \, \dot{z}_{\mathbf{q}i}. \tag{11}$$

Problem 4

Apply the classification of Table **8**-4.1, noticing the degeneracies and the groups of **q**. At Γ and $X(q = \pi/a)$ the normal modes are classified, besides **q**, by the irreducible representations of $\mathbf{C}_i$ (small representations) as gerade (symmetrical) and ungerade (antisymmetrical) with respect to the inversion at the centre of the cell. Thus, in principle, the bands at Γ and X are either symmetrical or antisymmetrical. (But see below.) The normal coordinate $\mathbf{z_q}(\mathbf{r})$ gets multiplied by $\exp(-\mathrm{i}\mathbf{q}\cdot\mathbf{t})$ under a translation by **t** (see eqn 3). Take **t** equal to **a** in order to see how the normal coordinate changes from cell to cell. For $\Gamma(q = 0)$ and $X(q = \pi/a)$, $\exp(-\mathrm{i}\mathbf{q}\cdot\mathbf{a})$ equals $+1$ and -1 respectively.

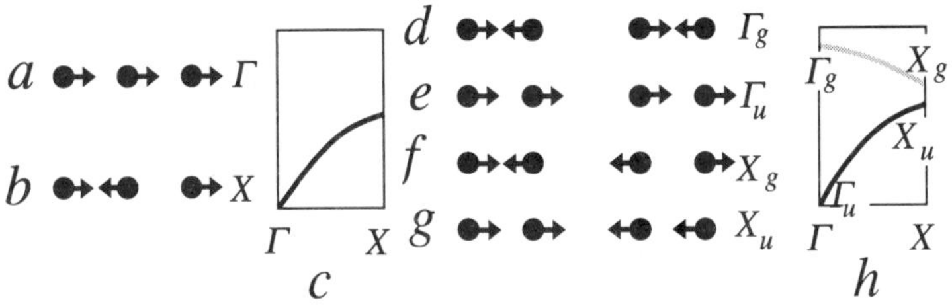

Fig. 15-11.1. Normal vibrations and phonon bands in strictly one-dimensional chains, simple (*a* to *c*) and with a basis (*d* to *h*). The black and grey curves in *h* are the acoustic and optic branches respectively. The vertical scales of the figures for the bands indicate frequencies.

In a strictly one-dimensional example normal coordinates that are gerade with respect to the centre of inversion cannot be constructed if there is no basis, as can be seen in Fig. 1. (Notice the difference with the Bloch functions, for which *s* type gerade cell functions can be chosen.) In comparing Fig. 1*a* with 1*b*, it can be seen at once that Fig. 1*a* is a crystal translation as a whole and that in going from *a* to *b* the energy must go up, as shown in *c*. In deciding on the qualitative shape of this picture notice the remark at the very end of § **11**-4. The case with a basis (inversion centre to be taken at the centre of the bond between the two atoms of the basis) is treated in the same way in Fig. 1 but now gerade and ungerade normal coordinates can be constructed. Assume that the bond between the atoms of the basis is the shortest one, as shown in the figure. Γ_u is a translation of the chain as a whole and it is the lowest energy state. In the gerade modes the atoms of the basis (shortest bond) are out of phase and thus these states are higher in energy (optic branch). In going from *e* to *g* in Fig. 1 along the low energy band (acoustic branch) the energy increases because the atoms of the longer bond become out of phase. In going from *d* to *f* the atoms on the long bond become in phase, which means that the energy must go down.

12 Problems 12-4

Problem 1

Call v the volume of the sphere, r its radius, Ω the volume of the primitive cell of the face-centred cubic lattice, v_{BZ} the volume of the Brillouin zone, and **N** the number of electrons required. Because v_{BZ} contains two electrons per atom,

$$\mathbf{N} = 2v/v_{BZ}. \tag{1}$$

The radius r is ΓL in Problem **9**-4.3:

9.9 $$r = \sqrt{(3)}\pi a^{-1} \qquad \Rightarrow \qquad v = 4\pi^4\sqrt{(3)}a^{-3}. \tag{2}$$

,4.8 $$\Omega = \tfrac{1}{4}a^3 \Rightarrow v_{BZ} = \Omega^{\#} = (2\pi)^3\Omega^{-1} = 32\pi^3 a^{-3} \Rightarrow \mathsf{N} = \frac{\sqrt{3}}{4}\pi = 1.36. \tag{3}$$

Notice an alternative way to get v_{BZ}, the volume of the primitive cell in reciprocal space, by forming the mixed triple product of the primitive vectors of the reciprocal lattice given in (**4**-7.4).

Problem 2

The character is invariant under a similarity transformation. Hence, when calculating the character of an operation, it can always be assumed that a similarity transformation has been applied which diagonalizes the matrix representative of that operation. The proof of (**12**-4.1) is immediate, since the character is the sum of the diagonal elements.

$$g\langle u_1, u_2, \ldots, u_m| = \langle u_1, u_2, \ldots, u_m| \begin{bmatrix} U_1 & & & \\ & U_2 & & \\ & & \cdots & \\ & & & U_m \end{bmatrix} \Rightarrow gu_i = U_i u_i. \tag{4}$$

The effect of g on the basis $\langle u| \otimes \langle v|$ of (**12**-4.2) is

,4 $$g\, u_i v_j = gu_i\, gv_j = U_i u_i\, V_j v_j = U_i V_j\, u_i v_j, \tag{5}$$

which shows that the corresponding matrix representative on this basis is diagonal:

5, **12**-4.1 $$\chi(g\,|\,u \otimes v) = \sum_{ij} U_i V_j = \sum_i U_i \sum_j V_j = \chi(g\,|\,u)\,\chi(g\,|\,v). \tag{6}$$

When $\langle u|$ and $\langle v|$ are identical the basis $\langle u| \otimes \langle v|$ on (R**12**-4.2) is redundant, since $u_i v_j$ and $u_j v_i$ are both equal to $u_i u_j$ and thus identical. The redundancies are eliminated in the new basis

12-4.4 $$\langle u_1, \ldots, u_m| \mathbin{\bar{\otimes}} \langle u_1, \ldots, u_m| =_{\text{def}} \langle u_i u_j|, \quad i = 1, \ldots, m; \quad j \geq i. \tag{7}$$

5 $$g\, u_i u_j = U_i U_j\, u_i u_j. \tag{8}$$

8 $$\chi(g\,|\,u \mathbin{\bar{\otimes}} u) = \sum_{i,j \geq i} U_i U_j = \sum_i U_i^2 + \sum_{i,j > i} U_i U_j. \tag{9}$$

6 $$\chi(g\,|\,u \otimes u) = \sum_{ij} U_i U_j = \sum_i U_i^2 + 2\sum_{i,j>i} U_i U_j. \tag{10}$$

9, 10 $$2\chi(g\,|\,u \mathbin{\bar{\otimes}} u) = \chi(g\,|\,u \otimes u) + \sum_i U_i^2. \tag{11}$$

Identify the summation in (11) as follows. From (**2**-4.6), the matrix representative of g^2 is the product of the matrix representative for g times itself. From (4) this is the diagonal matrix with elements U_i^2. Thus the summation is the character of g^2 in the basis $\langle u|$. Also, from (6), $\chi(g\,|\,u \otimes u)$ is $\{\chi(g\,|\,u)\}^2$. Thus (**12**-4.5) follows.

Problem 3

From **12**-2.11, with k_F equal to $\dfrac{\pi}{2a}$,

$$\text{matrix for } [E\,|\,\mathbf{t}\}: \begin{bmatrix} e^{-i\frac{\pi}{2a}t} & \\ & e^{i\frac{\pi}{2a}t} \end{bmatrix}, \quad \text{matrix for } \{E\,|\,\mathbf{t}\}^2: \begin{bmatrix} e^{-i\frac{\pi}{2a}2t} & \\ & e^{i\frac{\pi}{2a}2t} \end{bmatrix}. \tag{12}$$

12 $$\chi(t\,|\,\psi) = 2\cos(\pi t/2a), \qquad \chi(t^2\,|\,\psi) = 2\cos(\pi t/a)\,. \tag{13}$$

13 $$\{\chi(t\,|\,\psi)\}^2 = 4\cos^2(\pi t/2a) = 2\{1 + \cos(\pi t/a)\}\,. \tag{14}$$

13,14 | **12**-4.5 $$\chi(t\,|\,\psi\,\bar{\otimes}\,\psi) = 1 + 2\cos(\pi t/a)\,. \tag{15}$$

Since $\mathbf{q} = \pi/a$, the character of t corresponding to a representation like (**12**-2.11) with $\mathbf{q}$ substituted for $\mathbf{k}$, is $2\cos(\pi t/a)$. Since this character is contained in (15) the integral (**12**-2.10) does not vanish.

When the inversion is included the characters of the operations which contain it vanish from Problem **8**-8.9.

Problem 4

We show in Fig. 1 the normal vibration required, which is symmetrical with respect to lattice translations and thus belongs to Γ. It is easy to verify, first, that the inversion at the mid-point of the basis is a symmetry operation of the space group and, secondly that the normal vibration shown in Fig. 1 is gerade.

The transition probability integral which has to be worked out is, as in (**12**-2.10),

$$I = \int \psi^*_{\mathbf{k}_F} U_{\mathbf{q}} \psi_{\mathbf{k}_F}\, d\tau\,, \qquad \mathbf{k}_F = \pi/a\,, \qquad \mathbf{q} = \mathbf{0} \in \Gamma_g\,. \tag{16}$$

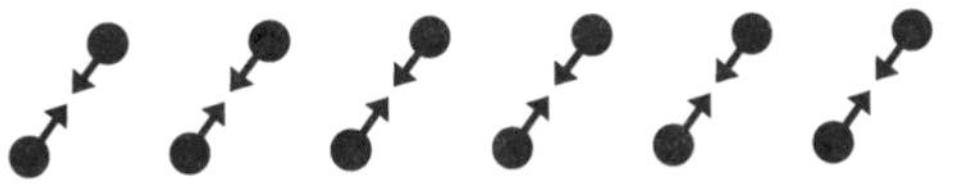

Γ_g

Fig. 15-12.1. The normal vibration required to drive the Peierls reconstruction sketched in Fig. **12**-3.2.

The Fermi level here is obtained from Fig. **12**-3.3. In principle, because $\mathbf{k}_F$ is at the Brillouin zone edge, $\mathbf{k}_F$ and $-\mathbf{k}_F$ are equivalent and we should have a one-dimensional representation. The argument given in § **12**-3, however, shows that there is an extra degeneracy introduced by the conjugator operator, so that we must have a two-dimensional basis, exactly as in (**12**-2.11). Equation (**12**-2.14) which leads to the direct product of $\psi^*_{\mathbf{k}_F}$ with $\psi_{\mathbf{k}_F}$ will have the same form, but since E_F is now π/a, (instead of $\pi/2a$), the exponentials in it will be $\exp(-\mathrm{i}2\pi t/a)$ and $\exp(\mathrm{i}2\pi t/a)$. The two diagonal elements equal to the unity remain but, when the redundancy in the basis is eliminated, only one of them will be left, just as before. The character of the representation to which $\mathbf{q}$ belongs is clearly unity, and it is thus contained in the direct product of $\psi^*_{\mathbf{k}_F}$ with $\psi_{\mathbf{k}_F}$. It follows that I should not vanish, unless the effect of the inversion symmetry forces it to do so. The group of the **k** vector $\mathbf{k}_F$ at the edge of the Brillouin zone contains the inversion, whereby the corresponding Bloch functions have to be classified by the two irreducible representations of the group C_i, that is, they must be gerade or ungerade. In either case, the product $\psi^*_{\mathbf{k}_F}\psi_{\mathbf{k}_F}$ must be gerade, and this is also the symmetry of $\psi_{\mathbf{q}}$, so that I does not vanish.

Problem 5

The screw axis corresponds to a binary rotation around the **c** axis accompanied by a translation by **w**, where **w** is $\frac{1}{2}\mathbf{c}$. The proof required is exactly like the one in § **12**-3, except for the following changes. In (**12**-3.2) take **k** on the top surface of the hcp Brillouin zone as $[00\frac{\pi}{c}]$. Therefore, because the first two components will not matter, take **t** as $[00c]$ or **c**. The operation γ will now be $\{C_2 \mid \frac{1}{2}\mathbf{c}\}$ with C_2 parallel to **c**. It is immediate that $C_2\mathbf{c}$ equals **c**, as required in (**12**-3.6) and that γ^2 equals $\{E \mid \mathbf{c}\}$ in (**12**-3.16). The rest of the proof in § **12**-3 requires only minor changes of notation. There is therefore an essential double degeneracy at the top surface of the hcp Brillouin zone and all bands stick together in pairs on that face. Because there is no energy discontinuity between the successive bands a double Brillouin zone must be used for the hcp bands along the $\mathbf{c}^{\#}$ (parallel to **c**) direction. The first 'band' incorporating the first and second bands will look like the band in Fig. **12**-3.3 (with c substituted for a) and the third and fourth bands will form a 'second' band related to the first as the second band is in Fig. **8**-4.2. Draw the corresponding figure. In the hcp structure there are two atoms per primitive cell so that we shall have 8 electrons in the primitive cell, whence we must fill the Brillouin zone four times, that is the first four bands, or the first two double bands, must be filled. The Fermi energy in your figure must be placed at the top of the 'second' (really fourth) band.

Three things happen in the transition from hcp to fcc. (i) The sticking together of the bands disappears. (ii) The period along the [111] direction (which replaces the former c direction) becomes $\frac{1}{2}c$, whence, with reference to the hcp value of c, the first Brillouin zone goes now from $-2\pi/c$ to $2\pi/c$. (Draw the new figure for the bands in the fcc structure, which should be identical to your previous figure, noting that your base line is now the entire extension of the first Brillouin zone along the [111] direction.) (iii) The primitive cell contains only one atom. Since we have to place four electrons, the first two bands must be full, which will put the Fermi energy at the top of the second band, with precisely the same value as before. It is thus clear that this change of phase cannot be explained by anything like a Peierls process. Other effects must be considered. For instance, because the close-packed layers are all registering in the fcc structure (whereas the atoms of the intermediate layer of the hcp structure sit cozily in the interstitial places of the other two) it is possible that on the intermediate layer sliding in the fcc structure the interplanar distance might expand (crystal relaxation) and this factor should contribute to the energy balance. This crystal relaxation would benefit the hcp structure and it is likely to be more important at low temperatures (because at high temperatures the interplanar distances are in any case larger) so that this effect might correlate in some cases with the stability of the hcp structure with respect to the fcc one in some fourvalent metals, like zirconium. Many other effects, however, are also important so that this type of argument must be taken with a large pinch of salt.

Problem 6

You must prove that the representation to which $\psi_i \psi_j$ belongs contains on reduction the totally symmetric representation at least once. This is what it means to 'transform like the totally symmetrical representation'. Their product may contain other terms which transform otherwise but such terms are not important since, for example, they cannot contribute anything to transition probability integrals. Call $\hat{G}$ the reducible representation in question. Because of the conservation of characters on reduction, the characters of this representation must be sums of characters of irreducible representations of the group ${}^i\hat{G}$, each repeated h_i times. It is this number that we must find (it must be different from zero when ${}^i\hat{G}$ is the totally symmetrical representation).

$$\chi(g \mid \hat{G}) = \sum_i h_i \chi(g \mid {}^i\hat{G}) \tag{17}$$

17

$$\sum_g \chi(g \mid {}^j\hat{G})^* \chi(g \mid \hat{G}) = \sum_{ig} h_i \chi(g \mid {}^j\hat{G})^* \chi(g \mid {}^i\hat{G}) \tag{18}$$

2-5.2 | 18

$$= \sum_i h_i n \delta_{ij} = h_j n. \tag{19}$$

Take ${}^j\hat{G}$ to be the totally symmetrical representation for which $\chi(g \mid {}^j\hat{G})$ equals unity for all g:

19

$$h_j = n^{-1} \sum_g \chi(g \mid \hat{G}) \tag{20}$$

The representation to which $\psi_i\psi_j$ belongs is the direct product representation for which the character is:

12-4.3 $$\chi(g\,|\,\hat{G}) = \chi(g\,|\,{}^i\hat{G})\chi(g\,|\,{}^j\hat{G}). \tag{21}$$

21|20 $$h_j = n^{-1}\sum_g \chi(g\,|\,{}^i\hat{G})\chi(g\,|\,{}^j\hat{G}). \tag{22}$$

Since from (2-5.2) the summation on (R22) equals zero unless ${}^i\hat{G}$ and ${}^j\hat{G}$ are complex conjugate and n in the latter case, it follows that it is only when ψ_i and ψ_j belong to conjugate representations that (R22) does not vanish, being unity otherwise.

13 Problems 13-4

Problem 1

13-1.10 $$h\varphi_i = hh_i\varphi_0 = h_j\varphi_0 = \varphi_i, \tag{1}$$

13-1.8 $$h_i\varphi_0 = \varphi_i. \tag{2}$$

1,2;1 $$h_j\varphi_0 = h_i\varphi_0 \quad\Rightarrow\quad h_i^{-1}h_j\varphi_0 = \varphi_0 \quad\Rightarrow\quad h_i^{-1}hh_i\varphi_0 = \varphi_0. \tag{3}$$

13-1.9 $$h_i^{-1}hh_i = E \quad\Rightarrow\quad h = h_ih_i^{-1} = E. \tag{4}$$

Problem 2

13-1.11,13-1.12 $$h_ih_{N-i} = h^ih^{N-i} = h^N = E \quad\Rightarrow\quad h_i^{-1} = h^{N-i}. \tag{5}$$

Problem 3

Follow the hint.

14 Problems 14-5

Problem 1

Call in this and in all the following problems σ_{v1}, σ_{v2}, σ_{v3}, the reflection planes through the mid-points of the bonds $a'a$, $c'c$, and $b'b$ respectively. C_3^+ and C_3^- are respectively counterclockwise and clockwise. Apply (2-5.4) to the orbital a for which purpose you must first find, from Fig. 14-2.1, its transform under all the group operations, as shown below:

E	C_3^+	C_3^-	σ_{v1}	σ_{v2}	σ_{v3}
a	c	b	a'	b'	c'

When you multiply the functions in the second row here by the characters in (**14**-5.1) and (**14**-5.2), respectively, you will get the first two rows of Table **14**-2.1.

Problem 2

$$E = \langle b - c + b' - c' | \mathbb{H} | b - c + b' - c' \rangle / \langle b - c + b' - c' | b - c + b' - c' \rangle. \tag{1}$$

14-2.4
14-2.5
$$\langle b - c + b' - c' | b - c + b' - c' \rangle =$$

$$= \langle b | b \rangle + \langle c | c \rangle + \langle b' | b' \rangle + \langle c' | c' \rangle = 4. \tag{2}$$

14-2.1
14-2.2, **14**-2.3
$$\langle b - c + b' - c' | \mathbb{H} | b - c + b' - c' \rangle =$$

$$= 4\langle b | \mathbb{H} | b \rangle + 4\langle b | \mathbb{H} | b' \rangle - 2\langle c | \mathbb{H} | b' \rangle = 4\alpha + 4\beta - 2\gamma. \tag{3}$$

Equations (2) and (3) into (1) yield the result.

Problem 3

Form a table of transforms of the orbitals under the required symmetry operations as follows:

	a	a'	b	b'	c	c'
$\mathbb{i}$	a'	a	c'	c	b'	b
σ_{v1}	a'	a	c'	c	b'	b

Problem 4

Since there are three primitive cells in the chain, there are three **k** vectors in the Brillouin zone, labelled as in (**7**-3.3) by κ equal to $-1, 0$, and 1. A drawing like **7**-3.1 will immediately yield the results required. Notice that X and $-X$ are not equivalent since they differ by a **k** vector with κ equal to 2, which is not a vector of the reciprocal lattice.

Problem 5

The hint is self-explanatory. The exponentials are the characters of the translation group. Translation groups with periodic boundary conditions are cyclic and since T is of order three it must be isomorphic in this case to the cyclic group $\mathbf{C}_3$ of Table **2**-7.1.

Problem 6

It is sufficient to follow the hints in order to generate the required orbitals, remembering that $\omega + \omega^*$ equals -1. As regards the orthogonality of the two orbitals stated you will find, after some simplification taking account of symmetry,

$$\langle b-c+b'-c'|b-c-b'+c'\rangle =$$
$$\langle b|b\rangle + \langle c|c\rangle - \langle b'|b'\rangle - \langle c'|c'\rangle - 2\langle b|c\rangle + 2\langle b'|c'\rangle = 0, \quad (4)$$

where, again, the fact has to be used that $\langle b|b\rangle$ and $\langle b'|b'\rangle$ are equal by symmetry, and other similar results.

Orthogonality of two molecular orbitals belonging to the same irreducible representation is achieved for the following reason. The two E bases are two (doubly degenerate) Bloch functions belonging to the same **k** and to the same small representation (remember that the group of X is the identity). Although they are not necessarily orthogonal, therefore, on account of space group symmetry, they can be made orthogonal through suitable choice of the cell functions. Notice that the cell function for the first pair labelled E in Table **14**-2.1 is symmetric with respect to inversion at all bond centres (bonding), whereas the cell function of the second pair is antisymmetric (antibonding). This confirms the statement (see Problem 7) that, although the bonding-antibonding property cannot be extended necessarily throughout the whole band, it is nevertheless possible to do so by continuity.

Problem 7

(i) The g and u symmetry (with respect to the inversion centre marked in Fig. **14**-2.1b or, what is the same to σ_{v1}) of the degenerate levels E at X is not significant because the inversion does not belong to the group of its **k** vector. (See Problem 4.) Notice, in fact, that each E pair has not a definite g, u symmetry, one function of the basis being g and the other u. (This is the trick that we used in order to make them orthogonal.) (ii) Inversion symmetry obtains for Γ: the gerade and ungerade functions are bonding and antibonding respectively and by continuity of the cell functions this property is assigned to the whole of each band. Notice that, actually, a and a', b and b', c and c', appear with the same sign in the first pair of E functions in Table **14**-2.1, and with different sign in the second pair. Thus the first and second pair are bonding and antibonding respectively.

Problem 8

Instead of (**14**-5.6) to (**14**-5.9) you will be landed with

$$(a+a') + \omega\,(b+b') + \omega^*(c+c'), \quad (5)$$
$$(a-a') + \omega\,(b-b') + \omega^*(c-c'), \quad (6)$$
$$(a+a') + \omega^*(b+b') + \omega\,(c+c'), \quad (7)$$
$$(a-a') + \omega^*(b-b') + \omega\,(c-c'). \quad (8)$$

Sums and differences of (5) and (7) or (6) and (8) will now produce nonsense. The method does not work because g, u symmetry is only valid for Γ.

Problem 9

On neglecting all interaction terms between non neighbouring orbitals,

F14-2.1c $$\langle a + a' | \mathbb{H} | a + a' \rangle = \langle a | \mathbb{H} | a \rangle + \langle a' | \mathbb{H} | a' \rangle = 2\alpha, \quad (9)$$

F14-2.1c $$\langle a + a' | a + a' \rangle = \langle a | a \rangle + \langle a' | a' \rangle = 2. \quad (10)$$

A factor of 2 can be taken out of the first row of the determinant.

Problem 10

Follow the hints.

References

Altmann, S. L. (1958). Equivalent functions: hybrids and Wannier functions. *Proceedings of the Cambridge Philosophical Society*, **54**, 197–206.

Altmann, S. L. (1974). The cellular method for metals. In *Orbital theories of molecules and solids* (ed. N. H. March), pp. 30–94. Clarendon Press, Oxford.

Altmann, S. L. (1977). *Induced representations in crystals and molecules. Point, space, and nonrigid molecule groups.* Academic Press, London.

Altmann, S. L. (1982). A note on the Peierls reconstruction of dislocation cores. *Journal of Physics C: Solid State Physics*, **15**, 907–11.

Altmann, S. L., Lapiccirella, A., Lodge, K. W., and Tomassini, N. (1982). A valence force field for the silicon crystal. *Journal of Physics C: Solid State Physics*, **15**, 5581–91.

Antončík, E. (1959). Approximate formulation of the orthogonalized plane-wave method. *Journal of the Physics and Chemistry of Solids*, **10**, 314–20.

Apostol, M. and Baldea, I. (1982). Electron–phonon coupling in one dimension. *Journal of Physics C: Solid State Physics*, **15**, 3319–31.

Arfken, G. (1985). *Mathematical methods for physicists*, (3rd edn). Academic Press, Orlando.

Ashcroft, N. W. and Mermin, N. D. (1976). *Solid state physics.* Holt, Rinehart and Winston, New York.

Atkins, P. W. (1983). *Molecular quantum mechanics*, (2nd edn). Oxford University Press.

Atkins, P. W. (1990). *Physical chemistry*, (4th edn). Oxford University Press.

Austin, B. J., Heine, V., and Sham, L. J. (1962). General theory of pseudopotentials. *Physical Review*, **127**, 276–82.

Bassani, F. and Celli, V. (1961). Energy-band structure of solids from a perturbation of the 'empty lattice'. *Journal of the Physics and Chemistry of Solids*, **20**, 64–75.

Blakemore, J. S. (1985). *Solid state physics*, (2nd edn). Cambridge University Press.

Boas, M. L. (1983). *Mathematical methods in the physical sciences*, (2nd edn). Wiley, New York.

Bouckaert, L. P., Smoluchowski, R., and Wigner, E. (1936). Theory of Brillouin zones and symmetry properties of wave functions in crystals. *Physical Review*, **50**, 58–67.

Božović, I. (1985). The analogue of the Jahn–Teller theorem for extended chain molecules. *Molecular Crystallography and Liquid Crystals*, **119**, 475–8.

Božović, I. B., Vujičić, M., and Herbut, F. (1978). Irreducible representations of the symmetry groups of polymer molecules. I. *Journal of Physics A: Mathematical and General Physics*, **11**, 2133–47.

Božović, I. B. and Vujičić, M. (1981). Irreducible representations of the symmetry groups of polymer molecules. II. *Journal of Physics A: Mathematical and General Physics*, **14**, 777–95.

Bradley, C. J. and Cracknell, A. P. (1972). *The mathematical theory of symmetry in solids. Representation theory for point groups and space groups.* Clarendon Press, Oxford.

Brillouin, L. (1946). *Wave propagation in periodic structures. Electric filters and crystal lattices.* McGraw Hill, New York.

Buerger, M. J. (1956). *Elementary crystallography. An introduction to the fundamental geometric features of crystals.* Wiley, New York.

Bullett, D. W. (1980). The renaissance and quantitative development of the tight-binding method. In *Solid State Physics*, Vol. **35**, (ed. H. Ehrenreich, F. Seitz, and D. Turnbull), pp. 129–214. Academic Press, New York.

Burns, G. and Glazer, A. M. (1978). *Space groups for solid state scientists.* Academic Press, New York.

Cochran, W. (1973). *The dynamics of atoms in crystals.* Arnold, London.

Cohen, M. L. and Heine, V. (1970). The fitting of pseudopotentials to experimental data and their subsequent application. In *Solid State Physics*, Vol. **24** (ed. H. Ehrenreich, F. Seitz, and D. Turnbull), pp. 37–248. Academic Press, New York.

Cornwell, J. F. (1984). *Group theory in physics*, two vols. Academic Press, London.

Cotton, F. A. (1971). *Chemical applications of group theory*, (2nd edn). Wiley-Interscience, New York.

Cottrell, A. (1988). *Introduction to the modern theory of metals.* Institute of Physics, London.

Cox, P. A. (1987). *The electronic structure and chemistry of solids.* Clarendon Press, Oxford.

Davison, S. G. and Levine, J. D. (1970). Surface states. In *Solid State Physics*, vol. **25** (ed. H. Ehrenreich, F. Seitz, and D. Turnbull), pp. 1–149. Academic Press, New York.

Dimmock, J. O. (1971). The calculation of electronic energy bands by the augmented plane wave method. In *Solid State Physics*, Vol. **26** (ed. H. Ehrenreich, F. Seitz, and D. Turnbull), pp. 103–274. Academic Press, New York.

Falicov, L. M. (1966). *Group theory and its physical applications.* University of Chicago Press.

García-Moliner, F. and Flores, F. (1979). *Introduction to the theory of solid surfaces.* Cambridge University Press.

Glazer, A. M. (1987). *The structure of crystals.* Adam Hilger, Bristol.

Gröbner, W. and Hofreiter, N. (1966). *Integraltafel. Zweiter Teil. Bestimmte Integrale*, (4th edn). Springer, Wien.

Hamermesh, M. (1962). *Group theory and its applications to physical problems.* Addison Wesley, Reading Mass.

Harris, R. A. and Falicov, L. M. (1969). Self-consistent theory of bond alternation in polyenes: normal state, charge-density waves, and spin-density waves. *Journal of Chemical Physics*, **51**, 5034–41.

Harrison, W. A. (1966). *Pseudopotentials in the theory of metals.* Benjamin, New York.

Harrison, W. A. (1980). *Electronic structure and the properties of solids.* Freeman, San Francisco.

Heine, V. (1960). *Group theory in quantum mechanics.* Pergamon Press, London.

Heine, V. (1970). The pseudopotential concept. In *Solid State Physics*, vol. **24** (ed. H. Ehrenreich, F. Seitz, and D. Turnbull), pp. 1–36. Academic Press, New York.

Heine, V. and Weaire, D. (1970). Pseudopotential theory of cohesion and structure. In *Solid State Physics*, vol. **24** (ed. H. Ehrenreich, F. Seitz, and D. Turnbull), pp. 249–463. Academic Press, New York.

Herring, C. (1940). A new method for calculating wave functions in crystals. *Physical Review*, **57,** 1169–1177.

Hoffmann, R. (1988). *Solids and surfaces: a chemist's view of bonding in extended structures.* VCH Publishers, New York.

Jansen, L. and Boon, M. (1967). *Theory of finite groups. Applications in physics. Symmetry groups of quantum mechanical systems.* North Holland, Amsterdam.

Janssen, T. (1973). *Crystallographic groups.* North-Holland, Amsterdam.

Jerome, D. and Schulz, H. J. (1982). Quasi-one dimensional conductors: the Peierls instability, pressure and fluctuation effects. In *Extended linear chain compounds*, Vol. 2 (ed. J. S. Miller), pp. 159–204. Plenum, New York.

Jones, H. (1960). *The theory of Brillouin zones and electronic states in crystals.* North-Holland, Amsterdam.

Keller, H. J. (1977). (Ed) *Chemistry and physics of one-dimensional metals.* Plenum, New York.

Kelly, A. and Groves, G. W. (1973). *Crystallography and crystal defects.* Longman, London.

Kittel, C. (1986). *Introduction to solid state physics*, (6th edn). Wiley, New York.

Knox, R. S. and Gold, A. (1964). *Symmetry in the solid state.* Benjamin, New York.

Koster, G. F. and Slater, J. C. (1954). Wave functions for impurity levels. *Physical Review*, **95,** 1167–76.

Kventsel, G. F. (1976). *Local electronic states in long polyene chains.* D. Sc. Thesis. Technion, Haifa.

Landau, L. D. and Lifshitz, E. M. (1977). *Quantum mechanics. Nonrelativistic theory*, (3rd edn). Pergamon Press, Oxford.

Lax, M. (1974). *Symmetry principles in solid state and molecular physics.* Wiley, New York.

Littlewood, P. B. and Heine, V. (1981). The effect of electron–electron interactions on the Peierls transition in metals with strong nesting of Fermi surfaces. *Journal of Physics C: Solid State Physics*, **14,** 2943–9.

Lodge, K. W., Altmann, S. L., Lapiccirella, A., and Tomassini, N. (1984). Core structure and electronic bands of the 90° partial dislocation in silicon. *Philosophical Magazine*, **B49,** 41–61.

Longuet-Higgins, H. C. and Salem, L. (1959). The alternation of bond lengths in long conjugated chain molecules. *Proceedings of the Royal Society*, **A251,** 172–85.

Loucks, L. J. (1967). *Augmented plane wave method*, Benjamin, Menlo Park, California.

Löwdin, P. O. (1950). On the non-orthogonality problem connected with the use of atomic wave functions in the theory of molecules and crystals. *Journal of Chemical Physics*, **18,** 365–78.

Löwdin, P. O. (1956). Quantum theory of cohesive properties of solids. *Advances in Physics*, **5,** 1–172.

Lukes, T. (1969). One-electron Green's functions in solid state physics. In *Solid state theory. Methods and applications* (ed. P. T. Landsberg), pp. 407–505. Wiley-Interscience, London.

McKie, D. and McKie, C. (1974). *Crystalline solids.* Nelson, London.

McWeeny, R. (1963). *Symmetry. An introduction to group theory and its applications.* Pergamon Press. Oxford.

McWeeny, R. (1979). *Coulson's Valence*, (3rd edn). Oxford University Press.

Mott, N. F. and Jones, H. (1936). *The theory of the properties of metals and alloys.* Clarendon Press, Oxford.

Ovchinnikov, A. A., Ukrainskiĭ, I. I., and Kventsel', G. V. (1973). Theory of one-dimensional Mott semiconductors and the structure of long molecules having conjugated bonds. *Soviet Physics Uspekhi*, **15,** 575–91.

Papaconstantopoulos, D. A. (1986). *Handbook of the band structure of elemental solids.* Plenum Press, New York.

Parry, D. E. and Thomas, J. M. (1975). Band structure of the one-dimensional metal polysulphur nitride. *Journal of Physics C: Solid State Physics*, **8,** L45–8.

Peierls, R. E. (1955). *Quantum theory of solids.* Clarendon Press, Oxford.

Phillips, J. C. and Kleinmann, L. (1959). New method for calculating wave functions in crystals and molecules. *Physical Review*, **116,** 287–94.

Rosenberg, H. M. (1988). *The solid state. An introduction to the physics of crystals for students of physics, materials science, and engineering*, (3rd edn). Oxford University Press.

Salem, L. (1966). *The molecular orbital theory of conjugated systems.* Benjamin, New York.

Slater, J. C. (1965). *Quantum theory of molecules and solids*, Vol. 2. *Symmetry and energy bands in crystals.* McGraw-Hill, New York.

Sneddon, I. N. (1961). *Fourier series.* Routledge & Kegan Paul, London.

Streitwolf, H.-W. (1971). *Group theory in solid-state physics.* Macdonald, London.

Sturge, M. D. (1967). The Jahn-Teller effect in solids. In *Solid State Physics*, vol. **20** (ed. F. Seitz, D. Turnbull, and H. Ehrenreich), pp. 91–211. Academic Press, New York.

Tinkham, M. (1964). *Group theory and quantum mechanics.* McGraw Hill, New York.

Tobin, N. (1960). Irreducible representations, symmetry coordinates, and the secular equation for line groups. *Journal of Molecular Spectroscopy*, **4,** 349–58.

van der Waerden, B. L. (1974). *Group theory and quantum mechanics. Die Grundlehren der mathematischen Wissenschaften in Einzeldarstellungen*, Vol. 214. Springer Verlag, Berlin.

Wannier, G. H. (1937). The structure of electronic excitation levels in insulating crystals. *Physical Review*, **52,** 191–97.

Weinberger, P. (1990). *Electron Scattering theory for ordered and disordered matter.* Clarendon Press, Oxford.

Whangbo, M.-H. (1982). Band structures of one-dimensional inorganic, organic, and polymeric conductors. In *Extended linear chain compounds*, vol. 2 (ed. J. S. Miller), pp. 127–58. Plenum Press, New York.

Whangbo, M.-H., Hoffmann, R., and Woodward, R. B. (1979). Conjugated one and two dimensional polymers. *Proceedings of the Royal Society*, **A366,** 23–46.

Wigner, E. P. (1959). *Group theory and its application to the quantum mechanics of atomic spectra*, (trans. J. J. Griffin). Academic Press, New York.

Ziman, J. M. (1960). *Electrons and phonons.* Clarendon Press, Oxford.

Ziman, J. M. (1963). *Electrons in metals. A short guide to the Fermi surface.* Taylor and Francis, London.

Ziman, J. M. (1964). *Principles of the theory of solids.* Cambridge University Press.

Ziman, J. M. (1971). The calculation of Bloch functions. In *Solid State Physics*, vol. **26** (ed. H. Ehrenreich, F. Seitz, and D. Turnbull), pp. 1–101. Academic Press, New York.

Index

Page references in **bold** are definitions. F, T, and P indicate Figure, Table, and Problem respectively. See Chapter **0** for the use of cross-references.

abelian group **42**
acceptor levels 159
acoustic bands 196
acoustic branches 196
active picture 28
alloys 198
  rigid band model 198
Altmann 45, 63, 119, 183, 195, 204, 212, 226
ammonia 214
amplitude 20
Antončík 183
Apostol 212
Arfken 85
Ashcroft 22, 63, 85, 165
Atkins 23, 45
augmented plane wave method 183
Austin 183

band 17, **95**
  see also band gap; band label; bands
Baldea 212
band gap 138
  Fermi level in it for semiconductors 159
  nearly free-electron method value 171 (**10**-1.33)
  states in it 227
  type I 173
  type II 173
band label 99
bands 17, 99
  antibonding 232
  bonding 232
  condition for two bands to touch or cross at given **k** 111, 116
  conduction 140, 158
  continuity over the Brillouin zone 112
  degeneracies 113, 115
  guesswork of trends 114
  inversion symmetry with respect to **k** 118 (**6**-5.6)
  periodicity of energy in reciprocal space 112 (**6**-4.1)
  relation to free atomic levels 100 F**5**-6.1
  valence 158
bases 35
  see also bases of space group representations; basis of a representation
bases of space group representations 103
  condition for same **k** to be repeated in same basis 111
  generated by **k** vectors 104
  group of the **k** vector **107**
  importance of use of bases of invariant subgroup T 104
  separate all **k** vectors of Brillouin zone in disjoint sets 104
  small representations **110**
basis of a crystal pattern **50**
basis of a representation **35**
  direct sum 35
  irreducible **37**
  reduction 36
Bassani 183
benzene 39
  energy levels 42 F**2**-6.2
binary rotations **53**
Blakemore 22, 165
Bloch functions **97** (**5**-6.3)
  asymptotic relation to atomic states 99
  asymptotic relation to plane waves 99
  definition depends on form of translation representative 98
  effect of space group operators 102 (**6**-1.1)
  as eigenfunctions of the translations 97 (**5**-6.2)
  form for the wrong translation representatives 100 P**5**-7.2
  as modulated plane waves 98 (**5**-6.9)
Bloch sums **174** (**10**-2.6)
  for a lattice with basis 246 P**14**-5.8
  for linear chain 175 (**10**-2.12)
  use in tight-binding method 174 (**10**-2.7)
Boas 85
body-centred cubic lattice 52 F**3**-3.2, 77, 84 P (**4**-11.4)
Boon 45
Born–von Karman boundary conditions 8
  in a periodic lattice 87
Bouckaert 119
boundary conditions
  Born–von Karman 8
  box **21** P**1**-9.6–P**1**-9.9, 249
  periodic 11
Božović 212
bra **79**

bra-ket **79** (**4**-9.3)
Bradley 63, 119
Bragg reflection 132
  its relation to group velocity 136
  see also Bragg scattering
Bragg scattering 193
brass alloy phases 199
  Hume–Rothery rules 199
  Jones theory 199
Bravais lattice **51**
  of crystal pattern **52**
  must be a lattice without basis 53
  synonymous with lattice and translational lattice 52
Brillouin 101
Brillouin zone **92**
  all operations of *P* are covering operations 93
  basic domain 115
  as centred primitive cell of the reciprocal lattice **92**
  folding back 147
  for one-dimensional chain 100 P**5**-7.5
  of higher order 141
    construction 143
  it has inversion symmetry from conjugator 118
  it has no geometrical symmetry of its own 93, 116
  number of states inside 94, 96
  open polyhedron property 93
  permitted **k** vectors inside label all energy eigenstates 94
  properties of faces 131
    cusps 134
    cusps, relation to group velocity 136
    energy derivatives 133 (**8**-2.8)
    k vectors 132
  quantization of **k** vectors inside 95 F**5**-5.1
  three-dimensional definition **143**
Buerger 63
Bullett 183, 226
Burns 63

$\mathbf{C}_3$ 42
  operations 42 F**2**-7.1
  representations 43 T**2**-7.1
$\mathbf{C}_{3v}$
  as the group of ammonia 213
  character table 44 T**2**-7.2
  operations 44 F**2**-7.2
$\mathbf{C}_{4v}$ 106
  character table 111 T**6**-3.2
$\mathbf{C}_{6v}$ 40
  operations 40 F**2**-6.1
  character table 41 T**2**-6.1
cell functions **98**
  asymptotic relation to atomic states 99
Celli 183
cellular method 183
centred primitive cell **50**, 92
character **37** (**2**-4.17)
circular frequency **19** (**1**-8.7)
circular wave number **19** (**1**-8.6)
class **24**
Cochran 197
Cohen 183
conduction band 140, 158
conductivity **184**
  relation to Fermi surface 185 (**11**-1.6)
conductor 140
configuration space **26**
configuration space operator **28**
conjugate element **24**
conjugate representations **251** P**2**-7.6
conjugator operator **32** (**2**-3.11)
  commutes with symmetry operations 33
  inverts **k** vector of Bloch functions 118 (**6**-5.1)
  it is non linear 33
Copper 161
  band structure 162 F**9**-2.1
  Brillouin zone 157 F**9**-1.3
    details 165 P**9**-4.2, P**9**-4.3
  crystal structure 162
  density of states 162 F**9**-2.1
  Fermi surface 163 F**9**-2.2, 164 F**9**-2.3
  free-electron energies 165 P**9**-4.4
core electrons 7
Cornwell 45
Cotton 45
Cottrell 22, 183, 200, 212
Coulson 240
covering operation **26**
Cox 22
Cracknell 63, 119
crystal structure **46**
crystal lattice 52, **64**
crystal pattern **46**
crystal vibrations 186
$\mathbf{C}_S$ 109
  character table 109 T**6**-3.1
current density 184
cyclic groups **42** P**2**-7.3, 215, 250 P**2**-7.3
  partition space into disjoint cells 214, 216
  properties 215
  their representations are faithful 216

d band 162
  in noble metal 164
  in transition metal 164
$\mathbf{D}_{6h}$ 40
dangling bond 230
de Haas–van Alphen effect 148
degeneracy **31**

density of states **14**
for free electrons 15 (**1**-5.7)
free-electron gas 16 F**1**-5.1
diamond crystal pattern 153, 154 F**9**-1.1
bond lengths 165 P**9**-4.1
Bravais lattice 154
point group 154
space group 155
Dimmock 183
direct product **25** (**2**-1.7), (**2**-1.8)
basis **204** (**12**-2.12)
character 210 (**12**-4.3)
matrix 204
representations 38
symmetrized 210
direct sum
of bases 35
of matrices 35
displacement vector 188
donor levels 159

electron density **14**
electron–phonon interactions 203
electron–phonon scattering 192
electrons 149
relation to holes 149
energy as a function of **k** 92, 112
contacts or crossings at given **k** 111
inversion symmetry in reciprocal space 118
periodicity in reciprocal space 112 (**6**-4.1)
symmetry in reciprocal space due to the star 115
energy eigenvalues **31** (**2**-3.3)
are labelled by **k** vector 92
energy level surfaces 135
discontinuities at the Brillouin zone edge 145 F**8**-7.1
equivalent functions 213, **215** (**13**-1.7)
close formula 219 (**13**-1.33)
construction 218 (**13**-1.20)
orthogonal for different cells 218
equivalent **k** vectors **91**
exchange integral **176** (**10**-2.15), 231 (**14**-2.2)
exchange interactions 166
extended zone scheme **143**, 141 F**8**-5.1
extrinsic semiconductivity 159
extrinsic semiconductor 140

face-centred cubic lattice 76 F**4**-7.1
Brillouin zone 131 F**8**-1.1
its reciprocal 76 F**4**-7.1, 84 P**4**-11.3
faithful representations **216**
Falicov 44, 212

Fermi energy 13, 146
for free electrons 16 (**1**-6.4)
Fermi sphere 14
Fermi surface 14
displacement under electric field 185 F (**11**-1.1)
experimental determination 148
extremal cross-sections 148
for simple square lattice 147 F**8**-7.2, 148 F**8**-7.3
fermions 13
first Brillouin zone, *see* Brillouin zone
Flores 247
folding back 147
forbidden energy gap 138
Fourier coefficients **81** (**4**-9.19)
Fourier series **78**, 81 (**4**-9.17)
for lattice with basis 82 (**4**-9.25)
fractional translation **55**
free-electron approximation 167
function products
transformation properties 204 (**12**-2.12), 266 P**12**-4.2, 269 P**12**-4.6
function space operator 28, **29** (**2**-2.7)
isomorphic to configuration space operators 30 (**2**-2.8)
wrong form 45
function transformation 28
wrong form 45
functional transformations 29

G
a group 24
a space group 56
García Moliner 247
general **k** vector **116**
bands cannot touch or cross at them 116
Glazer 63
glide reflection **55**
Gold 63, 119, 212
Green function **237**
explicit form 239 (**14**-3.30)
group **24**
closure **24**
conjugate element **24**
identity **24** (**2**-1.3)
inverse element **24** (**2**-1.4)
product of elements 24
associative condition **24** (**2**-1.2)
group of the **k** vector **107**
varies over star 110
group velocity **20** (**1**-8.13)
from energy gradient 135 (**8**-3.2)
it is normal to level surfaces of energy 135 F**8**-3.1
not always parallel to momentum 135
Groves 63, 85

Hamermesh 45
Hamiltonian 9
  vibronic 203 (**12**-2.8)
harmonic approximation 188
harmonic oscillations 186
harmonic oscillator 186
  eigenvalues 186 (**11**-2.3)
  quantum mechanical Hamiltonian 186 (**11**-2.1)
Harris 212
Harrison 183
hash **67**
Heine 44, 181, 183, 202, 211
Hermite functions 186
Herring 183
hexagonal close-packed structure 83 F**4**-9.1
  phase change 211 P**12**-4.5
  two dimensional (lattice) 73 F**4**-5.1
Hoffmann 22, 212, 247
hole states 149, 159
holes 149, 151
  effective mass 151
  velocity 151
holohedric point group **57**
Hückel approximation 177
Hume-Rothery 199
  rules 199

impurities and surface states 227
  Koster and Slater method 235
insulator 140, 149
interplanar distance 70
intrinsic semiconductivity 158
intrinsic semiconductor 140
invariant subgroup **25** (**2**-1.6)
inversion operator 103
  effect on Bloch functions 103 (**6**-1.9)
irreducibility test 38 (**2**-5.1)
irreducible representation **37**
  conjugate 211 P**12**-4.6
  dimension 38
  faithful **216**
  number 37, 38
  orthogonality relations 38 (**2**-5.2) (**2**-5.3)
  relation to degeneracy 37
  test for irreducibility 38 (**2**-5.1)

Jahn–Teller effect 205
Jansen 45
Janssen 63
Jerome 211
Jones 101, 199

k space 12
  quantization 12, 94
  volume per state 13 (**1**-3.25)
k vectors 11
  equivalent **91**
  group **107**
  quantization 11 (**1**-3.20), 94
  quasi-continuous distribution inside Brillouin zone 94
Kelly 63, 85
ket **79**
kinetic energy **9**, 188
  for vibrating crystal 188 (**11**-2.10)
    invariance under translations 189
Kittel 22, 165
KKR method 183
Kleinmann 183
Knox 63, 119, 212
Kohn–Korringa–Rostoker method 183
Koster and Slater method 235
  localized perturbation 239
Kronecker delta **38** (**2**-5.2), **80** (**4**-9.10)
Kventsel 212, 240

Landau 45, 203, 212
Lapiccirella 158, 195
lattice **49**
  synonymous with Bravais lattice 52
  with basis **49**
lattice planes **64**, 78
Lax 44
LCAO 177
Lennard-Jones 240
Levine 247
Lifshitz 45, 203, 212
linear chain 137
  alternation 200, 212
  bases and irreducible representations 152 (**8**-8.1)–(**8**-8.3)
  Brillouin zone 100 P**5**-7.5, 137 F**8**-4.1
  electron–phonon interactions 203
  energy bands 139 F**8**-4.2
    forbidden contacts 139 F**8**-4.3
    forbidden gaps 138
  N atoms, tight binding calculation 175
    eigenvalues 176 (**10**-2.26)
  Peierls instability 200
  reconstruction 200
  six atoms 120
    bands 127
    bases of the representations 125 T**7**-3.1
    Brillouin zone 126 F**7**-3.1
    classes 122 T**7**-1.1
    energy levels 177 T**10**-2.1, 177 F**10**-2.1
    irreducible representations 127 T**7**-3.3
    isomorphism with $\mathbf{C}_{6v}$ 122
    representations of invariant subgroup 124 T**7**-2.1
    small representations 138 T**8**-4.2
    stars and groups of **k** 137 T**8**-4.1

surface states 240, 244
tight-binding calculation 175
linear combination of atomic orbitals 173, 177
linear independence **31** (**2**-3.8)
linear operators **33**
linear variational method 167
Littlewood 202, 211
local potential 181
Lodge 158, 195
longitudinal oscillation 196
Longuet-Higgins 212
Loucks 183
Löwdin 225, 226
Löwdin orbitals **222**
close formula 224 (**13**-3.20)
Lukes 247

McKie 63
McWeeny 44, 240
matrix (representative) 34
direct sum 35
invariant 37
trace 37
matrix element **168**
matrix function **224**
Mermin 22, 63, 85, 165
metallic electrons 7
Miller indices **71** F**4**-4.2
properties 72
relation to reciprocal vectors 72 (**4**-4.11)
momentum 9, 10
eigenfunctions **9**, 10
operator **9**, 10
quantization 11
vector 84
not parallel to velocity 136
Mott 101

n-type semiconductors 159
nearly free-electron method 167
band gap 171 (**10**-1.33)
form of wave function 168 (**10**-1.10)
noble metal 161, 164
d band 164
s band 164
non-linear operator 33
non-symmorphic space group **62**
normal coordinates 187, **190**
as translation eigenvectors 191 (**11**-2.20)
normalization **79** (**4**-9.5)
normal mode of vibration 186, 191
longitudinal 196
transverse 196

one-dimensional chain
Brillouin zone 100 P**5**-7.5
see also linear chain
open surfaces **49**
operator
configuration space **28**
function space 28, **29** (**2**-2.7)
isomorphic to configuration space operators 30 (**2**-2.8)
wrong form 45
products **27**, 53
transformation 30 (**2**-2.13)
optic bands 196
optic branches 196
OPW, *see* orthogonalized plane waves method
orbital degeneracy 12
orbital state 13
orbits in **k** space 148
orthogonality **79** (**4**-9.4)
of irreducible representations 38
orthogonalized plane waves method 178
orthogonalized plane waves 179 (**10**-3.6)
wave function 179 (**10**-3.12)
orthonormal functions **80** (**4**-9.9)
orthonormality **79**
orthonormal vectors **65** (**4**-1.5) (**4**-1.6)
oscillator force constant 186
Ovchinnikov 212
overlap integral 168

P point group **57**
of space group **60**
p-type semiconductor 159
Papaconstantopoulos 162, 183
Parry 212
passive picture 28
never used in this book 28
pattern (crystal) **46**
Peierls 200, 211
Peierls instability 200
linear chain 200
lowering of Fermi energy 201 (**12**-2.1)
quasi linear chain 205
total energy balance 202
vibronic interactions 203
Peierls transition 201
period 19
periodic boundary conditions 8 F**1**-2.1
in a lattice 87
in a one-dimensional lattice 87 F**5**-1.1
order of the translation group 88 (**5**-1.3)
three-dimensional form 9
periodic functions 65
are periodic in the lattice 65
expansion in Fourier series 81 (**4**-9.17)
in any lattice 68 (**4**-2.12)

periodically repeated scheme **143**
periodicity 47
  it is fully determined by translation lattice 64
  its relation to translation symmetry 47
phase changes 199
  relation to Fermi surface 199
phase velocity 19 (**1**-8.8)
Phillips 183
phonon drag 194
phonons 192
phonon spectrum 195
  cusps 196
  for silicon 195 F**11**-4.1
plane group **53**
plane stacks **69**
plane waves **18**, 18 (**1**-8.2), 65, 77 (**4**-8.1)
  corresponding to lattice planes 78
  in the lattice 84 (**4**-10.2)
  momentum of 84 (**4**-10.3)
  normalized 79 (**4**-9.6), 84 (**4**-10.1)
  of the lattice 84 (**4**-10.1)
  orthogonal over crystal volume 100 P**5**-7.4
  orthonormality 80 (**4**-9.9)
  propagation vector 11, 77
  as translation eigenfunctions 100 P**5**-7.3
planes (crystal) 70
  equation in Miller indices 71 (**4**-4.7)
  equation in the normal 70 (**4**-4.2)
  of the lattice 78
point group **57**
  leaves Bravais lattice invariant 57, 58
  relation to space group 60
polyacetylene 206 F**12**-3.1
polyene chain 206
  degeneracy due to glide 207, 209
  double Brillouin zone 208, 209, F**12**-3.3
  glide plane 206
  Peierls instability 209
  reconstruction 210 F**12**-3.4
position vectors **64**
  always given in direct components 68, 73
  are expressed in terms of the crystal lattice 66
  as vectors *in* or *of* the lattice 68 (**4**-2.8) (**4**-2.9)
potential energy 9
  for vibrating crystal 188 (**11**-2.11)
    invariance under translations 189
potential field 166
  local 181
primitive cell **48**
  centred **50**, 92
  construction as maximal sets of translationally inequivalent points 50
  contains only one lattice point per cell 53
  as a maximal set of translationally inequivalent points 49
  open polyhedron property 49
  standard 51
  standard and centred have same extension 51
  with a basis **50**, 53
primitive vectors **48**
projection operator 38 (**2**-5.4)
  for plane waves 179 (**10**-3.10)
propagation vector **11**, 77
pseudo Hamiltonian 180 (**10**-4.6)
pseudomomentum 99
pseudopotential **180** (**10**-4.8)
pseudopotential method 180
  Austin–Heine–Sham pseudopotential **182** (**10**-4.15)
  form of wave equation 181 (**10**-4.9)
pseudo wave function **180** (**10**-4.2)

quasi linear chain 205
  see also polyene chain
quasi-continuous **94**

reciprocal lattice **74**
  blind to any basis in direct lattice 75
  cannot have a basis 75
  correct relation to the direct lattice 74
  entirely determined by translational lattice 74
  of face-centred cubic lattice 76 F**4**-7.1
  the **k** vectors in its primitive cell label all representations of T 91
reciprocal vectors **67**
  construction 69 (**4**-3.3) (**4**-3.5) (**4**-3.6)
  definition in crystallography **67** (**4**-2.4) (**4**-2.5)
  definition in solid-state theory **68** (**4**-2.6) (**4**-2.7)
  direction in terms of plane stack 72 (**4**-4.11)
  error to be avoided in getting moduli 74
  in the lattice **75** (**4**-6.3)
  modulus in terms of interplanar spacing 72 (**4**-4.12)
  of the lattice 75 (**4**-6.4)
  relation to plane stacks 72 (**4**-4.11) (**4**-4.12)
  used to denote directions 73
reconstruction of surface 235
reduced coordinates **189**
reduced zone scheme **142**
reduction of a representation 35
  number of times an irreducible is contained 269 (**15**-12.19)
reflections 53
relaxation time **185**
representation **35**
  conjugate 211 P**12**-4.6, **251** P**2**-7.6

irreducible 37
reduction 36
partition over irreducibles 38, 269 (**15**-12.19)
see also irreducible representation
resolvent **241**
rigid band model 198
Rosenberg 22, 165, 197

s band 162, 164
in noble metal 164
in transition metal 164
Salem 212
scattering
electron–phonon 192
of electron by rigid crystal 193
Schrödinger equation 9 (**1**-3.1)
Schrödinger group **30**
Schulz 211
screening 7
screw rotation **55**
secular determinant 168
Seitz operator **59** (**3**-6.2), 59 F**3**-6.1
inverse **60** (**3**-6.7)
product **60** (**3**-6.6)
self-consistent field 166
semiconductor 140, 149
extrinsic 140, 159
intrinsic 140, 158
n-type 159
position of Fermi energy in band gap 159
p-type 159
semimetal 149
set **23**
order **24**
product **24**
commutation 24
Sham 181, 183
Shockley states 229
linear chain example 230
silicon 153
band structure 158 F**9**-1.4
band trends (guess) 160
Brillouin zone 157 F**9**-1.3
features of small representations 157
origin of narrow gap 156
phonon spectrum 195 F**11**-4.1
silicon structure 153, 154, F**9**-1.1
Bravais lattice 154
orbital basis 155 F**9**-1.2
point group 154
space group 154
see also diamond crystal pattern
similarity 36 (**2**-4.16)
simple metals 153
simple square lattice **91** F**5**-3.1
bands 113 F**6**-4.1, 145 F**8**-7.1
Brillouin zone 91 F**5**-3.1, 106 F**6**-3.1
basic domain 116 F**6**-4.3
of higher order 144 F**8**-6.1
cell functions 114 F**6**-4.2
Fermi surface 147 F**8**-7.2, 148 F**8**-7.3
reciprocal lattice 91 F**5**-3.1
Slater 45, 247
small representations **110**
Smoluchowski 119
Sneddon 85
sodium 7
soft X-rays 17
space group operators **59** (**3**-6.2)
effect on Bloch functions 102 (**6**-1.1)
space group representations 102
bases 103
irreducibles and energy as a function of **k** **112**
space groups **53**
classification 62
compound symmetry operations 54
different factorizations 55
different settings 55
non-symmorphic **62**
point groups of **60**
symmorphic **62**
span
basis spans a representation 36
spin
always included when counting states 14
square lattice, simple, *see* simple square lattice
stacks of planes **69**
star **107**
degeneracies introduced 112, 115
step function 242 (**14**-4.10) (**14**-4.11), 242 F**14**-4.1
sticking together of bands 160, 171
Streitwolf 63, 119, 197
strict two-dimensionality **53**, 106
structure factor **82** (**4**-9.26)
for hexagonal close-packed structure 83
Sturge 212
subgroups **25**
surface states 227
in linear chain 240
Shockley 229
Tamm 229
symmetrized direct product 210
basis 210 (**12**-4.4)
character 210 (**12**-4.5)
symmetry operation **26**, **28**
active picture **28**
as a covering operation 26
passive picture (never used) **28**
product **27**
their set forms a group 28
in unit cell 55

symmetry operator **28**
symmorphic space groups **62**

T translation subgroup, *see* translation group
Tamm states 227
Thomas 212
tight-binding method 173
  Bloch sums 174
  form of wavefunction 173 (**10**-2.1)
  linear chain 175
time-reversal operator 33
Tinkham 44
Tobin 212
Tomassini 158, 195
total energy **16**
  for free electrons 17 (**1**-6.7)
totally symmetrical (trivial) representation 216
  condition for a function product to transform like it 269 **P12**-4.6
  functions that belong to it 269 **P12**-4.6
trace **37**
transition metal 164
  d band 164
  s band 164
transition probability integral **203** (**12**-2.9)
translation 27 F**2**-2.1
translation eigenfunctions **95** (**5**-5.1)
  as Bloch functions 97
  give all the representations of space group 96
  plane waves as particular case 100 **P5**-7.3
  provide energy eigenfunctions 96
  vary continuously over Brillouin zone 95
translation group **60**, 86
  abelian property 61
  direct product form 61 (**3**-8.5), 88 (**5**-1.6)
  factorized out of space group 55
  faithful property of representations 217
  form of the irreducible representations 90 (**5**-2.9)
  invariant of the space group 61
  number and dimensions of its representations 89
  number of classes 89
  order under periodic boundary conditions 88 (**5**-1.3)
translation subgroup, *see* translation group
translational lattice 49
  synonymous with lattice and Bravais lattice 52
translationally inequivalent points **50**
  construction of sets 50, 144
translations 49
  commute with Hamiltonian 95 (**5**-5.3)
  effect on potential and kinetic energies of a crystal 189
  eigenfunctions (Bloch functions) 97 (**5**-6.2)
  eigenvalue equation 97 (**5**-6.2)
  eigenvalues are not degenerate 95 (**5**-5.2)
  eigenvectors (normal coordinates) 191 (**11**-2.20)
  relation to unit cells 47
transverse oscillations 196
trigonal lattice 85
trivial representation **216**

Ukrainskiĭ 212
umklapp process 194
unit cell **47**
  may contain translationally equivalent points 48
  volume 79 (**4**-9.1)
unit vectors **47**

valence band 158
valence electrons 7
van der Waerden 45
vector *in* the lattice 60, 64, **66** (**4**-1.12), **75** (**4**-6.1)
vector *of* the lattice **60**, 64, **65** (**4**-1.7), **75** (**4**-6.2)
vectors
  in the lattice **75** (**4**-6.1)
  of the lattice **75** (**4**-6.2)
  see also position vectors
vibronic Hamiltonian **203** (**12**-2.8)
vibronic interaction 203
  transition probability integral 203 (**12**-2.9)
von Laue conditions 132 (**8**-2.1)

Wannier 225
Wannier functions **219**
  close formula 220 (**13**-2.2)
  for free electrons 221 (**13**-2.11), 221 F**13**-2.1
  localization 220 (**13**-2.4)
wave length 19
wave number, *see* circular wave number
wave packets 19, 20
wave velocity 19
waves 18
  superposition 20
Weaire 183
Weinberger 183
Whangbo 211, 212
Wigner 45, 119
Wigner-Seitz cells **51**
Woodward 212

Ziman 22, 183, 184, 197, 226, 247
zone schemes 141

GOSHEN COLLEGE - GOOD LIBRARY
3 9310 01035618 4